Wastewater Treatment, Plant Dynamics and Management in Constructed and Natural Wetlands

Jan Vymazal

Editor

Wastewater Treatment, Plant Dynamics and Management in Constructed and Natural Wetlands

 Springer

Editor
Dr. Jan Vymazal
ENKI, o.p.s.
and
Institute of Systems Biology and Ecology
Czech Academy of Sciences
Dukelská 145
379 01 Třeboň
Czech Republic

ISBN 978-1-4020-8234-4   e-ISBN 978-1-4020-8235-1

Library of Congress Control Number: 2008921925

Printed on acid-free paper

9 8 7 6 5 4 3 2 1

springer.com

Preface

At present, constructed wetlands for wastewater treatment are a widely used technology for treatment of various types of wastewaters. The International Water Association (then International Association on Water Pollution Research and Control) recognized wetlands as useful tools for wastewater treatment and established the series of biennial conferences on the use of wetland systems for water pollution control in 1988. In about 1993, we decided to organize a workshop on nutrient cycling in natural and constructed wetlands with the major idea to bring together researchers working on constructed and also natural wetlands. It was not our intention to compete with IWA conferences, but the workshop should rather complement the series on treatment wetlands by IWA. We believed that the exchange of information obtained from natural and constructed wetlands would be beneficial for all participants. And the time showed that we were correct.

The first workshop took place in 1995 at Třeboň in South Bohemia and most of the papers dealt with constructed wetlands. Over the years we extended the topics on natural wetlands (such as role of wetlands in the landscape or wetland restoration and creation) and during the 6th workshop held at Třeboň from May 30 to June 3, 2006, nearly half of 38 papers presented during the workshop dealt with natural wetlands. This workshop was attended by 39 participants from 19 countries from Europe, Asia, North and South Americas and Australia. The volume contains 29 peer-reviewed papers out of 38 papers which were presented during the workshop.

The organization of the workshop was partially supported by grants No. 206/06/0058 "Monitoring of Heavy Metals and Selected Risk Elements during Wastewater Treatment in Constructed Wetlands" from the Czech Science Foundation and Grant. No. 2B06023 "Development of Mass and Energy Flows Evaluation in Selected Ecosystems" from the Ministry of Education, Youth and Sport of the Czech Republic.

Praha   Jan Vymazal
August 2007

In Memoriam for Olga Urbanc-Berčič

Olga Urbanc-Berčič (1951–2007)

Olga Urbanc-Berčič was a biologist in the real sense of the word. She regarded her profession as a vocation which influenced her whole life. In 1975, after her diploma she got a post on the National Institute of Biology in Ljubljana in the laboratory for electronic microscopy. Some years later she joined the group researching freshwater and terrestrial ecosystems in the same institution. In 1988 she finished her Master's with a thesis titled "The use of *Eichhornia crassipes* and *Lemna minor* for wastewater treatment". In 2003 she successfully defended her Ph.D. thesis titled "The availability of nutrients in the rhizosphere of reed stands (*Phragmites australis*) in relation to water regime in the intermittent Lake Cerknica". Her service to her professional interests was totally unselfish. She was involved in many different projects, but most of all she liked the research dedicated to wetlands and aquatic plants. We were a perfect team for many years. I will never forget the fruitful time we spent in the field sampling and researching. The results of her research are summarised in numerous scientific and professional publications. Her studies of the role of water-level fluctuations in nutrient cycling led to a wider understanding of wetland functions. Her work additionally clarified the importance of macrophytes in aquatic systems. She was active in different non-governmental organisations, being the president of the Slovenian Ecological Society for many years. As a warm-hearted, generous, enthusiastic and positively oriented person she was a link among people and an efficient advocate of nature. On a cold, grey Wednesday in February, we accompanied her to her last home. Her death was a great loss for family, friends, colleagues and the community. We will miss her, but her work and her spirit will live with us forever.

Selected Bibliography

Olga Urbanc-Berčič authored more than 100 contributions in international and Slovenian research and popular journals, monographs and conference proceedings. The following list contains only a short selection of her publications.

Cimerman, A., Legiša, M., Urbanc-Berčič, O., & Berberovič, R. (1982). Morphology of connidia of citric acid producing Aspergillus niger strains by scanning electron microscopy. *Biol. Vestn.*, *30*(2), 23–31.

Urbanc-Berčič, O., & Gaberščik, A. (1989). The influence of temperature and light intensity on activity of water hyacinth (*Eichhornia crassipes* (Mart.) Solms.). *Aquat. Bot.*, *35*, 403–408.

Urbanc-Berčič, O., & Blejec, A. (1993). Aquatic macrophytes of lake Bled: Changes in species composition, distribution and production. *Hydrobiologia (Den Haag)*, *262*, 189–194.

Urbanc-Berčič, O. (1994). Investigation into the use of constructed reedbeds for municipal waste dump leachate treatment. *Wat. Sci. Tech.*, *29*(4), 289–294.

Urbanc-Berčič, O. (1995). Aquatic vegetation in two pre-alpine lakes of different trophic levels (Lake Bled and Lake Bohinj): Vegetation development from the aspect of bioindication. *Acta Bot. Gall.*, *142*, 563–570.

Urbanc-Berčič, O. (1995). Constructed wetlands for treatment of landfill leachates: Slovenian experience. In J. Vymazal (Ed.), *Nutrient cycling and retention in wetlands and their use for wastewater treatment* (pp. 15–23). Třeboň, Czech Republic: Institute of Botany; and Praha: Czech Republic: Ecology and Use of Wetlands.

Gaberščik, A., & Urbanc-Berčič, O. (1995). Monitoring approach to evaluate water quality of intermittent lake Cerknica. In: *Proc. 2nd International IAWQ Specialized Conf. and Symp. on Diffuse Pollution: Brno & Prague, Czech Republic, August 13–18, 1995, part 2,* pp. 191–196.

Urbanc-Berčič, O., & Gaberščik, A. (1995). Potential of the littoral area in lake Bled for reed stand extension. In R. Ramadori, R. Cingolani, & L. Cameroni, (Eds.), *Proc. Internat. Seminar Natural and Constructed Wetlands for Wastewater Treatment and Reuse: Experiences, Goals and Limits* (pp. 95–99). 26–28 October 1995. Perugia: Centro.

Urbanc-Berčič, O., & Griessler Bulc, T. (1995). Integrated constructed wetland for small communities. *Wat. Sci. Tech.*, *32*(3), 41–47.

Gaberščik, A., & Urbanc-Berčič, O. (1996). Monitoring approach to evaluate water quality of intermittent lake Cerknica. *Wat. Sci. Tech.*, *33*(4–5), 357–362.

Gaberščik, A., & Urbanc-Berčič, O. (1996). Lakes of the Triglav national park (Slovenia): Water chemistry and macrophytes. In A. Gaberščik, O. Urbanc-Berčič, & G. A. Janauer, (Eds.), *Proc. Internat. Workshop and 8th Macrophyte Group Meeting IAD-SIL* (pp. 23–28) September 1–4, 1996. Bohinj, Ljubljana, Slovenia: National Institute of Biology.

Urbanc-Berčič, O., & Gaberščik, A. (1996). The changes of aquatic vegetation in lake Bohinj from 1986 to 1995. In A. Gaberščik, O. Urbanc-Berčič, & G. A. Janauer (Eds.), *Proc. Internat. Workshop and 8th Macrophyte Group Meeting IAD-SIL* (pp. 69–72). September 1–4, 1996, Bohinj, Ljubljana, Slovenia: National Institute of Biology.

Urbanc-Berčič, O., & Kosi, G. (1997). Catalogue of limnoflora and limnofauna of Slovenia (Katalog limnoflore in limnofavne Slovenije). *Acta Biol. Slov., 41*, 149–156.

Urbanc-Berčič, O., & Gaberščik, A. (1997). Reed stands in constructed wetlands: "Edge effect" and photochemical efficiency of PS II in common reed. *Wat. Sci. Tech., 35*(5), 143–147.

Urbanc-Berčič, O. (1997). Constructed wetlands for the treatment of landfill leachates: The Slovenian experience. *Wetlands Ecol. Manag., 4*, 189–197.

Germ, M., Gaberščik, A., & Urbanc-Berčič, O. (1997). Environmental approach to the status of the river ecosystem. In M. Roš (Ed.), *Proc. 1st Internat. Conf. Environmental Restoration* (pp. 269–274) July 6–9, 1997. Cankarjev dom, Ljubljana, Slovenia: Slovenian Water Pollution Control Association.

Gaberščik, A., Urbanc-Berčič, O., Brancelj, A., & Šiško, M. (1997). Mountain lakes – remote, but endangered. In M. Roš (Ed.), *Proc. 1st Internat. Conf. Environmental Restoration* (pp. 452–456) July 6–9, 1997. Cankarjev dom, Ljubljana, Slovenia: Slovenian Water Pollution Control Association.

Urbanc-Berčič, O., Bulc, T., & Vrhovšek, D. (1998). Slovenia. In J. Vymazal, H. Brix, P. F. Cooper, M. B. Green, & R. Haberl, (Eds.), *Constructed wetlands for wastewater treatment in Europe* (pp. 241–250). Leiden, The Netherlands: Backhuys Publishers.

Brancelj, A., Gorjanc, N., Jačimovič, R., Jeran, Z., Šiško, M., & Urbanc-Berčič, O. (1999). Analysis of sediment from Lovrenška jezera (lakes) in Pohorje (Analiza sedimenta iz Lovrenškega jezera na Pohorju). *Geogr. Zb., 39*, 7–28. http://www.zrc-sazu.si/giam/zbornik/brancelj_39.pdf.

Germ, M., Gaberščik, A., & Urbanc-Berčič, O. (1999). Aquatic macrophytes in the rivers Sava, Kolpa and Krka (Vodni makrofiti v rekah Savi, Kolpi in Krki). *Ichthyos (Ljublj.), 16*, 23–34.

Urbanc-Berčič, O., & Gaberščik, A. (1999). Seasonal changes of potential respiration of root systems in common reed (Phragmites australis) grown on the constructed wetland for landfill leachate treatment. In J. Vymazal, (Ed.), *Nutrient cycling and retention in natural and constructed wetlands* (pp. 121–126). Leiden, The Netherlands: Backhuys Publishers.

Germ, M., Gaberščik, A., & Urbanc-Berčič, O. (2000). The wider environmental assessment of river ecosystems (Širša okoljska ocena rečnega ekosistema). *Acta Biol. Slov., 43*, 13–19.

Gaberščik, A., Urbanc-Berčič, O., & Martinčič, A. (2000). The influence of water level fluctuation on the production of reed stands (*Phragmites australis*) on intermittent lake Cerkniško jezero. In S. Cristofor, A. Sârbu, & M. Adamecsu, (Eds.), *Proc. Internat. Workshop and 10th Macrophyte Group Meeting IAD-SIL* (pp. 29–33). August 24–28, 1998. Danube Delta, Bucureşti, Romania: Editura Universitâţii din Bucureşti.

Germ, M., Gaberščik, A., & Urbanc-Berčič, O. (2000). The distribution of aquatic macrophytes in the rivers Sava, Kolpa and Krka (Slovenia). In S. Cristofor, A. Sârbu, & M. Adamecsu, (Eds.), *Proc. Internat. Workshop and 10th Macrophyte Group Meeting IAD-SIL* (pp. 34–40). August 24–28, 1998. Danube Delta, Bucureşti, Romania: Editura Universitâţii din Bucureşti.

Urbanc-Berčič, O., & Gaberščik, A. (2001). The influence of water table fluctuations on nutrient dynamics in the rhizosphere of common reed (*Phragmites australis*). *Wat. Sci. Tech., 44*(11–12), 245–250.

Gaberščik, A., & Urbanc-Berčič, O. (2001).Reed dominated intermittent lake Cerkniško jezero as a sink for nutrients. In J. Vymazal (Ed.), *Transformations of Nutrients in Natural and Constructed Wetlands* (pp. 225–234). Leiden, The Netherlands: Backhuys Publishers.

Urbanc-Berčič, O., Gaberščik, A., Šiško, M., & Brancelj, A. (2002). Aquatic macrophytes of the mountain lake Krnsko jezero, Slovenia (Vodni makrofiti Krnskega jezera, Slovenija). *Acta Biol. Slov., 45*, 25–34.

Urbanc-Berčič, O. (2003). Charophytes of Slovenia, their ecological characteristics and importance in aquatic ecosystems (Parožnice (Characeae) Slovenije, njihove ekološke značilnosti ter pomen v vodnih ekosistemih). *Hladnikia (Ljubl.), 15/16*, 17–22.

Gaberščik, A., Urbanc-Berčič, O., Kržič, N., Kosi, G., & Brancelj, A. (2003). The intermittent lake Cerknica: Various faces of the same ecosystem. *Lakes Reserv.*, *8*, 159–168.

Urbanc-Berčič, O., & Gaberščik, A. (2003). Microbial activity in the rhizosphere of common reed (Phragmites Australis) in the intermittent lake Cerkniško jezero. In J. Vymazal (Ed.), *Wetlands: Nutrients, metals and mass cycling* (pp. 179–190). Leiden, The Netherlands: Backhuys Publishers.

Urbanc-Berčič, O., & Gaberščik, A. (2004). The relationship of the processes in the rhizosphere of common reed *Phragmites australis*, (Cav.) Trin. ex Steudel to water fluctuation. *Int. Rev. Hydrobiol.*, *89*, 500–507.

Germ, M., Urbanc-Berčič, O., Gaberščik, A., & Janauer, G.A. (2004). Distribution and abundance of macrophytes in the river Krka. In I. Teodorivič, S. Radulovič, & J. Bloesch (Eds.), *Limnological Reports* (pp. 433–440). Novi Sad, Serbia: International Association for Danube Research – IAD.

Kuhar, U., Gaberščik, A., Germ, M., & Urbanc-Berčič, O. (2004). Macrophytes and ecological status of three streams in the river Drava plain. In I. Teodorivič, S. Radulovič, & J. Bloesch (Eds.), *Limnological reports* (pp. 441–447). Leiden, The Netherlands: International Association for Danube Research – IAD.

Germ, M., Urbanc-Berčič, O., & Kocjan Ačko, D. (2005). The response of sunflower to acute disturbance in water availability(Odziv sončnic na akutno pomanjkanje vode). *Acta Agric. Slov.*, *85*, 135–141.

Urbanc-Berčič, O., Kržič, N., Rudolf, M., Gaberščik, A., & Germ, M. (2005) The effect of water level fluctuations on macrophyte occurrence and abundance in the intermittent Lake Cerknica. In J. Vymazal (Ed.), *Natural and constructed wetlands: Nutrients, metals and management* (pp. 312–320). Leiden, The Netherlands: Backhuys Publishers.

Kržič, N., Germ, M., Urbanc-Berčič, O., Kuhar, U., Janauer, G.A., & Gaberščik, A. (2007) The quality of the aquatic environment and macrophytes of karstic watercourses. *Plant Ecol.* (Dordrecht), 192(1): 107–118.

Germ, M., Kreft, I., Stibilj, V., & Urbanc-Berčič, O. (2007) Combined effect of selenium and drought on photosynthesis and mitochondrial respiration in potato. *Plant Physiol. Biochem.* (Paris), 45(2): 162–167.

Ljubljana   Alenka Gaberščik
June 2007

Contents

Contributors

Edit Ágoston-Szabó
Institute of Ecology and Botany, Hungarian Danube Research Station, H-2163
Vácrátót, Hungary

Jani Anttila
Peatland Ecology Group, University of Helsinki, Department of Forest Ecology,
Helsinki, Finland

Hiroyuki Araki
Institute of Lowland Technology, Saga University, Saga, Japan

Sofia Kallner Bastviken
IFM-Biology, Linköping University, SE-581 83 Linköping, Sweden

Herbert John Bavor
Centre for Water and Environmental Technology – Water Research Laboratory,
University of Western Sydney – Hawkesbury, Locked Bag 1797, Penrith South
DC, NSW 1797, Australia

Simon Bird
University of Vermont, Department of Plant and Soil Science, Hills Agricultural
Building, 105 Carrigan Drive, Burlington, VT 05405, USA

Jakub Brom
Laboratory of Applied Ecology, Faculty of Agriculture, University of South
Bohemia, Studentská 13, České Budějovice, CZ-370 05, Czech Republic;
ENKI o.p.s., Dukelská 145, Třeboň, CZ-379 01, Czech Republic

Aracelly Caselles-Osorio
Environmental Engineering Division; Hydraulics, Maritime and Environmental
Engineering Department; Technical University of Catalonia, Jordi Girona, 1-3,
08034-Barcelona, Spain; Department of Biology, Atlantic University, km 7
Highway Old Colombia Port, Barranquilla, Colombia

David Cooper
ARM Ltd, Rydal House, Colton Road, Rugeley, Staffordshire, WS15 3HF,
United Kingdom

Paul Cooper
ARM Ltd, Rydal House, Colton Road, Rugeley, Staffordshire, WS15 3HF,
United Kingdom; Independent Consultant, PFC Consulting, The Ladder House,
Cheap Street, Chedworth, Cheltenham, GL54 4AB, United Kingdom

Christopher Craft
School of Public and Environmental Affairs, Indiana University, Bloomington IN
47405-1701, USA

Mária Dinka
Institute of Ecology and Botany, Hungarian Danube Research Station, H-2163
Vácrátót, Hungary

Aleksandra Drizo
University of Vermont, Department of Plant and Soil Science, Hills Agricultural
Building, 105 Carrigan Drive, Burlington, VT 05405, USA

Jiří Dušek
University of South Bohemia, Faculty of Biological Sciences, Branišovská 31,
370 05 české Budějovice, Czech Republic

Phil M. Fermor
Middlemarch Environmental Ltd, Triumph House, Birmingham Road, Allesley,
Coventry CV5 9AZ, United Kingdom

Masanori Fujita
Deanery, Kochi National College of Technology, 200-1 Monobe Otsu,
Namgoku, Kochi 783-8508, Japan

Alenka Gaberščik
Department of Biology, Biotechnical Faculty, University of Ljubljana,
Večna pot 111, Ljubljana, Slovenia

Joan García
Environmental Engineering Division; Hydraulics, Maritime and Environmental
Engineering Department; Technical University of Catalonia, Jordi Girona, 1-3,
08034-Barcelona, Spain

Mateja Germ
National Institute of Biology, Večna pot 111, 1000 Ljubljana, Slovenia

Andreas Graber
University of Applied Sciences Waedenswil, Institute of Natural Resource
Sciences, Section Ecological Engineering, Gruental, CH - 8820 Waedenswil,
Switzerland

Paul Griffin
Severn Trent Water Ltd., Technology and Development, Avon House, Coventry,
CV3 6PR, United Kingdom

Raimund Haberl
Institute of Sanitary Engineering and Water Pollution Control, University of
Natural Resources and Applied Life Sciences, Vienna, Muthgasse 18, A-1190
Vienna (BOKU), Austria

Peter D. Hedges
School of Engineering and Applied Science, Aston University, Aston Triangle,
Birmingham, B4 7ET, United Kingdom

Brigita Horvat
Department of Biology, Biotechnical Faculty, University of Ljubljana, Večna pot
111, Ljubljana, Slovenia

Michihiko Ike
Deptartment of Environmental Engineering, Graduate School of Engineering,
Osaka University, 2-1 Yamadaoka, Suita, Osaka 565-0871, Japan

Ranka Junge
University of Applied Sciences Waedenswil, Institute of Natural Resource
Sciences, Section Ecological Engineering, Gruental, CH - 8820 Waedenswil,
Switzerland

Pille Kängsepp
Institute of Molecular and Cell Biology, University of Tartu, Riia 23, Tartu,
51010, Estonia; School of Pure and Applied Natural Sciences Kalmar University,
Kalmar 39182, Sweden

Suwasa Kantawanichkul
Department of Environmental Engineering, Chiang Mai University, Chiang Mai
50202, Thailand

Marjut Karsisto
Finnish Forest Research Institute, Vantaa Research Unit, Finland

Veikko Kitunen
Finnish Forest Research Institute, Vantaa Research Unit, Finland

Margit Kõiv
Institute of Geography, University of Tartu, Vanemuise 46, Tartu, 51014, Estonia

Mait Kriipsalu
Institute of Forestry & Rural Engineering, Estonian University of Life Sciences,
Kreutzwaldi 64, Tartu, 51014, Estonia

Lenka Kröpfelová
ENKI, o.p.s., Dukelská 145, 379 01 Třeboň, Czech Republic

Raija Laiho
Peatland Ecology Group, Department of Forest Ecology, University of Helsinki,
Finland

Gijs Du Laing
Laboratory for Analytical Chemistry and Applied Ecochemistry, Department of
Applied Analytical and Physical Chemistry, Ghent University, Coupure Links
653, B-9000 Gent, Belgium

Günter Langergraber
Institute of Sanitary Engineering and Water Pollution Control, University of
Natural Resources and Applied Life Sciences, Vienna, Muthgasse 18, A-1190
Vienna (BOKU), Austria

Klaus Leroch
ÖKOREAL GmbH, Carl Reichert-Gasse 28, A-1170 Vienna, Austria

Els Lesage
Laboratory for Analytical Chemistry and Applied Ecochemistry, Department of
Applied Analytical and Physical Chemistry, Ghent University, Coupure Links
653, B-9000 Ghent, Belgium

Ülo Mander
Institute of Geography, University of Tartu, Vanemuise 46, Tartu, 51014, Estonia

Fabio Masi
IRIDRA Srl, via Lorenzo il Magnifico 70, Florence, 50129, Italy

Troy Meyers
Mathematics Department, Luther College, Decorah, Iowa, USA

Kari Minkkinen
Peatland Ecology Group, University of Helsinki, Department of Forest Ecology,
Helsinki, Finland

Annelies M.K. van de Moortel
Laboratory of Analytical Chemistry and Applied Ecochemistry, Department of
Applied Analytical and Physical Chemistry, Ghent University, Coupure Links
653, B-9000 Ghent, Belgium

Charity Mundia
Laboratory of Analytical Chemistry and Applied Ecochemistry, Ghent University,
Coupure Links 653, 9000 Ghent, Belgium

Jaime Nivala
North American Wetland Engineering LLC, 4444 Centerville Road, Suite 140,
White Bear Lake, Minnesota 55127, USA

Hanna Obarska-Pempkowiak
Gdansk University of Technology, Faculty of Civil and Environmental
Engineering, Narutowicza 11/12, 80-952 Gdansk, Poland

Sean O'Hogain
School of Civil, Structural and Building Services Engineering, Dublin Institute
of Technology, Bolton Street, Dublin 1, Ireland

Amanda Parker
U.S. Environmental Protection Agency, Washington, DC, USA

Niels De Pauw
Laboratory of Environmental Toxicology and Aquatic Ecology, Ghent University,
J. Plateaustraat 22, 9000 Ghent, Belgium

Libor Pechar
Laboratory of Applied Ecology, Faculty of Agriculture, University of South
Bohemia, Studentská 13, České Budějovice, CZ-370 05, Czech Republic;
ENKI o.p.s., Dukelská 145, Třeboň, CZ-379 01, Czech Republic;
Institute of System Biology and Ecology, Academy of Sciences of the Czech
Republic, Dukelská 145, Třeboň, CZ-379 01, Czech Republic

Timo Penttilä
Finnish Forest Research Institute, Vantaa Research Unit, Vantaa, Finland

Silvana Perdomo
Limnosistemas, Avda. Costanera Mz 205 S2, El Pinar, Canelones, Uruguay

Kiattisak Pingkul
Department of Environmental Engineering, Chiang Mai University, Chiang Mai
50202, Thailand

Jan Pokorný
ENKI o.p.s., Dukelská 145, Třeboň, CZ-379 01, Czech Republic;
Institute of System Biology and Ecology, Academy of Sciences of the Czech
Republic, Dukelská 145, Třeboň, CZ-379 01, Czech Republic

Christoph Prandtstetten
ÖKOREAL GmbH, Carl Reichert-Gasse 28, A-1170 Vienna, Austria

Alexander Pressl
Institute of Sanitary Engineering and Water Pollution Control, University of
Natural Resources and Applied Life Sciences, Vienna, Muthgasse 18, A-1190
Vienna (BOKU), Austria

Jan Procházka
Laboratory of Applied Ecology, Faculty of Agriculture, University of South
Bohemia, Studentská 13, České Budějovice, CZ-370 05, Czech Republic

Jaume Puigagut
Environmental Engineering Division; Hydraulics, Maritime and Environmental
Engineering Department; Technical University of Catalonia, Jordi Girona, 1-3,
08034-Barcelona, Spain

Katy E. Read
Middlemarch Environmental Ltd, Triumph House, Birmingham Road, Allesley,
Coventry CV5 9AZ, United Kingdom

Roland Rohrhofer
ÖKOREAL GmbH, Carl Reichert-Gasse 28, A-1170 Vienna, Austria

Donald Ross
University of Vermont, Department of Plant and Soil Science, Hills Agricultural
Building, 105 Carrigan Drive, Burlington, VT 05405, USA

Diederik P.L. Rousseau
Department of Environmental Resources, UNESCO-IHE, P.O.Box 3015, 2601
DA Delft, The Netherlands

Eric Seitz
University of Vermont, Department of Plant and Soil Science, Hills Agricultural
Building, 105 Carrigan Drive, Burlington, VT 05405, USA

Kirsten Sleytr
Institute of Sanitary Engineering and Water Pollution Control, University of
Natural Resources and Applied Life Sciences, Vienna, Muthgasse 18, A-1190
Vienna (BOKU), Austria

Nina Šraj-Kržič
Department of Biology, Biotechnical Faculty, University of Ljubljana, Večna pot
111, 1000 Ljubljana, Slovenia

Otto R. Stein
Center for Biofilm Engineering, Montana State University, Bozeman, MT 59717,
USA; Department of Civil Engineering, Montana State University, Bozeman,
MT 59717, USA

Bo Stenberg
Department of Soil Sciences, SLU, Skara, Sweden

Jana Štíchová
Department of Applied Chemistry and Chemistry Teaching, Faculty of
Agriculture, University of South Bohemia, Studentská 13, České Budějovice,
CZ-370 05, Czech Republic

Paul J. Sturman
Center for Biofilm Engineering, Montana State University, Bozeman, MT 59717,
USA

Christer Svedin
IFM-Biology, Linköping University, SE-581 83 Linköping, Sweden

Filip M.G. Tack
Laboratory for Analytical Chemistry and Applied Ecochemistry, Department of
Applied Analytical and Physical Chemistry, Ghent University, Coupure Links
653, B-9000 Ghent, Belgium

Tarja Tapanila
Finnish Forest Research Institute, Vantaa Research Centre, Finland

Masafumi Tateda
Department of Environmental Technology, College of Technology, Toyama
Prefectural University, 5180 Kurokawa, Kosugi-machi, Imizu-Gun, Toyama, Japan

Karin S. Tonderski
IFM-Biology, Linköping University, SE-581 83 Linköping, Sweden

Eamon Twohig
University of Vermont, Department of Plant and Soil Science, Hills Agricultural
Building, 105 Carrigan Drive, Burlington, VT 05405, USA

Olga Urbanc-Berčič
National Institute of Biology, Večna pot 111, 1000 Ljubljana, Slovenia

Nuria Vaello
Environmental Engineering Division; Hydraulics, Maritime and Environmental
Engineering Department; Technical University of Catalonia, Jordi Girona, 1-3,
08034-Barcelona, Spain

Petra Vávřová
Peatland Ecology Group, University of Helsinki, Department of Forest Ecology,
Helsinki, Finland; Finnish Forest Research Institute, Vantaa Research Unit,
Vantaa, Finland

Marc G. Verloo
Laboratory of Analytical Chemistry and Applied Ecochemistry, Department of
Applied Analytical and Physical Chemistry, Ghent University, Coupure Links
653, B-9000 Ghent, Belgium

Jan Vymazal
ENKI, o.p.s., Dukelská 145, 379 82 Třeboň, Czech Republic; Institute of Systems
Biology and Ecology, Dukelská 145, 379 01 Třeboň, Czech Republic

Scott Wallace
North American Wetland Engineering LLC, 4444 Centerville Road, Suite 140,
White Bear Lake, Minnesota 55127, USA

Chad Washburn
School of Public and Environmental Affairs, Indiana University, Bloomington IN
47405-1701, USA

Michael Thomas Waters
SMEC International, P.O. Box 1052, North Sydney, NSW 2060 Australia

David Weber
Vermont Agency of Agriculture Food & Markets. 116 State Street, Drawer 20
Montpelier, VT 05620–2901, USA

Ewa Wojciechowska
Gdansk University of Technology, Faculty of Civil and Environmental
Engineering, Narutowicza 11/12, 80-952 Gdansk, Poland

Chapter 1
Reed Stand Conditions at Selected Wetlands in Slovenia and Hungary

Mária Dinka[1], Edit Ágoston-Szabó[1], Olga Urbanc-Berčič[2], Mateja Germ[2], Nina Šraj-Kržič[3], and Alenka Gaberščik[3](✉)

Abstract We determined the characteristics of reed stands at an intermittent lake in Slovenia and degraded and vital reed stands in Hungary. The disturbance in reed performance was measured through growth analysis, amino acid analysis in basal culm internodes, and photochemical efficiency of photosystem II (PSII) in leaves. Morphological parameters indicated higher disturbance in the development of degraded and intermittent reed stands in comparison to vital reed stands. Similarly, total free amino acid contents in basal culm internodes reflected temporary stress response in degraded and intermittent reed stands. On the other hand, potential photochemical efficiency showed undisturbed energy harvesting of all reed stands, even though actual photochemical efficiency revealed temporary disturbance of PSII. The most unfavourable condition for reed development seems to be degraded reed stand of Kis-Balaton wetland and littoral reed stand of intermittent Lake Cerknica.

Keywords Free amino acids, reed biometry, photochemical efficiency of PSII, *Phragmites australis*

1.1 Introduction

Phragmites australis (Cav.) Trin. ex Steud. (common reed) is the most widely distributed angiosperm, characteristic species of the ecotone between terrestrial and aquatic environments in freshwater to brackish ecosystems (van der Putten, 1997;

[1] Institute of Ecology and Botany, Hungarian Danube Research Station, H-2163 Vácrátót, Hungary

[2] National Institute of Biology, Večna pot 111, 1000 Ljubljana, Slovenia

[3] Department of Biology, Biotechnical Faculty, University of Ljubljana, Večna pot 111, 1000 Ljubljana, Slovenia

(✉) Corresponding author: e-mail: alenka.gaberscik@bf.uni-lj.si

J. Vymazal (ed.) *Wastewater Treatment, Plant Dynamics and Management in Constructed and Natural Wetlands*,
© Springer Science + Business Media B.V. 2008

Cronk & Fennessy, 2001; Mauchamp & Méthy, 2004). *P. australis* may be temporarily exposed to complete submersion or to drought ranging from few days to several months (Mauchamp & Méthy, 2004). It acclimatises to deep water and water deficit with phenotypic plasticity (Vretare *et al.*, 2001; Pagter *et al.*, 2005). Deep water may affect the performance of *P. australis* by constraining oxygen supply to the below-ground parts of the plant (White & Ganf, 2002). Under such conditions, reed allocates more assimilates to stem weight, and produces fewer but taller stems, maintaining positive carbon balance (Dinka & Szeglet, 1999) and effective gas exchange between emerged and below-ground parts (Vretare *et al.*, 2001).

Despite high functional plasticity of *P. australis*, reed stands throughout Europe experienced severe decline in last decades (Ostendorp, 1989). Previous studies have shown that different environmental factors may contribute to the decreasing vitality of the reed stands (Ostendorp, 1989; van der Putten, 1997): changes in water level (Dienst *et al.*, 2004), reduced oxygen supply to roots and rhizomes (Armstrong & Armstrong, 1990; Brix *et al.*, 1992), internal eutrophication (e.g. high ammonium concentration), etc. These stress factors affect metabolic pool of whole plant, which may be reflected by changes in amino acid patterns in basal culm internodes (Haldemann & Brändle, 1988; Kohl *et al.*, 1998; Rolletschek *et al.*, 1999; Koppitz, 2004). Plants subjected to stress often show accumulation of specific free amino acids and/or reduced protein synthesis (Marschner, 1995; Rabe, 1990; Smolders *et al.*, 2000; Koppitz, 2004), and decreased photochemical efficiency of PSII due to photo-inhibition (Schrieber *et al.*, 1995).

The aim of this study was to determine the characteristics of selected reed stands in Slovenia and Hungary. Localities differ in vitality of reed stands and to a great extent in water regimes. We hypothesised that different reed stands will experience different levels of disturbance, as measured through growth analysis, amino acid analysis, and photochemical efficiency. We assumed that reed stands of the intermittent lake in Slovenia and degraded reed stands in Hungary will be more disturbed in comparison to vital reed stands in Hungary.

1.2   Methods

1.2.1   Area Description

The survey of reed stand conditions was performed at selected wetlands of Slovenia (Lake Cerknica) and Hungary (Lake Fertő and Kis-Balaton wetland of Lake Balaton) in growth periods 2004 and 2005.

Lake Cerknica is *locus typicus* for intermittent lakes, appearing at the bottom of the karstic valley of Cerkniško polje (38 km^2). Due to floods in spring and autumn, the valley changes into a lake (20–25 km^2). Floods last on average 260 days a year and the dry period usually starts in late spring (Krajnc, 2002). The lake was designated for the Ramsar List in 2006.

Table 1.1 Reed stands characteristics at Lake Cerknica (Slovenia), and Lake Fertő and Kis-Balaton wetland (Hungary), surveyed in 2004 and 2005

Lake	Location		Characteristics
Cerknica, SLO	Zadnji Kraj 1	CE 1	Littoral reed stands, nutrient-poor, variable water regime (0–2.5 m throughout a year)
45°45'N, 14°20'E	Zadnji Kraj 2	CE 2	
	Gorenje jezero	CE 3	Ecotonal reed, variable water regime, but efficient water supply
Fertő, HU	Fertőrákos	FE 1	Homogeneous, vital reed stand in shallow water (0–0.3 m)
(Neusiedler See)	Nádas 3	FE 3	Clumped distribution, loose, degraded reed stand (0.3–0.5 m)
47°42'N, 16°46'E	Herlakni 5	FE 5	Homogeneous, loose, vital reed stand in deep water (0.8–1.2 m)
Kis-Balaton, HU	Ingói berek 1	KB 1	Vital reed stand in deep water (0.5–0.8 m)
46°50'N 17°44'E	Ingói berek 2	KB 2	Degraded reed stand in shallow water (0.3–0.5 m)

Lake Fertő (Neusiedler See) is the largest sodic lake in Europe (309 km²), declared as a biosphere reserve by UNESCO in 1977/79. It is a eutrophic steppe lake, situated on the Hungarian–Austrian border (Löffler, 1979). The water is permanent, but extremely shallow (mean depth 1.1 m, maximal depth 1.8 m), with regulated outflow. As a consequence of shallowness, 54% of the whole lake and 85% of the Hungarian part is covered by reed.

Kis-Balaton is 81 km² large wetland, located SW of Lake Balaton (594 km²). Large parts were drained due to agriculture in the beginning of the 20th century. Later the re-establishment of the Kis-Balaton wetland was implemented. The extended area was given the classification of Landscape Protected Area, and was designated for the Ramsar List in 1989.

All three wetlands are dominated by reed stands. Different sampling sites were selected with respect to nutrient conditions, water regime, and reed vitality (Table 1.1). Hungarian locations were nutrient-rich and with permanent water (Dinka, 1993; Pomogyi, 1993; Tátrai et al., 2000; Dinka et al., 2004), while Slovenian locations were nutrient-poor and with variable water regime (Šraj-Kržič et al., 2006). Growth seasons 2004 and 2005 differed with regard to precipitation pattern and consequently water regime (Fig. 1.1).

1.2.2 Growth Analyses

Shoot density was measured within four squares (0.25 m²). Randomly harvested shoots (n = 8–12) were used for measurements of shoot height, shoot diameter, shoot dry mass, and specific leaf area (Dykyjová et al., 1973; Květ, 1971). The dry weight of samples was estimated after 24 h of drying at 105°C (Sterimatic ST-11, Instrumentaria, Zagreb). The leaf area was measured using area meter (Delta-T

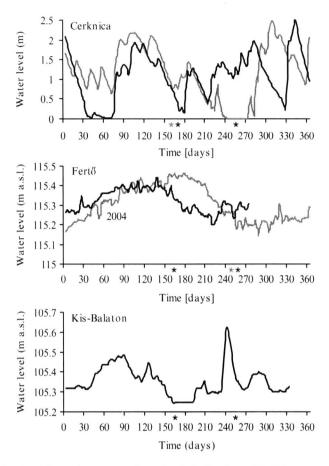

Fig. 1.1 Water level fluctuations at Lake Cerknica, Lake Fertő, and Kis-Balaton wetland in 2004 (—) and 2005 (—). Asterisks indicate sampling time in 2004 (*) and 2005 (*)

Devices Ltd., Cambridge, England). Specific leaf area was calculated as the ratio between leaf area and leaf dry weight (cm^2 g^{-1}).

1.2.3 Analysis of Amino Acids

For the analysis of amino acids in basal culm internodes of randomly harvested primary culms ($n = 3$–6) we followed the method of Koppitz (2004). Samples were frozen in liquid N$_2$, transferred to the laboratory, stored ($-20°$C), pulverised under liquid N$_2$, and divided into two subsamples. Powdered samples (250 mg) were extracted three times with 3 ml of ethanol (80% v/v) at room temperature. Combined fractions were sonicated in an ultrasound bath (10 min), evaporated under liquid N$_2$, and the remaining moisture eliminated by freeze-drying. Dry samples were dissolved in 1 ml of ethanol (80% v/v). Amino acids were derivatised with 9-fluorenylmethoxycarbonyl chloride/1-aminoadamantane (FMOC/ADAM), detected using high performance liquid chromatography (HPLC) (thermo Separation P200 as pump, gradient elution, GromSil 250 × 4 mm column) and UV150 detector at 263 nm, and separated with Na-acetate buffer and acetonitrile/water. Standard mixture of 20 amino acids was used for identification and quantification of samples. The contents of amino acids were calculated per dry weight (μmol/g).

1.2.4   Measurements of Photochemical Efficiency

Chlorophyll *a* (Chl *a*) fluorescence of PSII is an indicator of photosynthetic electron transport in intact leaves and therefore reflects changes in primary processes of photosynthesis (Schrieber *et al.*, 1995). To estimate the disturbance to the light harvesting of PSII we monitored Chl *a* fluorescence (modulated fluorometer OS-500, OPTI-SCIENCES, Tyngsboro, MA, USA). Measurements were carried out on fully developed leaves ($n = 5$–12) on clear days at noontime, when photosynthetic photon flux density (PPFD) exceeded 1,200 μmol m^{-2} s^{-1}. The potential photochemical efficiency (F_v/F_m) was determined after dark-acclimation (15 min) using saturating pulses of white light (PPFD $\approx$ 8,000 μmol m^{-2} s^{-1}, duration 0.8 s). Actual photochemical efficiency (Y) was measured under ambient light using saturating pulses of white light (PPFD $\approx$ 9,000 μmol m^{-2} s^{-1}, duration 0.8 s). It gives the information on energy conversion in PSII (Björkman & Demmig-Adams, 1995; Schrieber *et al.*, 1995).

1.2.5   Statistical Analyses

The significance of differences between sampling sites and sampling times was tested using the analysis of variance (one-way ANOVA) for parametrical data, and Mann–Whitney U test for non-parametrical data. Relationships between two parameters were tested using Spearman's rank-order correlation. Statistical analyses were preformed using SPSS for Windows 13.0.

1.3   Results

1.3.1   Growth Parameters

Low reed density was determined at Lake Cerknica (in average 60 reeds m^{-2}), degraded reed stand FE3 and deepwater, vital reed stand of FE5 at Lake Fertő (in average 20 and 85 shoots m^{-2}, respectively). Temporal changes in reed density were not determined. On the contrary, high density of vital reed stand FE1 declined significantly from 110 shoots m^{-2} in 2004 to 51 shoots m^{-2} in 2005. The density of degraded reed stand KB2 (ranging from 35 to 90 shoots m^{-2}) differed significantly from vital reed stand KB1 at Kis-Balaton wetland (ranging from 150 to 200 shoots m^{-2}). Reed stands of Lake Cerknica and degraded reed stands FE3 and KB2 had the lowest basal diameter (ranging from 3 to 7 mm), followed by Lake Fertő and Kis-Balaton vital reed stands (ranging from 7 to 11 mm).

Table 1.2 shows shoot height and dry mass and specific leaf area of reeds from Lake Cerknica, Lake Fertő, and Kis-Balaton wetland, measured in June and September 2005. Significantly smaller reeds with lower dry mass were character-istic of degraded reed stands of FE3 and KB2 compared to vital reed stands of

Table 1.2 Shoot height and dry mass and specific leaf area of reed stands at Lake Cerknica, Lake Fertő, and Kis-Balaton wetland, measured in 2005. Data represent arithmetic mean ± SD, n = 8–12. One-way ANOVA; letters indicate differences between sampling sites ($p \leq 0.05$), and asterisks indicate differences between sampling time

	June 2005				September 2005				
Height (cm)									
CE1	196	±	15	b	194	±	33	b	ns
CE2	224	±	14	a	238	±	51	a	ns
CE3	179	±	8	c	195	±	10	b	*
FE1	238	±	15	a	270	±	27	a	**
FE3	152	±	15	b	159	±	11	b	ns
FE5	256	±	39	a	235	±	25	a	ns
KB1	335	±	48	a	228	±	26	a	***
KB2	137	±	23	b	158	±	20	b	ns
Shoot dry mass (g)									
CE1	12	±	4	B	10	±	5	b	ns
CE2	19	±	6	A	18	±	6	a	ns
CE3	13	±	2	B	13	±	3	a	ns
FE1	30	±	7	A	51	±	15	a	***
FE3	8	±	2	C	11	±	2	c	**
FE5	22	±	7	B	23	±	8	b	ns
KB1	63	±	23	A	16	±	8	a	***
KB2	9	±	4	B	11	±	4	b	ns
Specific leaf area (cm^2 g^{-1})									
CE1	615	±	42	A	1,056	±	137	a	***
CE2	508	±	64	B	1,007	±	109	a	***
CE3	409	±	33	C	767	±	55	b	***
FE1	72	±	7	B	85	±	12	b	**
FE3	75	±	9	B	111	±	7	a	**
FE5	108	±	10	a	77	±	14	b	***
KB1	107	±	5		112	±	24		ns
KB2	121	±	21		104	±	16		ns

ns 'not significant', * $p \leq 0.05$, ** $p \leq 0.01$, *** $p \leq 0.001$

Lake Fertő and Kis-Balaton wetland. Reeds from Lake Cerknica were of intermediate height and dry mass, which did not differ significantly between June and September. Specific leaf area of reeds from Lake Cerknica and Lake Fertő increased in time significantly.

1.3.2   Free Amino Acid Content

The highest content of total amino acids in basal culm internodes (Fig. 1.2) was detected at CE1 and degraded reed stand KB2 (17–22 µmol g^{-1}), followed by other sampling sites (3.5–12 µmol g^{-1}). Principal amino acids at all sampling sites were alanine (Ala), arginine (Arg), asparagine (Asn), γ-amino-butyric acid (Gaba), glutamine (Gln), and serine (Ser). The remaining 14 amino acids were presented as "other amino acids". The accumulation of Ala+Gaba+Ser ranged between 22% and 47% in reeds of Lake Cerknica and Lake Fertő. The percentage increased significantly from June to September in reeds of Kis-Balaton wetland (increase from 14% to 38%) and Lake Cerknica (increase from 31% to 43%). Additionally, high accumulation of Arg+Asn+Gln was detected at all sampling sites. The percentage declined significantly from spring to autumn in reeds of Lake Cerknica (decline from 25–50% to 12–27%) and Kis-Balaton wetland (decline from 57% to 25%), while relatively constant values were characteristic of reed stands at Lake Fertő (23–50%).

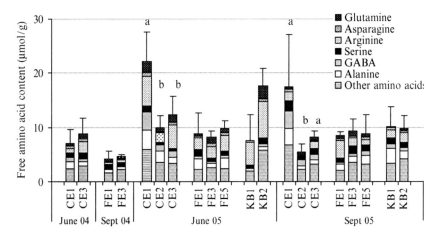

Fig. 1.2   Free amino acids in basal culm internodes in reeds at Lake Cerknica (CE), Lake Fertő (FE), and Kis-Balaton wetland (KB), sampled in 2004 and 2005. Data represent arithmetic mean ± SD, n = 3–6. Mann–Whitney U test; letters indicate differences between sampling sites ($p \leq 0.05$)

1.3.3 Photochemical Efficiency

The potential (F_v/F_m) and actual (Y) photochemical efficiency of PSII of reed stands of Lake Cerknica, Lake Fertő, and Kis-Balaton wetland are presented in Fig. 1.3. F_v/F_m was close to the value 0.8 in reeds of all the lakes. Significant changes in F_v/F_m in time were calculated ($p \leq 0.05$), with the highest values in June 2005. Y ranged between 0.3 and 0.5 throughout both seasons 2004 and 2005. There were no major differences between locations in the Lake Fertő, while locations at Lake Cerknica and Kis-Balaton wetland differed significantly. Reed stands of Lake Cerknica and degraded reed stands FE3 and KB2 showed notable decline in Y from June to September ($p \leq 0.05$). Spearman's rank-order correlation did not show significant relationship between photochemical efficiency and total amino acid content.

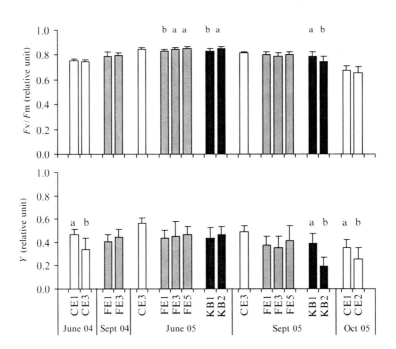

Fig. 1.3 Potential (F_v/F_m) and actual photochemical efficiency of photosystem II (Y) of reeds at Lake Cerknica (CE), Lake Fertő (FE), and Kis-Balaton wetland (KB), measured in 2004 and 2005. Data represent arithmetic mean ± SD, n = 5–12. One-way ANOVA; letters indicate differences between sampling sites ($p \leq 0.05$)

1.4 Discussion

Our study revealed some characteristics of vital and degraded reed stands of wetlands with permanent water regime (Hungary) and reed stands of intermittent wetlands (Slovenia).

Reed stands differed in morphological characteristics (Table 1.2), which might be attributed to the differences in environmental conditions (Dienst *et al.*, 2004; Brix *et al.*, 1992). Vital reed stands of Lake Fertő (FE1) and Kis-Balaton wetland (KB1) were denser, with better developed shoots than degraded, as already reported in the case of reeds from Lake Fertő (Dinka & Szeglet, 2001). The density of reed stands at Lake Fertő was decreasing, as also evident from the long-term database (Dinka, 2006). At vital reed stands of Lake Fertő, shoots were well developed, while at the degraded site shoot height, dry mass, and basal diameter revealed weaker reeds. We presume that plants were affected by low water level due to increased dissolved solids (conductivity up to $4,000\,\mu S\ cm^{-1}$). Reed stands from intermittent Lake Cerknica showed intermediate growth characteristics. Despite low density, shoots were relatively well developed, which reveals great functional plasticity of *P. australis* under variable water regime (Vretare *et al.*, 2001; White & Ganf, 2002; Gaberščik *et al.*, 2003).

Similarly the analysis of free amino acids in basal culm internodes (Fig. 1.2) revealed the presence of disturbance in some reed stands (CE1 and KB2). It is widely accepted that stress induces the production of free amino acids (Gzik, 1996; Šircelj *et al.*, 1999; Hartzendorf & Rolletschek, 2001; Koppitz, 2004), which reflect the conditions during the growth period. In the intermittent Lake Cerknica the growth period in 2005 was outstanding, since water level was relatively high during the whole summer. Consequently, plants revealed significantly higher total amino acid content, which could be the result of the oxygen shortage in the soil (Koppitz, 2004). In basal culm internodes at littoral reed CE1 we determined the highest content of total free amino acids due to large fractions of Ala+Gaba+Ser, which were also recorded at other sampling sites. Ala, Gaba, and Ser are reported as indicators of hypoxia and anaerobic metabolism (Haldemann & Brändle, 1988; Kohl *et al.*, 1998; Rolletschek *et al.*, 1999; Sánchez *et al.*, 1998; Koppitz *et al.*, 2004). In reeds of all sampling sites also relatively high fractions of Arg+Asn+Gln were detected, which indicated high NH_4^+ concentration in the environment, resulting in inhibition of protein synthesis (Smolders *et al.*, 2000). Asn is the main storage and transport compound of the intermediate N metabolism in *P. australis*. Therefore, the synthesis of specific N-efficient soluble amino acids like Asn and Arg prevents the accumulation of toxic free ammonium in the cells (Haldemann & Brändle, 1988; Rolletschek *et al.*, 1999).

Besides the content of free amino acids in basal culm internodes, the photochemical efficiency of PSII in leaves also gives an insight in plant performance under stress. The potential photochemical efficiency of PSII (F_v/F_m) of unstressed leaves of many species and ecotypes ranges from 0.80 to 0.83 (Schrieber *et al.*, 1995). The measurements of F_v/F_m of reed showed that the majority of plants exhibited normal energy harvesting, with F_v/F_m close to the optimal value. That indicated good physiological status of the reeds, which was also found by Mészáros *et al.* (2003). Decreases in F_v/F_m in intermittent Lake Cerknica in October 2005 indicated that PSII reaction centres had been damaged. This could be due to high water level through the season which suppresses oxidative processes in the reed roots (White & Ganf, 2002), or late season, when the senescence starts. Mauchamp & Méthy (2004) reported that F_v/F_m of reed was affected by submergence and exhibited varying recovery levels depending on duration and degree of submergence. Actual photochemical efficiency of PSII (Y) was generally lower than F_v/F_m, which was due to temporary stress during the midday depression. The effects of short-term photoinhibition were found to be reversible (Mauchamp & Méthy, 2004; Šraj-Kržič & Gaberščik, 2005). F_v/F_m and Y of degraded reed stands of Lake Fertő and Kis-Balaton wetland declined significantly from June to September, which reflected the temporary disturbance in the functioning of PSII. Similarly, Mészáros *et al.* (2003) reported decline in F_v/F_m in degraded reed stands of Lake Fertő.

1.5 Conclusions

This study revealed some functional characteristics of different reed stands (Lake Cerknica, Lake Fertő, and Kis-Balaton wetland). Biometric parameters indicated that degraded (FE1 and KB2) and intermittent reed stands (CE) were more disturbed in their development than vital reed stands. Similarly, total free amino acid contents reflected temporary stress response in some sampling sites (CE1 and KB2). Photochemical efficiency showed normal energy harvesting of all reed stands throughout the season. The most unfavourable condition for reed development seems to be reed stands of intermittent Lake Cerknica (littoral reed stand CE1) and Kis-Balaton wetland (degraded reed stand KB2).

Acknowledgements This research was financed by the National Office for Research and Technology, Hungary, and by the Ministry of Education, Science and Sport, Republic of Slovenia, through the bilateral project (OMFB-00455/2005) and OM-00371/2002 project. The authors thank Dr. P. Pomogyi for valuable suggestions and G. Horváth, M. Rudolf, and the personnel of the West-Transdanubian Environmental & Water Directorate (Keszthely), the Fertő -Hanság National Park Directorate, and the Fertőrákos Hydrometeorological Station for fieldwork assistance.

References

Armstrong, J., & Armstrong, W. (1990). Light-enhanced convective through flow increases oxygenation in rhizomes and rhizosphere of *Phragmites australis* (Cav.) Trin. Ex. Steud. *New Phytologist, 114*, 121–128.

Björkman O. & Demmig-Adams B. (1995). Regulation of phostosynthetic light energy capture, conversion, and dissipation in leaves of higher plants. In: E.D. Shulze & M.M. Caldwell, (Eds.) *Ecophysiology Photosynthesis.* (pp. 17–48). Berlin, Heidelberg, New York: Springer.

Brix, H., Sorell, B.K., & Orr, P.T. (1992). Internal pressurisation and convective gas flow in some emergent macrophytes. *Limnology and Oceanography, 37*, 1420–1433.

Cronk, J.K., & Fennessy, M.S. (2001). *Wetland plants – biology and ecology*. Boca Raton, FL: Lewis.

Dienst, M., Schmieder, K., & Ostendorp, W. (2004). Effects of water level variations on the dynamics of the reed belts of Lake Constance. *Limnologica, 34*, 29–36.

Dinka, M., (1993). Über die regionalen wasserchemischen Verschiedenheiten des ungarischen Seeteiles im Neusiedler See. *Biologisches Forschungsinstitut für Burgenland Berich, 79*, 31–39.

Dinka, M., & Szeglet, P. (1999). Carbohydrate and nutrient content in rhizomes of *Phragmites australis* from different habitats of Lake Fertő/Neusiedler See. *Limnologica, 29*, 47–59.

Dinka, M., & Szeglet, P. (2001). Some characteristics of reed (*Phragmites australis* /Cav./Trin ex Steudel) that indicate different health between vigorous and die-back stands. *Verhandlungen. Internationale Vereinigung Limnologie, 27*, 3364–3369.

Dinka, M., Ágoston-Szabó, E., Berczik, Á., & Kutrucz, G. (2004). Influence of water level fluctuation on the spatial dynamic of the water chemistry at Lake Fertő/Neusiedler See. *Limnologica, 34*, 48–56.

Dykyjová, D., Hejny, S., & Květ, J. (1973). Proposal for international comparative investigations of production by stands of reed (*Phragmites communis*). *Folia Geobotanica Phytotaxa (Praha), 8*, 435–442.

Gaberščik, A., Urbanc-Berčič, O., Kržič, N., Kosi, G., & Brancelj, A. (2003). The intermittent lake Cerknica: Various faces of the same ecosystem. *Lakes Reservoirs: Research and Management, 8*, 159–168.

Gzik, A. (1996). Accumulation of proline and pattern of α-amino acids in sugar beet plants in response to osmotic, water and salt stress. *Environmental and Experimental Botany, 36*, 29–38.

Haldemann, C., & Brändle, R. (1988). Amino acid composition in rhizomes of wetland species in their natural habitat and under anoxia. *Flora, 180*, 407–411.

Hartzendorf, T., & Rolletschek, H. (2001). Effects of NaCl-salinity on amino acid and carbohydrate contents of *Phragmites australis*. *Aquatic Botany, 69*, 195–208.

Kohl, J.G., Woitke, P., Kühl, H., Dewender, M., & König, G. (1998). Seasonal changes in dissolved amino acids and sugars in basal culm internodes as physiological indicators of the C/N-balance of at littoral sites of different trophic status. *Aquatic Botany, 60*, 221–240.

Koppitz, H. (2004). Effects of flooding on the amino acid and carbohydrate patterns of *Phragmites Australis*. *Limnologica, 34*, 37–47.

Koppitz, H., Dewender, M., Ostendorp, W., & Schmieder, K. (2004). Amino acids as indicators of physiological stress in common reed *Phragmites australis* affected by an extreme flood. *Aquatic Botany, 79*, 277–294.

Krajnc, A. (2002). Hidrološke značilnosti/Hydrology. In A. Gaberščik (Ed.), *Jezero, ki izginja – Monografija o Cerkniškem jezeru/The vanishing lake – Monograph on Lake Cerknica* (pp. 27–37). Ljubljana: Društvo ekologov Slovenije.

Květ, J. (1971). Growth analysis approach to the production ecology of reed swamp plant communities. *Hydrobiologia (Bucuresti), 12*, 15–40.

Löffler, H. (1979). *Neusiedler See: The limnology of a shallow lake in Central Europe* (pp. 1–543). The Hague, Boston, London: Funk Publishers.

Marschner, H. (1995). *The mineral nutrition of higher plants*. London: Academic Press.

Mauchamp, A., & Méthy, M. (2004). Submergence-induced damage of photosynthetic apparatus in *Phragmites australis*. *Environmental and Experimental Botany, 51*, 227–235.

Mészáros, I., Veres, S., Dinka, M., & Lakatos, G. (2003). Variations in leaf pigment content and photosynthetic activity of *Phragmites australis* in healthy and die-back reed stands of Lake Fertő/Neusiedlersee. *Hydrobiologia, 506–509*, 681–686.

Ostendorp, W. (1989). "Die-back" of reeds in Europe – a critical review of literature. *Aquatic Botany, 35*, 5–26.

Pagter, M., Baragato, C., & Brix, H. (2005). Tolerance and physiological responses of *Phragmites australis* to water deficit. *Aquatic Botany, 81*, 285–299.

Pomogyi, P. (1993). Nutrient retention by the Kis-Balaton water protection system. *Hydrobiologia*, *251*, 309–320.

Rabe, E. (1990). Stress physiology: The functional significance of the accumulation of nitrogen containing compounds. *Journal of Horticultural Science*, *65*, 231–243.

Rolletschek, H., Hartzendorf, T., Rolletschek, A., & Kohl, J.G. (1999). Biometric variation in *Phragmites australis* affecting convective ventilation and amino acid metabolism. *Aquatic Botany*, *64*, 291–302.

Sánchez, F.J., Manzanares, M., Andres, E.F., Tenorio, J.L., & Ayerbe, L. (1998). Turgor maintenance, osmotic adjustment and soluble sugar and proline accumulation in 49 pea cultivars in response to water stress. *Field Crops Research*, *59*, 225–235.

Schrieber, U., Bilger, W., & Neubauer, C. (1995). Chlorophyll fluorescence as a nonintrusive indicator for rapid assessment of in vivo photosynthesis. In E.D. Schulze & M.M. Caldwell (Eds.), *Ecophysiology of photosynthesis* (pp. 49–61). Berlin, Heidelberg, New York: Springer.

Šircelj, H., Batič, F., & Štampar, F. (1999). Effects of drought stress on pigment, ascorbic acid and free amino acid content in leaves of two apple tree cultivars. *Phyton*, *39*, 97–100.

Smolders, A.J.P., van Riel, M.C., & Roelofs, J.G.M. (2000). Accumulation of free amino acids as an early indication for physiological stress (nitrogen overload) due to elevated ammonium levels in vital *Stratiotes aloides* L. stands. *Archiv für Hydrobiologie*, *150*, 169–175.

Šraj-Kržič, N., & Gaberščik, A. (2005). Photochemical efficiency of amphibious plants in an intermittent lake. *Aquatic Botany*, *83*, 281–288.

Šraj-Kržič, N., Pongrac, P., Klemenc, M., Kladnik, A., Regvar, M., & Gaberščik, A. (2006). Mycorrhizal colonisation in plants from intermittent aquatic habitats. *Aquatic Botany*, *85*, 331–336.

Tátrai, I., Mátyás, K., Korponai, J., Paulovits, G., & Pomogyi, P. (2000). The role of the Kis-Balaton water protection system in the control of water quality of Lake Balaton. *Ecological Engineering*, *16*, 73–78.

van der Putten, W.H. (1997). Die-back of *Phragmites australis* in European wetlands: An overview of the European Research Programme on reed die-back and progression. *Aquatic Botany*, *59*, 263–275.

Vretare, V., Weisner, S.E.B., Strand, J.A., & Granéli, W. (2001). Phenotypic plasticity in *Phragmites australis* as a functional response to water depth. *Aquatic Botany*, *69*, 127–145.

White, S.D., & Ganf, G.G. (2002). A comparison of the morphology, gas space anatomy and potential for internal aeration in *Phragmites australis* under variable and static water regimes. *Aquatic Botany*, *73*, 115–127.

Chapter 2
Water Quality and Macrophyte Community Changes in the Komarnik Accumulation Lake (Slovenia)

Brigita Horvat[1], Olga Urbanc Berčič[2], and Alenka Gaberščik[1] (✉)

Abstract The Komarnik accumulation lake was built to retain high waters in the Pesnica valley. Nowadays it is used as an unfertilised fishpond. In order to estimate the human impacts we have monitored changes in macrophyte community for 5 years and changes in water chemistry during two vegetation periods. The values of chemical parameters indicated the input of nutrients and different ions entering the system through the run-off from the surrounding areas and by the tributary. At low water level during summer period, oxygen was lacking in the whole water column. The bottom and the water column of the lake were completely colonised by macrophytes comprising 17 species of different growth forms, among which *Trapa natans* and *Ceratophyllum demersum* prevailed. The Komarnik accumulation lake revealed to be a resilient system, since inter-annual changes of water level affected only species abundance and not species composition.

Keywords Accumulation lake, Komarnik lake, macrophytes, water chemistry

2.1 Introduction

Macrophytes are essential elements in the structure and function of freshwater ecosystems (Baattrup-Pedersen & Riis, 1999). As primary producers they play an important role in mineral transformation and cycling, presenting the link between sediment, water, and in some aspects also atmosphere (Cronin *et al.*, 2006). They provide habitat and shelter for numerous organisms whose condition is indicated

[1] Department of Biology, Biotechnical Faculty, University of Ljubljana, Večna pot 111, Ljubljana, Slovenia

[2] National Institute of Biology, Večna pot 111, Ljubljana, Slovenia

(✉) Corresponding author: e-mail: alenka.gaberscik@bf.uni-lj.si

J. Vymazal (ed.) *Wastewater Treatment, Plant Dynamics and Management in Constructed and Natural Wetlands*,
© Springer Science + Business Media B.V. 2008

indirectly by the condition of the macrophytes (Gaberščik, 1997). The deterioration of physical environment and the eutrophication of water bodies result in changes in macrophyte distribution, decline in macrophyte species richness, and greater abundance of more resistant species (Preston, 1995; Sand-Jensen *et al.*, 2000; Germ & Gaberščik, 2003).

After mass regulation of rivers in Europe in the previous century, different artificial water bodies and their surroundings have become an important refuge for numerous organisms. This is also the case in the Komarnik accumulation lake, built during the melioration of the Pesnica river in the 1960s. Many plant and animal species found there are listed on the Red List of endangered species. As a relatively small and diverse waterbody, the Komarnik lake could also serve as an indicator of short-term as well as long-term changes in the landscape due to different impacts, i.e. human activities and global changes, as it was established for some other waterbodies (Mckee *et al.*, 2002). We hypothesised that permanent human impacts (input of nutrients and drying) influence the biocoenosis and shape the macrophyte species composition and abundance in this shallow waterbody. For that reason we monitored changes in macrophyte community for 5 years and changes in water chemistry during two vegetation periods.

2.2   Material and Methods

2.2.1   Site Description

The Komarnik accumulation lake is one out of four lakes built along the Pesnica river. Its main purpose was to retain high waters (Fig. 2.1). Nowadays it is used as an unfertilised fishpond. It covers an area of 30 ha, and the maximum water depth is 2 m (Fliser *et al.*, 1985). The lake is surrounded by wetland vegetation (prevailing species are *Alisma* spp., *Caltha palustris, Carex* spp., *Equisetum palustre*, *Eleocharis palustris, Galium palustre, Glyceria maxima, Iris pseudacorus, Juncus effusus, Lycopus europaeus, Lysimachia vulgaris, Lythrum salicaria, Mentha aquatica, Phragmites australis, Polygonum amphibium, Rorippa amphibia, Schoenoplectus lacustris, Sparganium erectum, Typha latifolia*, and *Typha angustifolia*) transiting to the lowland forest (*Robori-Carpinetum* type) on the eastern side and to agricultural areas on the west. The accumulation is conditioned by the tributary Partinjski potok. The water is maintained at the same level by an overflow as long as the water supply from inflow Partinjski potok is higher than evapotranspiration from the lake. During drought the riverbed dries out and the water level decreases gradually; therefore, the water level during the peak season might vary a lot (Fig. 2.2a, b). In autumn (usually in October) when the vegetation period ends, the lake is usually dried out for approximately 2 weeks to enable fishermen to collect the fish.

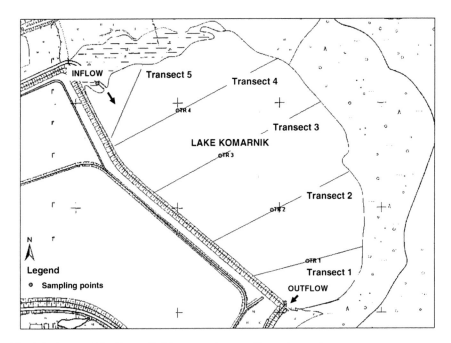

Fig. 2.1 A map of the Komarnik lake presenting sampling points and transects

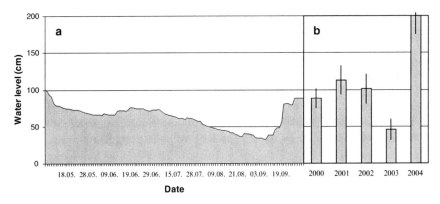

Fig. 2.2 (a) Water-level fluctuations in the Komarnik lake at the measuring station in the vegetation period 2001. (b) The average values of water level at transect 1 in the peak season from 2000 to 2004

2.2.2 Physical and Chemical Analyses

Water quality was monitored at six sampling points (inflow, outflow, and four sites in the lake) during the whole vegetation periods 2000 and 2001 (Fig. 2.1). The analysis of physical parameters (electric conductivity, temperature) and chemical parameters (pH, oxygen concentration, total phosphorus and nitrogen, silica dioxide, different ions, i.e. orthophosphate, nitrite, nitrate, ammonium, sodium, potassium, calcium, magnesium, chlorine and sulphate) was performed using standard methods (APHA, 1996).

2.2.3   Macrophyte Survey

The survey of macrophytes in the whole lake along five transects was carried out monthly during vegetation period 2001. The inter-annual changes from 2000 to 2004 were estimated on transect 1 only, which was the most representative of species composition. The distribution and abundance of macrophytes were assessed from a boat using a rake. The relative abundance was estimated on five belt transects, each divided into 20 m reaches, using a five-degree scale presenting mass index (MI) (Kohler & Janauer, 1995): 1 = very rare, 2 = infrequent, 3 = common, 4 = frequent, and 5 = abundant, predominant. For estimation of the quantitative significance of certain species at the time of sampling we calculated real biomass (MI^3). MI^3 is related to the mass index with a function $f(x) = x^3$, which is based on empirical data (Kohler & Janauer, 1995; Schneider & Melzer, 2003). Plants were identified using the following keys: Casper and Krausch (1980), Preston (1995), and Martinčič *et al.* (1999).

2.2.4   Statistical Analysis

The significance of differences in water chemistry parameters between inflow, lake, and outflow was tested. Analysis of variance was performed using one-way ANOVA for parametrical data and Mann–Whitney U test for non-parametrical data (SPSS for Windows 13.0). The inter-annual differences in macrophyte composition and abundance at transect 1 were analysed according to Bray–Curtis index.

2.3   Results

The extent of water level fluctuations in the Komarnik lake was quite high. In the vegetation period 2001 the water level fluctuated for about 70 cm. The lowest level was detected at the end of August (Fig. 2.2a). Measurements during peak season at transect 1 also revealed notable inter-annual fluctuations (Fig. 2.2b). Different chemical parameters revealed the input of nutrients and different ions entering the system with the run-off from the surrounding areas and by the tributary. The electric conductivity ranged from approximately 200 to 400 μS/cm, with the average in the lake being significantly lower than in the inflow. These hold true also for the concentrations of the some ions (Table 2.1) which exhibited great variability during vegetation period. Significant decrease of Ca^{2+} and K^+ was obtained during the intensive growth of macrophytes (data not shown), while there was constant supply at the inflow. Hypoxic conditions at the bottom were detected in early morning hours during the whole season (Fig. 2.3a). At low water level during summer period oxygen was lacking in the whole water column (Fig. 2.3b), which resulted in the presence of NH_4^+ (Table 2.1). Total N was rather low, the average concentration in the lake being 2.9 mg l^{-1}. Soluble P was not detected, while total P was variable, being significantly lower in the lake than in the inflow (Table 2.1).

The area of the lake was overgrown by macrophytes of different growth forms. Natant species covered the whole surface while the submersed species filled up the water column. In total we identified 17 species, among which *Trapa natans* and *Ceratophyllum demersum* prevailed. Monitoring of seasonal dynamic of species revealed that *C. demersum* reached its peak production in June, while *T. natans* increased its abundance until August (Fig. 2.4). Towards the end of the vegetation period favourable conditions for free-floating macrophytes of genus *Utricularia*, namely *Utricularia vulgaris*, and *U. australis* occurred. In 2003 and 2004 they had appeared already in July.

In spite of inter-annual water level variations high plant diversity in the Komarnik lake was detected. The abundance of species somehow differed between different seasons but the species composition was only slightly different. Species *Myriophyllum spicatum* and *Potamogenton crispus* were present in very low abundance. The former did not appear in the vegetation period 2001 and the latter was

Table 2.1 Concentrations of different ions, as well as total P and N in inflow, the Komarnik lake and outflow water during vegetation periods 2000 and 2001. Data are presented as average and standard deviation (SD). The significance of differences ($p \leq 0.05$) between inflow and other locations is indicated by asterisk (*), while between lake and other locations is indicated by dot (•)

Parameter	Unit	Inflow		Lake		Outflow	
		average	SD	average	SD	average	SD
pH		7.9	0.3	8.1*	0.4	7.8*	0.3
El. cond.	µS cm^{-1}	368	107	288*	70	339	61
Ca^{2+}	mg l^{-1}	28.2	8.8	25.1	9.0	29.0	7.3
Mg^{2+}	mg l^{-1}	11.7	2.9	12.2	3.4	12.2	3.1
Na^+	mg l^{-1}	8.8	2.9	6.7*	2.1	7.0•	1.7
K^+	mg l^{-1}	2.8	1.6	1.8	1.3	2.3	1.0
Cl^-	mg l^{-1}	10.1	2.5	9.6	3.0	10.3	3.4
SO_4^{2-}	mg l^{-1}	11.6	7.2	9.6	6.7	8.4	6.9
NO_3^-	mg l^{-1}	0.2	0.2	0.1	0.3	0.1	0.2
NO_2^-	mg l^{-1}	0.02	0.03	0.02	0.05	0.01	0.00
NH_4^+	mg l^{-1}	0.26	0.31	0.15	0.09	0.21	0.16
N total	mg l^{-1}	2.9	0.5	2.9	1.2	3.0	0.7
P total	µg l^{-1}	131	71	69*	30	113	78
SiO_2	mg l^{-1}	11.4	3.3	5.5*	3.1	7.5*	3.4
COD	mg l^{-1}	7.9	2.7	8.2	1.9	8.0	1.5

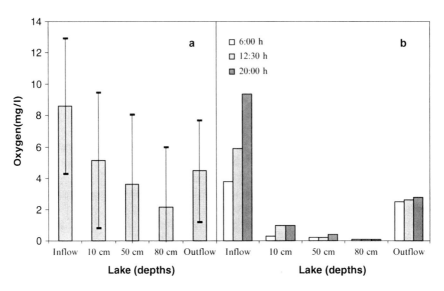

Fig. 2.3 (a) The average oxygen conditions in inflow, lake (on different depths), and outflow water in the vegetation periods 2000 and 2001. (b) The example of diurnal changes of oxygen concentration during peak season 2001

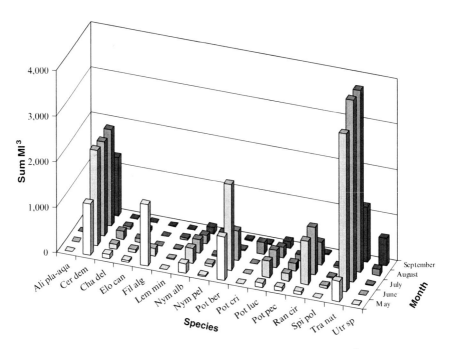

Fig. 2.4 The quantitative significance of single species expressed as sum of MI³ (real biomass) for single species in all transects in the vegetation period 2001. Ali pla-aqa = *Alisma plantago-aquatica*, Cer dem = *Ceratophyllum demersum*, Cha del = *Chara delicatula*, Elo can = *Elodea canadensis*, Fil alg = filamentous algae, Lem min = *Lemna minor*, Nym alb = *Nymphaea alba*, Nym pel = *Nymphoides peltata*, Pot ber = *Potamogeton bertholdii*, Pot cri = *P. crispus*, Pot luc = *P. lucens*, Pot pec = *P. pectinatus*, Ran cir = *Rananculs circinatus*, Spi pol = *Spirodela polyrhiza*, Tra nat = *Trapa natans*, Utr sp = *Utricularia* sp

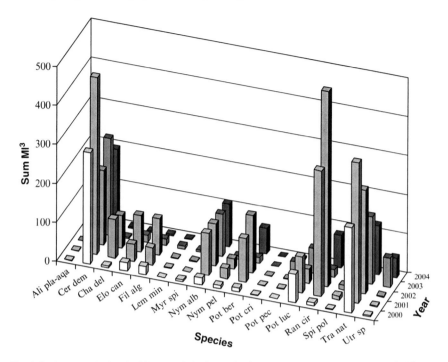

Fig. 2.5 The quantitative significance of single species in transect 1 expressed as a sum of MI³ (real biomass) from 2000 to 2004. Myr spi = *Myriophyllum spicatum*; for other abbreviations see Fig. 2.4

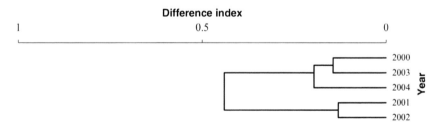

Fig. 2.6 The difference index of macrophyte species composition and abundance in transect 1 in different years

absent in the vegetation period 2004. During the extremely dry conditions in peak season 2003 the water level at transect 1 dropped to less than 50 cm and the surface of the fishpond decreased significantly. The monitoring revealed that this decrease reflected in lower plant species abundance in the following year (Fig. 2.5). The comparison of years revealed two clusters, with 2001 and 2002 in the first cluster, and others in a separate cluster (Fig. 2.6).

2.4   Discussion

The bottom and the water column of the Komarnik lake were completely colonised by macrophytes (Fig. 2.4). The water was transparent, since macrophytes maintain clear water by a variety of mechanisms, e.g. stabilising sediments and promoting zooplankton communities (Cronin *et al.*, 2006). Abundant population of filamentous algae appeared in spring only before macrophytes started their growth cycle (Fig. 2.5). The comparison of shallow lakes revealed that they can exhibit two possible states: a clear water state where the system is dominated by macrophytes and a turbid water state where the system is dominated by algae (Rip *et al.*, 2006). In the former state, macrophytes present a crucial element in the processes of energy through flow and matter cycling (Cronin *et al.*, 2006). Abundant macrophyte vegetation can also influence other organisms, e.g. increase the richness of zoobenthic community (Mastrantuono & Mancinelli, 1999). Some researches also pointed out several mechanisms involved in the impacts of macrophytes on the planktonic food web (van Donk & van de Bund, 2002).

From our study it is evident that macrophytes contributed a lot to the nutrient balance in the lake, since the majority of measured ions, as well as total P and N, were lower in the lake in comparison to inflow and outflow. This is in accordance with the results of other authors, claiming that a large proportion of metabolism occurs in macrophyte beds (Piezynska, 1993; Gessner, 2000; Marion & Paillisson, 2003). Organisms in densely vegetated lakes often experience oxygen limitations, which was also a case in the Komarnik lake, where the oxygen concentrations were low during the whole vegetation period. The presence of *U. vulgaris* and *U. australis* reflected the decrease in nutrient concentrations in lake water during intensive macrophyte growth. As a carnivorous species they could be competitively successful when nutrients are exhausted from the water column (Ulanowicz, 1995).

The highest abundance of macrophytes in the Komarnik lake occurred from the end of June to the end of August. The maximum production of *T. natans*, *Nymphoides peltata*, and *Nymphaea alba* coincided with the production of these species in shallow lakes in western France, where the former two reached the highest abundance in July and August, while *N. alba* exhibited the highest biomass at the beginning of the vegetation period (Marion & Paillisson, 2003).

Changes of water level in the Komarnik lake affected mainly species abundance and not species composition as it was also reported in other studies (Urbanc-Berčič *et al.*, 2005). The results of Riis and Hawes (2002), who studied macrophytes in shallow New Zealand lakes, showed that species richness in the lakes with interannual water level fluctuations is lower than in those with intra-annual changes. Pallisson and Marion (2006) found out that only small deviations in spring water level might control the above-ground biomass of *N. alba*. In our study, the abundance remained the same in 5 subsequent years, because the decrease in water level usually occurred in the middle of the growth season. Long-term monitoring of the dynamic of macrophyte populations in shallow eutrophic lakes revealed considerable changes in vegetated areas that might differ in two subsequent years for 50% (Scheffer *et al.*, 2003). That was not the case in the Komarnik fishpond, which supported resilient macrophyte community in spite of its small size. The results of Rooney and Kalff (2000) showed that changes in macrophyte community can be attributed to increased water temperatures, which increased biomass in macrophytes of five shallow lakes in Eastern Townships in Quebec (Canada). The temperatures in the Komarnik lake were rather variable ranging from $9.9°C$ to $30.4°C$ during vegetation period.

The Komarnik accumulation lake is a diverse and resilient waterbody, since inter-annual changes of water level affected only species abundance and not species composition. In spite of its small size, it is a precious system contributing a lot to the landscape heterogeneity and biodiversity.

Acknowledgements This research was financed by the Ministry of Education, Science and Sport, Republic of Slovenia, through the programmes "Biology of plants" (P1-0212). The authors thank Nina Šraj who critically revised the manuscript.

References

APHA (1996). Standard methods for the examination of water and wastewater, 19th edition. Washington DC, USA.

Baattrup-Pedersen, A., & Riis, T. (1999). Macrophyte diversity and composition in relation to substratum characteristics in regulated and unregulated Danish streams. *Freshwater Biology, 42*, 375–385.

Casper, S.J., & Krausch, H.D. (1980). *Süsswasserflora von Mitteleuropa. Pteridophyta und Antophyta.* Jena, Stuttgart, Germany: VEB Gustav Fischer.

Cronin, G., Lewis, W.M. Jr., & Schiehser, M.A. (2006). Influence of freshwater macrophytes on the littoral ecosystem structure and function of a young Colorado reservoir. *Aquatic Botany, 85*, 37–43.

Gaberščik, A. (1997). Makrofiti in kvaliteta voda./Aquatic macrophytes and water quality. *Acta Biologica Slovenica, 41*, 141–148 (in Slovenian with English summary).

Germ, M., & Gaberščik, A. (2003). Comparison of aerial and submerged leaves in two amphibious species, *Myosotis scorpioides* and *Ranunculus trichophyllus*. *Photosynthetica, 41*, 91–96.

Gessner, M.O. (2000). Breakdown and nutrient dynamics of submerged *Phragmites* shoots in the littoral zone of a temperate hardwater lake. *Aquatic Botany, 66*, 9–20.

Fliser, B., Kosi, G., Vrhovšek, D., Bricelj, M., Brancelj, A., Zidar, K., Poga ar, A., Brumen, S., Lapajne, S., Lobnik, S., & Glavič, B. (1985). *Ugotavljanje stanja kakovosti jezer (naravnih in umetnih) na območju povodja Drave in Mure.* Maribor, Slovenia: Vodnogospodarsko podjetje Maribor, Inštitut za biologijo.

Kohler, A., & Janauer, G. (1995). Zur Methodik der Untersuchung von aquatischen Makrophyten in Fließgewässern. In C. Steinberg, H. Bernhardt, & H. Klapper (Eds.), *Handbuch Angewandte Limnologie* (VIII-1.1.3.). Landsberg/ Lech, Klapper: Ecomed Verlag.

Marion, L., & Paillisson, J.-M. (2003). A mass balance assessment of the contribution of floating-leaved macrophytes in nutrient stocks in an eutrophic macrophyte dominated lake. *Aquatic Botany, 75*, 249–260.

Martinčič A., Wraber, T., Jogan, N., Ravnik, V., Podobnik, A., Turk, B., & Vreš, B. (1999). *Mala flora Slovenije. Ključ za določanje praprotnic in semenk.* Ljubljana, Slovenia: Tehniška založba Slovenije.

Mastrantuono, L., & Mancinelli, T. (1999). Long-term changes of zoobenthic fauna and submerged vegetation in the shallow lake Monterosi (Italy). *Limnologica – Ecology and Management of Inland Waters, 29*, 105–119.

Mckee, D., Hatton, K., Eaton, J.W., Atkinson, D., Atherton, A., & Harvey, I. (2002). Effects of simulated climate warming on macrophytes in freshwater microcosm communities. *Aquatic Botany, 74*, 71–83.

Pallisson, J.L., & Marion, L. (2006). Can small water level fluctuations affect the biomass of *Nymphea Alba* in large lakes? *Aquatic Botany, 84*, 259–266.

Piezynska, E. (1993). Detritus and nutrient dynamics in the shore zone of lakes: A review. *Hydrobiologia, 251*, 49–58.

Preston, C.D. (1995). *Pondweeds of Great Britain and Ireland*. London: Botanical Society of the British Isles.

Riis, T., & Hawes, I. (2002). Relationships between water level fluctuations and vegetation diversity in shallow water of New Zealand lakes. *Aquatic Botany, 74*, 133–148.

Rip, W.J., Rawee, N., & de Jong, A. (2006). Alternation between clear, high-vegetation and turbid, low-vegetation states in a shallow lake: The role of birds. *Aquatic Botany, 85*, 184–190.

Rooney, N., & Kalff, J. (2000). Inter-annual variation in submerged macrophyte community biomass and distribution: The influence of temperature and lake morphometry. *Aquatic Botany, 68*, 321–335.

Sand-Jensen, K., Riis, T., Vestergaard, O., & Larsen, S.E. (2000). Macrophyte decline in Danish lakes and streams over the past 100 years. *Journal of Ecology, 88*, 1030–1040.

Scheffer, M., de Redelijkheid, M.R., & Noppert, F. (2003). Distribution and dynamics of submerged vegetation in a chain of shallow eutrophic lakes. *Aquatic Botany, 42*, 199–216.

Schneider, S., & Melzer, A. (2003). The trophic index of macrophytes (TIM) – a new tool for indicating the trophic state of running waters. *International Review of Hydrobiology, 88*, 49–67.

Ulanowicz, R.E. (1995). *Utricularia*'s secret: The advantage of positive feedback in oligotrophic environments. *Ecological Modelling, 79*, 49–57.

Urbanc-Berčič O., Kržič, N., Rudolf, M., Gaberščik, A., & Germ, M. (2005). The effect of water level fluctuation on macrophyte occurrence and abundance in the intermittent Lake Cerknica. In J. Vymazal (Ed.), *Natural and constructed wetlands: Nutrients, metals and management* (pp. 312–320). Leiden, The Netherlands: Backhuys Publishers.

van Donk, E., & van de Bund, W.J. (2002). Impact of submerged macrophytes including charophytes on phyto- and zooplankton communities: Alellopathy versus other mechanisms. *Aquatic Botany, 72*, 261–274.

Chapter 3
Latitudinal Trends in Organic Carbon Accumulation in Temperate Freshwater Peatlands

Christopher Craft[1](✉), **Chad Washburn**[1], **and Amanda Parker**[2]

Abstract The 30-year rate of organic carbon (C) accumulation, based on cesium-137 (^{137}Cs), was measured in 15 freshwater peatlands across a latitudinal gradient from southern Florida (26°N) to northern Minnesota (47°N) to identify relationships between climate (temperature) and C accumulation. Organic C accumulation was inversely related to mean annual air temperature (MAAT, °C) in acidic peatlands (pH < 5) (C, g m^{-2} year^{-1} = 199 − 7.94 × MAAT; r^2 = 0.64, $p \leq 0.01$), with greatest accumulation in the coldest climate. There was a weak but non-significant relationship between C accumulation and MAAT in circumneutral peatlands (pH > 5) (r^2 = 0.41, $p \leq 0.17$). A regression model that incorporated both temperature and precipitation (rain factor, f = mean annual precipitation in cm/MAAT) was no more effective in predicting organic C accumulation than one with temperature alone (r^2 = 0.57 for acidic peatlands, r^2 = 0.36 for circumneutral peatlands). Across all sites, circumneutral peatlands sequestered less C (49 ± 11 g m^{-2} year^{-1}) than acidic peatlands (88 ± 20 g m^{-2} year^{-1}) regardless of temperature. Our findings suggest that, like terrestrial ecosystems, organic C accumulation in freshwater peatlands is linked to climate through the effects of temperature. Local factors such as pH, hydroperiod and nutrient enrichment also should be considered when assessing the potential of freshwater wetlands to sequester C.

Keywords Cesium-137 (^{137}Cs), climate change, Histosol, precipitation, temperature

[1] School of Public and Environmental Affairs, Indiana University, Bloomington IN 47405-1701, USA

[2] U.S. Environmental Protection Agency, Washington, DC, USA

(✉) Corresponding author: e-mail: ccraft@indiana.edu

J. Vymazal (ed.) *Wastewater Treatment, Plant Dynamics and Management in Constructed and Natural Wetlands,*
© Springer Science + Business Media B.V. 2008

3.1   Introduction

Histosols, peat-accumulating wetland soils, occupy 1.3% of the total land area but store 21–29% of the world's soil organic carbon (C) (Eswaran *et al.*, 1993, 1995; Batjes, 1996). Anaerobic conditions resulting from inundation and soil saturation lead to incomplete decomposition of plant detritus and accumulation of C as soil organic matter (Ponnamperuma, 1972); accumulation of organic matter in the soil over time results in the formation of soils with significant accumulation of peat.

In terrestrial ecosystems, accumulation of litter and soil organic C is determined primarily by climate, especially temperature (Post *et al.*, 1982; Burke *et al.*, 1989; Raich & Schlesinger, 1992; Schimel *et al.*, 1994; Couteaux *et al.*, 1995) with greater accumulation in cooler and moister climates than warmer, drier climates (Schlesinger, 1990; Vogt *et al.*, 1986; Chadwick *et al.*, 1994). The importance of climate to organic C accumulation in peatlands has not been determined, although the fact that 50% of peatlands are in northern latitudes (Matthews & Fung, 1987; Aselmann & Crutzen, 1989; Franzen, 1994) suggests that peat accumulation is regulated, in part, by temperature.

Long-term rates of soil organic C accumulation, based on cesium-137 (^{137}Cs), were measured in 15 freshwater peatlands across a latitudinal gradient from southern Florida (26°N) to northern Minnesota (47°N) to identify whether, like terrestrial soils, trends exist in organic C accumulation that are linked to climate.

3.2   Methods

Soil cores were collected from 15 temperate peatlands of the eastern continental USA between 26°N (south Florida) and 47°N (Minnesota). Peatlands represented a range of latitudes and vegetation types (bog, fen, marsh, forest) and were sampled between 1994 and 2000 (Table 3.1). Our sampling sites included large temperate peatlands such as the Great Dismal Swamp (VA), Okefenokee Swamp (GA) and the Everglades (FL) in the southern USA and bogs and fens in the northern USA (MN, MI). Mean annual air temperature (MAAT) of the locations ranged from 3.9°C to 23.9°C and was strongly related to latitude ($r^2 = 0.98$).

One to three undisturbed peatlands were sampled at each location. Within each peatland, one to three cores, each 30–50 cm in length, were collected using a piston corer 8.5 cm in diameter. A total of 31 cores were collected. Our sampling criteria was based on wetlands that were underlain by organic soils (i.e. Histosols or histic epipedons) greater than 30 cm thick and containing >12% organic C (USDA, 1999), and with no obvious evidence of recent natural disturbances (fire) and human disturbances like ditches, levees or logging.

Cores were sectioned into 2 cm depth increments and analyzed for ^{137}Cs, bulk density, organic C and pH. Cesium-137 was measured by gamma spectrometry of the 661.62 keV photopeak (Craft & Richardson, 1998). The ^{137}Cs maximum in each

Table 3.1 Sampling location, predominant vegetation, climate attributes, peat characteristics and organic C accumulation of freshwater peatlands

State	Vegetation	Latitude (°N)	MAAT[a] (°C)	pH[b]	Accretion[c] (mm year⁻¹)	Bulk density (g cm⁻³)[d]	Organic C (%)[d]	Accumulation (g C m⁻² year⁻¹)	
MN (1,1)[e]	Bog	47.0	3.0	3.5	10.3	0.04	42.8	198	—
MI (1,1)	Cedar[f]	46.0	5.0	5.5	1.5	0.15	42.1	95	—
MI (1,1)	Bog	45.6	5.0	4.0	6.4	0.05	44.7	132	—
MI (2,1)	Fen	45.4	5.0	6.0	0.9	0.09	44.1	38	(28–47)[g]
IN (2,2–3)	Marsh	41.5	10.0	5.6	1.5	0.15	29.8	61	(15–109)
NJ (2,1–2)	AWC[h]	40.0	12.0	4.0	1.8	0.12	47.4	97	(67–136)
VA (1,3)	AWC[h]	36.6	5.0	3.8	1.6	0.14	47.8	105	(90–142)
NC (1,3)	AWC[h]	35.8	16.0	3.9	3.5	0.05	47.0	90	(66–109)
NC (1,1)	Pocosin[i]	34.8	17.0	3.8	0.3	0.08	48.3	13	—
GA (3,1)	Cypress	31.3	18.0	5.5	0.7	0.32	16.5	36	(15–56)
GA (1,2)	Marsh[j]	30.8	20.0	4.0	0.8	0.07	47.6	24	(9–39)
FL (1,2)	AWC[h]	30.2	19.0	3.9	0.3	0.13	48.4	17	(16–18)
FL (1,1)	Cypress	29.6	19.0	3.9	2.1	0.12	48.6	122	—
FL (1,1)	Fen[k]	26.5	23.0	5.5	0.8	0.05	51.0	19	19
FL (2,1)	Fen[l]	25.7	23.0	6.0	1.0	0.12	46.0	46	(37–56)

[a] Mean annual air temperature (MAAT) was determined from 1960 to 2000 data from the nearest NOAA National Weather Service reporting station (http://www.ncdc.noaa.gov).

[b] Soil pH (0–10 cm depth) was measured in 10:1 distilled water to soil solution ratio using a hydrogen ion electrode.

[c] ¹³⁷Cs accretion rate.

[d] Averaged over 0–10 cm depth.

[e] Number of peatlands sampled, number of cores collected within each peatland.

[f] Northern cedar swamp forest.

[g] Range of organic C accumulation.

[h] Atlantic white cedar forest (Pine Barrens NJ, Dismal Swamp VA, Alligator River NC, Apalachicola FL).

[i] Evergreen shrub bog.

[j] Okefenokee swamp (Only one interpretable ¹³⁷Cs profile was obtained).

[k] Everglades (Loxahatchee peat and Everglades peat, respectively).

core, corresponding to the 1964 period of maximum deposition of ¹³⁷Cs from above-ground nuclear weapons testing, was used to determine the 30-year rate of vertical accretion (Ritchie & McHenry, 1990). Accretion was calculated using the midpoint of the increment containing the ¹³⁷Cs peak (e.g. 5 cm if the peak was located in the 4–6 cm depth increment). Only cores that contained interpretable ¹³⁷Cs profiles, such as shown in Fig. 3.1, were used in the statistical analyses.

Cesium-137 is a powerful marker because it is strongly adsorbed onto clay and organic particles, its uptake by vegetation is low and its diffusion is usually limited (Ritchie & McHenry, 1990). Even when diffusion does occur, such movement likely will not change the position of the ¹³⁷Cs peak (Ritchie & McHenry, 1990). Cesium-137 is more mobile under acidic than circumneutral conditions (Appleby *et al.*, 1991), so that, in some acidic peat, it is not effective for determining vertical accretion (Oldfield *et al.*, 1995). Our ombrotrophic peat cores, however, have well-defined

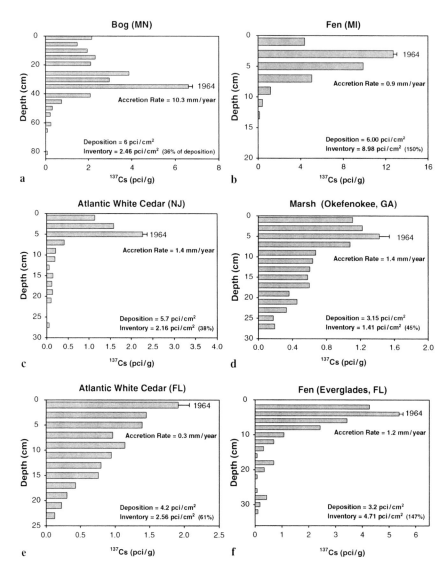

Fig. 3.1 ^{137}Cs profiles, cumulative atmospheric deposition of ^{137}Cs and ^{137}Cs inventories (0–30 cm depth) in peat cores collected from (a) bog in Minnesota, (b) fen in Michigan, (c) Atlantic white cedar forest in New Jersey, (d) Okefenokee marsh in Georgia, (e) Atlantic white cedar forest in Florida, and (f) Everglades fen in Florida

peaks with inventories comparable to atmospheric deposition (see Figs. 3.1a–f) suggesting that ^{137}Cs is a relatively immobile and reliable marker horizon in our acidic peat soils.

Cumulative atmospheric deposition of ^{137}Cs and ^{137}Cs inventories in peat (0–30 cm depth) were calculated for representative cores collected from bogs (MN), fens (MI,

FL Everglades) marshes (GA Okenfenokee), and Atlantic white cedar forests (NJ, FL). For each core, atmospheric deposition of ^{137}Cs was extrapolated from cumulative ^{137}Cs deposition (1953–1972) at Chicago, Illinois (MN Bog, MI fen), Upper Hudson River, New York (NJ AWC), and Columbia, South Carolina (GA Okefenokee marsh, FL AWC and Everglades) (Gustafson *et al.*, 1965; Eisenbud, 1973; McHenry & Ritchie, 1975) using the relationship between ^{137}Cs deposition and latitude (Davis, 1963). Calculation of ^{137}Cs inventories revealed that our peat soils contained 36% (MN bog) to 150% (MI fen) of the cumulative atmospheric deposition of [137]Cs (Fig. 3.1a–f). Cesium-137 inventories in acidic peatlands were smaller as compared to circumneutral peatlands. Relative to atmospheric inputs, higher latitude peatlands contained proportionally less ^{137}Cs than peatlands at lower latitudes.

Bulk density was measured by weighing depth increments that were dried at 70°C. Depth increments were ground and analyzed for organic C by dry combustion (Perkin-Elmer CHN analyzer). Organic C accumulation was calculated using ^{137}Cs-based accretion rates and bulk density and C content averaged across depth increments above and including the ^{137}Cs marker layer.

3.3   Results and Discussion

Organic C accumulation of temperate freshwater peatlands increased with increasing latitude ($r^2 = 0.35$, $p \leq 0.02$) and decreasing MAAT ($r^2 = 0.38$, $p \leq 0.02$). An even stronger relationship emerged when peatlands were separated on the basis of pH. Organic C accumulation of acidic (pH < 5) peatlands increased with latitude ($r^2 = 0.55$, $p \leq 0.03$) and with decreasing MAAT (Fig. 3.2a). Organic C accumulation of circumneutral (pH > 5) peatlands exhibited a weak but non-significant relationship with latitude and MAAT (Fig. 3.2a). A regression model based on separate slopes for acidic and circumneutral peatlands explained more variation ($r^2 = 0.67$) than the model that included all peatlands ($r^2 = 0.38$). Similar to temperature, peatland C accumulation increased with rain factor, f (f = mean annual precipitation/ mean annual temperate; Lang cited in Eggelsmann 1976), for acidic peatlands but not for circumneutral peatlands (Fig. 3.2b). However, temperature alone explained more variation in peatland C accumulation than rain factor.

Organic C accumulation in soil represents the balance between net primary production (NPP), which adds C to the soil, and decomposition, which converts litter and soil organic matter to CO_2. Accumulation of C in soil is affected more by decomposition rate than by NPP (Cebrian & Duarte, 1995; Schlesinger, 1997), and temperature is a primary factor controlling the rate of decomposition in terrestrial (Raich & Schlesinger, 1992; Couteaux *et al.*, 1995; Kirschbaum, 1995) and peatland soils (Updegraff *et al.*, 2001). Our findings support the idea that, like terrestrial soils, C accumulation in freshwater peatlands is controlled primarily by temperature, not precipitation.

Variation in peatland C accumulation was driven mostly by differences in accretion rate rather than by differences in soil bulk density and organic C content (Table 3.1).

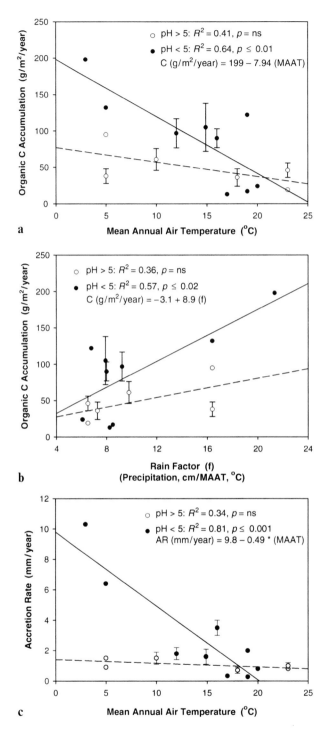

Fig. 3.2 (a) Relationship between organic C accumulation and MAAT of acidic (pH < 5) versus circumneutral (pH > 5) peatlands. (b) Relationship between organic C accumulation and rain factor (*f*) of acidic (pH < 5) versus circumneutral (pH > 5) peatlands. (c) Relationship between accretion rate and MAAT of acidic (pH < 5) versus circumneutral (pH > 5) peatlands. Bars denote one standard error for locations where multiple cores were collected and analyzed

Accretion rate (in mm/year) was inversely related to MAAT across all sites ($r^2 = 0.36$). And, like organic C accumulation, accretion was more strongly related to MAAT in acidic versus circumneutral peatlands (Fig. 3.2c). Rain factor explained comparable amounts of variation in accretion rates in acidic peatlands ($r^2 = 0.89$, $p < 0.001$) but not in circumneutral peatlands ($r^2 = 0.21$).

Mean organic C accumulation (86 ± 21 g C m^{-2} year^{-1}) was two times greater in acidic than circumneutral peatlands (49 ± 11 g C m^{-2} year^{-1}) in spite of very low C accumulation (<20 g C m^{-2} year^{-1}) in several acidic peatlands in the southeastern USA (NC, GA, FL). Optimum pH for most bacteria is in the range of pH 6–8 (Atlas & Bartha, 1987) and, as a result, decomposition of organic matter proceeds faster in circumneutral than acidic wetland soils (Schlesinger, 1990). Enhanced organic C accumulation in acidic peatlands may be attributed to lower decomposition and C mineralisation caused, in part, by low pH (DeLaune et al., 1981; Benner et al., 1985; Farrish & Grigal, 1988; Bridgham et al., 1991, 1998; Updegraff et al., 1995; Verhoeven & Toth, 1995). There was no consistent difference in surface soil (0–10 cm) bulk density or organic C concentration (%) with latitude or pH (Table 3.1).

In conclusion, organic C accumulation in temperate freshwater peatlands is controlled both by climate through the effects of temperature and local factors such as pH. While predicting future rates of C sequestration by freshwater peatlands one should consider the varying effects of global (i.e. temperature) and local (i.e. pH) factors on organic C accumulation.

Acknowledgements This research was funded in part through contract OW-0986-NAEX from the Office of Water and Wetlands, U.S. Environmental Protection Agency, Washington DC. We are grateful to Scott Bridgham (MN), Chev Kellogg (IN), Rob Atkinson and Pattie Duttry (VA/NC AWC), Bill Casey (FL cypress), John Sacco (NJ AWC) and Jerry Williams (Okefenokee) for their help in collecting peat cores.

References

Appleby, P.G., Richardson, N., & Dolan, P.J. (1991).[241]Am dating of lake sediments. *Hydrobiologia*, *214*, 35–42.

Aselmann, J., & Crutzen, P.J. (1989). Global distribution of natural freshwater wetlands and rice paddies, their net primary productivity, seasonality and possible methane emissions. *Journal of Atmospheric Chemistry, 8*, 307–358.

Atlas, R.M., & Bartha, R. (1987). *Microbial ecology: Fundamentals and applications* (2nd ed.). Reading, MA: Benjamin/Cummings.

Batjes, N.H. (1996). Total carbon and nitrogen in soils of the world. *European Journal of Soil Science, 47*, 151–163.

Benner, R., Moran, M.A., & Hodson, R.E. (1985). Effects of pH and plant source on lignocellulose biodegradation rates in two wetland ecosystems: The Okefenokee Swamp and a Georgia salt marsh. *Limnology and Oceanography, 30*, 489–499.

Bridgham, S.D., Richardson, C.J., Maltby, E., & Faulkner, S.P. (1991). Cellulose decay in natural and disturbed peatlands in North Carolina. *Journal of Environmental Quality, 20*, 695–701.

Bridgham, S.D., Updegraff, K., & Pastor, J. (1998). Carbon, nitrogen and phosphorus mineralization in northern peatlands. *Ecological Applications, 79*, 1545–1561.

Burke, I.C., Yonker, C.M., Parton, W.J., Cole, C.V., Flach, K., & Schimel, D.S. (1989). Texture, climate and cultivation effects on soil organic matter content in U.S. grasslands. *Soil Science Society of America Journal, 53,* 800–805.

Cebrian, J., & Duarte, C.M. (1995). Plant growth-rate dependence of detrital carbon storage in ecosystems. *Science, 268,* 1606–1608.

Chadwick, O.A., Kelley, E.F., Merritts, D.M., & Amundson, R.G. (1994). Carbon dioxide consumption during soil development. *Biogeochemistry, 24,* 115–127.

Couteaux, M.M., Bottner, P., & Berg, B. (1995). Litter decomposition, climate and litter quality. *Trends in Ecology and Evolution, 10,* 63–66.

Craft, C.B., & Richardson, C.J. (1998). Recent and long-term organic soil accretion and nutrient accumulation in the Everglades. *Soil Science Society of America Journal, 62,* 834–843.

Davis, J.J. (1963). Cesium and its relationships to potassium in ecology. In V. Schultz & A.W. Klement, Jr. (Eds.), *Radioecology* (pp. 539–556). New York: Reinhold.

DeLaune, R.D., Reddy, C.N., & Patrick, W.H., Jr. (1981). Organic matter decomposition in soil as affected by pH and redox condition. *Soil Biology and Biochemistry, 13,* 533–534.

Eggelsmann, R. (1976). Peat consumption under influence of climate, soil condition and utilization. In *Proceedings of the 5th International Peat Conference* (Vol. I, pp. 233–247). Poznan, Poland: International Peat Society.

Eisenbud, M. (1973). *Environmental radioactivity.* New York: Academic Press.

Eswaran, H., Van Den Berg, E., & Reich, P. (1993). Organic carbon in soils of the world. *Soil Science Society of America Journal, 57,* 192–194.

Eswaran, H., Van den Berg, E., & Kimble, J. (1995). Global soil carbon resources. In I.R. Lal, J. Kimble, E. Levine, & B.A. Stewart (Eds.), *Soils and global change* (pp. 27–43). Boca Raton, FL: CRC/Lewis.

Farrish, K.W., & Grigal, D.F. (1988). Decomposition in an ombrotrophic bog and a minerotrophic fen. *Soil Science, 145,* 353–358.

Franzen, L.G. (1994). Are wetlands the key to the ice-age cycle enigma? *Ambio, 23,* 300–308.

Gustafson, P.F., Brar, S.S., & Muniak, S.E. (1965). Fission product deposition and dietary levels in the Chicago area. In A.W. Klement, Jr. (Ed.), *Proceedings 2nd Conference on Radioactive Fallout from Nuclear Weapons Tests,* Nov. 3–6, 1964, Germantown, MD (pp. 783–790). Washington DC: U.S. Atomic Energy Commission, Division of Technical Information.

Kirschbaum, M.U.F. (1995). The temperature dependence of soil organic matter decomposition, and the effect of global warming on soil organic C storage. *Soil Biology and Biochemistry, 27,* 753–760.

Matthews, E., & Fung, I. (1987). Methane emission from natural wetlands: Global distribution, area and environmental characteristics of sources. *Global Biogeochemical Cycles, 1,* 61–86.

McHenry, J.R., & Ritchie, J.C. (1975). Redistribution of cesium-137 in southeastern watersheds. In F.G. Howell, J.B. Gentry, & M.H. Smith (Eds.), *Mineral cycling in southeastern ecosystems* (pp. 455–462). Washington, DC: Energy Research and Development Administration (ERDA) Symposium Series. CONF 74-740513. U.S. ERDA.

Oldfield, F., Richardson, N., & Appleby, P.G. (1995). Radiometric dating of recent ombrotrophic peat and evidence for changes in mass balance. *The Holocene, 5,* 141–148.

Ponnamperuma, F.N. (1972). The chemistry of submerged soils. *Advances in Agronomy, 24,* 29–96.

Post, W.M., Emanuel, W.R., Zinke, P.J., & Stangenberger, A.G. (1982). Soil carbon pools and world life zones. *Nature, 298,* 156–159.

Raich, J.W., & Schlesinger, W.H., Jr. (1992). The global carbon dioxide flux in soil respiration and its relationship to vegetation and climate. *Tellus, 44B,* 81–99.

Ritchie, J.C., & McHenry, J.R. (1990). Application of radioactive fallout cesium-137 for measuring soil erosion and sediment accumulation rates and patterns: A review. *Journal of Environmental Quality, 19,* 215–233.

Schimel, D.S., Braswell, B.H., Holland, E.A., McKeown, R.,Ojima, D.S., Painter, T.H., Parton, W.J., & Townsend, A.R. (1994). Climatic, edaphic and biotic controls over storage and turnover of carbon in soils. *Global Biogeochemical Cycles, 8,* 279–293.

Schlesinger, W.H., Jr. (1990). Evidence from chronosequence studies for a low carbon-storage potential of soils. *Nature, 348*, 232–234.

Schlesinger, W.H. (1997). *Biogeochemistry: An analysis of global change.* New York: Academic.

Updegraff, K., Pastor, J., Bridgham, S.D., & Johnston, C.A. (1995). Environment and substrate controls over carbon and nitrogen mineralization in northern wetlands. *Ecological Applications, 5*, 151–163.

Updegraff, K., Bridgham, S.D., Pastor, J., Weishampel, P., & Harth, C. (2001). Response of CO_2 and CH_4 emissions from peatlands to warming and water table manipulations. *Ecological Applications, 11*, 311–326.

USDA (United States Department of Agriculture), Natural Resources Conservation Service (1999). *Soil taxonomy: Agricultural handbook no. 436.* Washington, DC: US Government Printing Office.

Verhoeven, J.T.A., & Toth, E. (1995). Decomposition of *Carex* and *Sphagnum* litter in fens: Effect of litter quality and inhibition by living tissue homogenates. *Soil Biology and Biochemistry, 27*, 271–275.

Vogt, K.A., Grier, C.C., & Vogt, D.J. (1986). Production, turnover and nutrient dynamics of above- and below-ground detritus of world forests. *Advances in Ecological Research, 15*, 303–409.

Chapter 4
Buffering Performance in a Papyrus-Dominated Wetland System of the Kenyan Portion of the Lake Victoria Basin

Herbert John Bavor[1]([✉]) and Michael Thomas Waters[2]

Abstracts The study was aimed at assessing buffering processes of the Eldoret Chepkoilel wetland in the Kenyan portion of the Lake Victoria Basin. Changing patterns of use and activities in the region of the wetland have exerted enormous pressure on land and water resources, including increased nutrient loads, increased erosion and thus increased sediment loadings to receiving waters. The Chepkoilel wetland was located a few kilometres north of Eldoret town. The wetland was sited in a shallow trough-like valley that lies at an elevation of between 2,110 and 2,140 m. The dominant wetland vegetation included a central band of dense papyrus (*Cyperus papyrus*) with narrow fringing stands of shorter emergent macrophytes, *Cyperus* spp. (*C. rotundus, C. triandra, C. laevigatus*). The section of the wetland described in the investigations was approximately 10 km long. The wetland received inputs from surrounding agricultural land, a flower farm and also a small domestic sewage treatment plant. The wetland received a permanent stream inflow from the Misikuri river. The Chepkoilel campus of Moi University was located to the west of the wetland, and effluent from the campus sewage treatment plant also discharged into the wetland. The monitoring data indicated significant nutrient-processing performance, reflecting a relatively long water-residence time within the system. Problems and approaches to monitoring the system are discussed.

Keywords Buffering Kenya, Lake Victoria, nutrients, papyrus, runoff, surface flow

4.1 Introduction

Wetlands within the Kenyan part of the Lake Victoria catchment include a series of highland wetlands in high rainfall areas as well as large wetlands adjacent to Lake Victoria. These wetlands of the Lake Victoria Basin are extremely important to

[1] Centre for Water and Environmental Technology – Water Research Laboratory, University of Western Sydney – Hawkesbury, Locked Bag 1797, Penrith South DC, NSW 1797, Australia

[2] SMEC International, P.O. Box 1052, North Sydney, NSW 2060 Australia

([✉]) Corresponding author: e-mail: j.bavor@uws.edu.au

J. Vymazal (ed.) *Wastewater Treatment, Plant Dynamics and Management in Constructed and Natural Wetlands,*
© Springer Science + Business Media B.V. 2008

the water-quality cycles that occur in the basin's waterways, specifically through roles such as:

- Reducing and delaying highly polluted flood flows entering the lake
- Trapping suspended sediments and nutrients
- Cycling nutrients between sediments, biota and the water column

Wetlands in the African context may also be considered a very appropriate means of improving water quality compared with other options as they have additional benefits including that they

- Form a rich resource base for the local communities
- Have low input requirements in terms of energy, building/engineering infrastructure and local maintenance for sustainability
- Support ecosystem functions that would be lost when land-use development occurs, for example by providing habitat for migrating, rare and endangered species
- Are resilient to a wide range of water level and flow fluctuations

Few investigations have addressed pollutant-removal performance for such wetland systems in the Basin (Kansiime & Nalubega, 1999; Mwanuzi *et al.*, 2003). The primary aim of the study described here was to assess buffering processes of the Eldoret Chepkoilel wetland in the Kenyan portion of the Lake Victoria Basin, as part of the Lake Victoria Environment Management Program (LVEMP) Wetland Component. Changing patterns of use and activities in the region of the wetland have exerted enormous pressure on land and water resources (Ntiba *et al.*, 2001), including increased nutrient loads and increased erosion, and thus, increased sediment loadings to receiving waters.

4.2   Methods

Chepkoilel is a major wetland running south–north, a few kilometres north of the city of Eldoret. It is a permanent, riverine wetland with a high length to width ratio, about 10 km long and about 700 m wide at the widest point. The wetland is sited in a shallow trough-like valley that lies at an elevation between 2,110 and 2,140 m above sea level. Wetland vegetation was dominated by a central band of dense common papyrus (*Cyperus papyrus*) flanked on either side by shorter emergent vegetation dominated by *Cyperus* spp. (*C. rotundus, C. triandra, C. laevigatus*). Shallow and ephemerally flooded sections of the wetland have been encroached by agricultural activities where tomatoes, cabbages and kale are the main crops. In these areas, ditches have been dug either to drain water away or to detain water for irrigating these crops.

    The local community used the name 'Chepkoilel' for both the wetland in this area and for the stream passing through it both upstream and downstream of the wetland; however, the wetland is also referred to in places as 'Murula', while the stream is officially recorded as the Misikuri river. The 560 ha Chepkoilel wetland has a catchment area of 21,000 ha with major inflows to the wetland contributed by the Sergoit–Misikuri river system. The catchment drained areas of mild slopes ranging up to 2,160 m

Table 4.1 Eldoret Chepkoilel wetland monitoring stations. UTM Zone 36N

Station name	Description	Easting (UTM km)	Northing (UTM km)
IE1	Chepkoilel US Bridge	36,758,817	61,544
IE2	Flower Farm Weir	36,758,576	61,992
IE3	Moi Chepkoilel STP	36,757,674	64,621
OE4	Downstream transect midpoint	36,759,651	70,316

above sea level. Due to the extent of agricultural activities and limited natural vegetation cover remaining in the catchment, a considerable amount of surface runoff was therefore expected to be generated during rainfall events.

Four monitoring stations were established in the wetland, reflecting the major sources of inflow and potential nutrient loads entering the wetland. These stations, with georeferenced location, are described in Table 4.1.

Field measurements carried out at the sites on a monthly basis from June 2004 to February 2005 and also during rainfall events covered the following parameters:

1. Depths and flow rates for the various streams were determined using standardised hydrology techniques to quantify total pollution loadings into the wetland.
2. Temperature, electrical conductivity, dissolved oxygen, pH and turbidity were determined using a Hydrolab Data Sonde 4A or YSI 660 Sonde in the field.
3. Suspended sediments, nitrogen and phosphorus were measured by taking samples for analysis using standard methods.

A sample collection was carried out in coincidence with hydrologic assessment of flow rates for the streams. All water samples were collected in pre-rinsed plastic bottles for the above parameters and held in cooler boxes for transport to the laboratory for analyses.

The base modelling framework utilised in the performance estimation was the LAVINKS-WEB model described by Bavor and Waters (2007). This model has been most widely used in Australia, but is written generally enough so that other applications across a range of climatic and landscape conditions are possible. The model has also been used successfully in countries with tropical climates such as Thailand.

4.3 Results and Discussion

Water-quality monitoring was carried out through both wet and dry periods to allow characterisation of inflow and outflow loads at both wetlands. Using this data, relationships were determined between inflow loads for total suspended solids (TSS), total phosphorus (TP), total nitrogen (TN) and daily rainfall, such that modelling of the wetlands performance could be performed and compared with outflow loads.

Intensive monitoring over storm events indicated that the wetland system could be a net exporter of sediments and nutrients during storm events; however, the effect was limited and did not take place in all storm events. It was demonstrated that the wetland represented a net 'sink' under low flow and moderate flow conditions, but potentially exported accreted nutrients and sediments during large storm events. Such behaviour is not uncommonly reported for both wetland and pond systems (Davies *et al.*, 2001).

However, the net impact over both low- and high-flow periods was of a net removal of pollutants by the Chepkoilel system. It is recognised that longer-term monitoring should be undertaken to more fully understand the buffering process and to ensure that management strategies are in place to promote sustainability of the system.

Monthly rainfall, ranging from less than 30 mm to greater than 170 mm per month, inflow/outflow characteristics for the inlet (total), mid-transect and outlet and a schematic of the system are shown below in Figs. 4.1 and 4.2.

In Table 4.2, water-quality data is summarised for the key parameters of interest – TSS, TP and TN respectively for the Chepkoilel wetland. The high nutrient

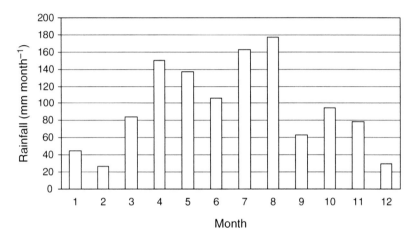

Fig. 4.1 Average monthly rainfall at Eldoret, 1984–2004, including historical data prior to the investigation period

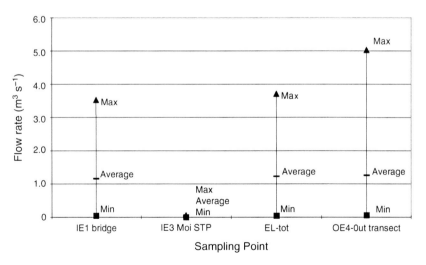

Fig. 4.2 Eldoret Chepkoilel wetland measured flow rates summary, showing inlet flows (IE 1 and 2), combined inlet flows (EL-tot) and the system outlet flow (OE4-out)

Table 4.2 Summary of buffering capacity effectiveness: Chepkoilel wetland, Eldoret, Kenya

	Inlet water quality				Outlet water quality				
	SS	TP	TN	TDS	SS	TP	TN	TDS	N/P ratio
Maximum	90.0	0.649	2.877	0.177	28.77	0.213	1.07	0.09	108
Average	10.6	0.077	0.414	0.051	5.65	0.028	0.12	0.05	5
Minimum	0.0	0.002	0.008	0.006	0.00	0.001	0.00	0.02	0.024
	mg l^{-1}	mg l^{-1}	mg l^{-1}	mg l^{-1}	mg l^{-1}	mg l^{-1}	mg l^{-1}	mg l^{-1}	

Load in	discharge	TSS	TP	TN	TDS
Average	115.4	2,533.4	18.9	92.1	6.2
	m^3 × 10^3 day^{-1}	kg day^{-1}	kg day^{-1}	kg day^{-1}	kg day^{-1}

Load out	discharge	TSS	TP	TN	TDS
Average	115	997	5.94	27.0	5.69
	m^3 × 10^3 day^{-1}	kg day^{-1}	kg day^{-1}	kg day^{-1}	kg day^{-1}

Loading rates	discharge	TSS	TP	TN	TDS
Average	0.20	44	0.33	1.60	0.11
	m day^{-1}	kg ha^{-1} day^{-1}	kg ha^{-1} day^{-1}	kg ha^{-1} day^{-1}	kg ha^{-1} day^{-1}

Load removed	TSS	TP	TN	TDS
Average	1,536	12.9690	65	0.486
	kg day^{-1}	kg day^{-1}	kg day^{-1}	kg day^{-1}

Areal removal rate (kg ha^{-1} day^{-1})	TSS	TP	TN	TDS
	26.74	0.23	1.13	0.01

Relative reduction (%) (load removed/incoming load)	TSS	TP	TN	TDS
Average	61%	69%	71%	8%

retention is suggested to reflect a long detention time for the wetland system, of the order of 12 days as estimated from preliminary NaCl tracer studies. Additionally, the specific hydrologic profile and the ratio of overflow to underflow water column dynamics of the papyrus dominated system are suggested to have contributed to the performance. The characteristics of nutrient and disinfection performance in papyrus wetland systems have also been noted by Kansiime and van Bruggen (2001).

4.4 Conclusions

The buffering capacity of the Chepkoilel wetland was determined using monitoring and historical data in the wetland model described above. This modelling was performed for a 7-year period. The results indicated that the wetland buffered suspended sediments and total phosphorus by approximately 60% and total nitrogen by approximately 70%. The effective pollutant-buffering performance determined for this upland wetland system strongly supports the vital need for wetland preservation in the Kenyan portion of Lake Victoria in ensuring sustainability of the basin. Steps should be taken to prevent possible destruction of wetlands in the future and to encourage the restoration and management of degraded wetlands in the basin to ensure they continue to play this vital role.

Acknowledgements The support of the Lake Victoria Environment Management Program (LVEMP): Wetland Component is gratefully acknowledged. Additionally, the support and encouragement of Mr. Stephen Katua as LVEMP Kenyan Wetlands Component coordinator; Mr. Stanley Ambasa as buffering capacity coordinator; and Ms. Rose Angweya, LVEMP Water Quality Component laboratory manager at Kisumu is acknowledged.

References

Bavor, H.J., & Waters, M.T. (2008). Pollutant transformation performance and model development in African wetland systems: Large catchment extrapolation. In: Wastewater Treatment, Plant Dynamics and Management in Constructed and Natural Wetlands. Ed. J. Vymazal. Ecological Studies Series, Springer-Verlag New York. This Volume, Chapter 28.

Davies, C.M., Sakadevan, K., & Bavor, H.J. (2001). Removal of stormwater-associated nutrients and bacteria in constructed wetland and water pollution control pond systems. In J. Vymazal (Ed.), *Transformations of nutrients in natural and constructed wetlands* (pp. 483–495). Leiden, The Netherlands: Backhuys.

Kansiime, F., & Nalubega, M. (1999). *Wastewater treatment by a natural wetland: The Nakivubo Swamp, Uganda – processes and implications*. Rotterdam, The Netherlands: A.A. Balkema.

Kansiime, F., & van Bruggen, J.J.A. (2001). Distribution and retention of faecal coliforms in the Nakivubo wetland in Kampala, Uganda. *Water Science & Technology, 44*(11–12), 199–206.

Mwanuzi, F., Aalderink, H., & Mdamo, L. (2003). Simulation of pollution buffering capacity of wetlands fringing Lake Victoria. *Environment International, 29*, 95–103.

Ntiba, M.J., Kudoja, W.M., & Mukasa, C.T. (2001). Management issues in the Lake Victoria watershed. *Lakes Reservoirs Research and Management, 6*, 211–216.

Chapter 5
Changes in Concentrations of Dissolved Solids in Precipitation and Discharged Water from Drained Pasture, Natural Wetland and Spruce Forest During 1999–2006 in Šumava Mountains, Czech Republic

Jan Procházka[1]($\boxtimes$), Jakub Brom[1,2], Libor Pechar[1,2,3], Jana Štíchová[4], and Jan Pokorný[2,3]

Abstract Changes in the chemistry of bulk precipitation and stream water between 1999 and 2006 are described for small drained, wetland and forest catchments located within Šumava Mountains (Bohemian Forest) in south-west of the Czech Republic. The chapter is focused on interpretation of hydrochemical trends in selected time periods, especially on effect of deposition changes and landscape management. In bulk precipitation, the annual mean pH increased from 4.8 to 5.4, concentrations of base cations were relatively low and constant over time, nitrate moderately increased, conductivity and alkalinity decreased, while sulphate decreased (from 16.8 mg l^{-1} to 7.2 mg l^{-1}) during the 7 year monitoring period. Catchments with wetland and forest appear to be more resistant to recovery in terms of water conductivity, alkalinity and base cations in comparison with the relatively fast response from the drained pasture. High nitrate, bicarbonate and base cations concentrations in the stream water discharged from the pasture indicate continuous acidification of the catchment, and they result most probably from fast mineralisation (oxidation) of soil organic matter.

Keywords Stream water chemistry, bulk precipitation chemistry, discharge of dissolved solids, base cations, nitrate and sulphate

[1]Laboratory of Applied Ecology, Faculty of Agriculture, University of South Bohemia, Studentská 13, České Budějovice, CZ-370 05, Czech Republic

[2]ENKI o.p.s., Dukelská 145, Třeboň, CZ-379 01, Czech Republic

[3]Institute of System Biology and Ecology, Academy of Sciences of the Czech Republic, Dukelská 145, Třeboň, CZ-379 01, Czech Republic

[4]Department of Applied Chemistry and Chemistry Teaching, Faculty of Agriculture, University of South Bohemia, Studentská 13, České Budějovice, CZ-370 05, Czech Republic

($\boxtimes$) Corresponding author: e-mail: prochaz@zf.jcu.cz

J. Vymazal (ed.) *Wastewater Treatment, Plant Dynamics and Management in Constructed and Natural Wetlands*,
© Springer Science + Business Media B.V. 2008

5.1   Introduction

The present agricultural management has a severe impact on the quality of surface and ground waters. Leaching of matter – including nutrients such as nitrogen and phosphorus but even more importantly of base cations – deprives soil of its fertility, and surface and ground waters of their quality. The impact of management and the distribution of agricultural land and natural ecosystems can be observed as differences in the dissipation of incoming solar energy (overheating of agricultural land), as well as in the water quality and discharge patterns.

According to the holistic approach (Ripl, 1995, 2003), well-functioning landscape is characterised by the high ability of vegetation cover to dissipate solar energy, i.e. attenuate high solar-energy pulses received by the Earth everyday, the ability of biocoenosis to retain high water content within its domain and reduce the outflow of both dissolved (leaching) and solid (erosion) matter. The 'least-loss' efficiency relates to the long-term sustainability of the whole catchments, sub-catchments, individual habitats or forms of land use. Minimal matter losses through the water discharged from a given catchment is both an ecological as well as economic interest. This concept of landscape efficiency based on matter losses and on solar-energy dissipation was tested in three small catchments in Šumava Mountains (Bohemian Forest, Czech Republic) of very low population density (c. 2.5 inhabitant km^{-2}) (Procházka *et al.*, 2001). A previous study showed that different land use in small catchments resulted in different matter losses and different temperature distributions (Procházka *et al.*, 2001, 2006).

Since 1999, research of the three catchments has also focused on the impact of acid deposition on the acidification of soil and freshwater. The research has led to the interpretation of hydrochemical trends by the choice of the time series, especially in relation to deposition changes and landscape management. This chapter presents a more detailed description of the relationships among concentrations of soluble solids in precipitation waters on one side and in the water discharged from the drained pasture during the period 1999 to 2006 on the other.

5.2   Material and Methods

Selected catchments of Mlýnský (1), Horský (2) and Bukový (3) streams represent three different types of land cover: (1) semi-intensive drained pasture, (2) wetlands, forest and mowed meadows, and (3) forests. The selection of the experimental sites for long-term monitoring was done in 1997 on the basis of results of the previous research in the area of Svatotomášská Highlands at the Šumava Mountains in the border region between Austria and the Czech Republic (Fig. 5.1). The altitude range of the studied catchments is between 780 and 1,026 m with an average annual rainfall between c. 950 and 1,050 mm and with a mean annual air temperature c. 5°C. The geology of all the catchments is primarily granite. The three catchments have similar area (200–260 ha) and exposition (SE), but they differ markedly in land use and management (Procházka *et al.*, 2001).

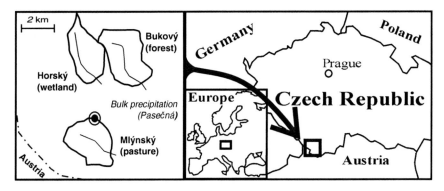

Fig. 5.1 Location of study area containing long-term sampling sites

According to "CORINE land cover technical guide" (Bossard *et al.*, 2000) the catchments are classified as follows:

- Mlýnský pasture: class 2 – Agricultural areas 2.3 – Pastures 2.3.1 – Pastures
- Horský wetland: class 4 – Wetlands 4.1 – Inland wetlands 4.1.1 – Inland marshes and class 2 – Agricultural areas 2.4 – Heterogeneous agricultural areas 2.4.3 – Land principally occupied by agriculture, with significant areas of natural vegetation
- Bukový forest: class 3 – Forest and semi-natural areas 3.1 – Forest 3.1.2 – Coniferous forest

Vegetation type of Mlýnský pasture belongs to communities of alliance *Cynosurion*, with dominating *Agrostis capillaris* L., *Festuca pratensis* Huds., *Poa pratensis* L. and *Trifolium repens* L. Vegetation types of Horský wetlands belong mostly to herbal and shrubby communities of alliance *Calthion* with dominating *Scirpus sylvaticus* L., *Filipendula ulmaria* L. Maxim. and *Lysimachia vulgaris* L. and alliance *Salicion cinereae* with dominating *S. aurita* L. A large area of the Horský potok catchment is covered by spruce forest. Forest segments of the Bukový catchment are mostly formed by spruce forest (*Picea abies* (L.) Karsten), and semi-natural areas are mostly herbal wetlands. For more details on vegetation and primary production see Hakrová (2003) and Procházka *et al.* (2001).

In each catchment, 58–70 soil probes characteristic for the variety of vegetation types were taken in 1998 (Hakrová & Procházka, 1998). A spade of long narrow steel plate approximately 20 cm wide and 80 cm long was used for sampling in order to reach the horizon C located at least at 40 cm depth. The core was divided according to structure, colour and humus content into 2–3 parts: A – surface (humic), B – middle, and C – lower (soil-forming substrate, parent material). More than 400 soil samples were taken in this way. Simultaneously, samples for estimation of the soil volume weight were taken. The total soil amount and consequently the elutable pool of matter in the soil was calculated for each respective catchment. Organic matter content in the soil was estimated by loss on ignition, 5 g of homogenised soil sample was exposed to 550°C for 5 h. Elutable anions and ammonium (NH_4^+) were estimated spectrophotometrically in water extract (10 g of soil in 20 ml of distilled water) using Tecator flow injection analyser (Fiastar 5000) after 5 min extraction, shaking and filtration; the main base cations were estimated by absorption spectrometry with spectrophotometer Varian Spectr AA-640 (Procházka *et al.*, 2001).

Since November 1999, daily precipitation has been monitored, and integrated samples of precipitation collected weekly for chemical analysis of atmospheric deposition at precipitation gage station Pasečná (Mlýnský catchment) (Fig. 5.1). Since 1998, regular sampling of stream water and flow-rate measurements has been done in 3-week intervals at the closing profiles. Water temperature and conductivity were also measured directly in situ with a Multiline P4 instrument (WTW, Germany). Since 1999, automatic monitoring stations have been working in each of the catchments. They continuously recorded water level (by pressure sensor), conductivity and temperature of the discharged water. In the collected samples, pH and alkalinity values (by potentiometric titration with 0.1 M HCl) were measured. Cations of calcium (Ca^{2+}), magnesium (Mg^{2+}), potassium (K^+) and sodium (Na^+) were determined by atomic absorption spectrometric (AAS) method with Varian SpectrAA-640 instrument. Other major ions Cl^- and SO_4^{2-} and inorganic forms of N were determined by FIA on Tecator FIA-Star instrument. Precipitation data and water chemistry data were subjected to statistical analyses in order to evaluate the differences between the catchments. Statistical evaluation was performed using descriptive statistics and one-way analysis of variance (ANOVA).

Water and element budgets for hydrological years 2000–2004 were calculated from the data of total precipitation and total discharged water monitored at the concluding profiles and from the results of chemical analyses. Samples for chemical analyses were taken in 3-week intervals. Water discharge was monitored continuously. For detailed results see Procházka (2004) and Procházka *et al.* (2006).

5.3   Results

5.3.1   *Chemical Composition of the Soil*

Organic matter content, nutrients and K^+, Na^+, Mg^{2+}, Ca^{2+} were estimated in the soil samples of all three catchments. Lowest concentration of nutrients and other substances was found in the soil of the Mlýnský catchment (pasture); relatively higher concentration of Na is explained by presence of cattle. The results are shown in Table 5.1.

5.3.2   *Bulk Precipitation Chemistry*

The chemical composition of bulk precipitation at Pasečná is summarised in Table 5.2 as annual rainfall-volume weighted means and standard deviation for the period

Table 5.1 Average concentrations of nutrients and elements in soil samples in three monitored catchments

	Organic matter %	Cl⁻ mg kg⁻¹	total N mg kg⁻¹	PO_4^{3-}–P mg kg⁻¹	SO_4^{2-} mg kg⁻¹	Ca^{2+} mg kg⁻¹	Mg^{2+} mg kg⁻¹	Na^+ mg kg⁻¹	K^+ mg kg⁻¹
Mlýnský	12.3	13.5	11.9	0.7	76.1	8.6	2.0	5.6	3.2
Horský	27.8	19.1	13.2	2.7	156.8	12.1	3.0	3.2	8.6
Bukový	25.4	14.9	12.2	0.9	133.1	9.1	2.2	3.1	6.9

1999–2006 inclusive. Bulk precipitation is typically dilute, mean conductivity 44 μS cm⁻¹, and moderately acidic with a mean pH of 5.4. The dominant acids in precipitation are sulphate (7.6 mg SO_4^{2-} l⁻¹) and nitrate (2.9 mg NO_3^- l⁻¹).

5.3.3 Stream Water Chemistry

The total content of dissolved matter expressed as conductivity, as well as average concentrations of bicarbonates (alkalinity), nitrates (NO_3^-), Ca^{2+}, Mg^{2+}, Na^+, K^+ and chlorides (Cl⁻), were significantly higher in the water of the Mlýnský (drained pasture) stream than in the water of the Horský (wetland) and Bukový (forest) streams. Results of the 'one-way ANOVA' proved a difference among the catchments on 5% and 1% significance levels, respectively (Table 5.2).

5.3.4 Long-Term Changes in Precipitation and Stream Water Chemistry

The atmospheric deposits were estimated from chemical analysis of bulk precipitations. An evident decline in sulphate (SO_4^{2-}) concentration has been found for the period since 1999. In 2006, the sulphate concentration (7.2 mg l⁻¹) was approximately 2.3 times lower than in 1999 (16.8 mg l⁻¹). The decline has not been linear and large fluctuations have occurred over the 7-year period with the lowest concentration recorded in 2001 (6.1 mg l⁻¹) and 2004 (4.1 mg l⁻¹). These fluctuations cannot be attributed solely to variability in annual rainfall amounts. Annual concentrations of NO_3^- do not show similar trends as those of SO_4^{2-}, although again there are large year-to-year fluctuations, which are not directly correlated with rainfall amount (Fig. 5.2).

Water quality of the three monitored streams shows several types of trends during the time period of 7 years (Figs. 5.3–5.6). The trends in all the streams were approximated by simple linear regression models (significance at p-level = 0.05).

Table 5.2 Mean bulk precipitation chemistry and stream water chemistry during the years 1999–2006 (conductivity in μS cm⁻¹, Alkalinity in mEq l⁻¹, the others in mg l⁻¹)

Parameter	Precipitation N	Mean	SD	Mlýnský stream N	Mean	SD	Horský stream N	Mean	SD	Bukový stream N	Mean	SD	ANOVA F; *p*-level
Conductivity	293	**44.0**	30.0	115	**91.3**	17.3	118	**42.5**	7.7	116	**36.5**	7.4	**820.25; **
pH	303	**5.4**	0.6	122	**6.4**	0.3	122	**6.1**	0.4	121	**6.0**	0.6	**20.96; **
Alkalinity	272	**0.6**	6.1	121	**0.5**	0.1	121	**0.2**	0.1	118	**0.1**	9.3	**230.67; **
NO₃⁻	297	**2.9**	6.0	124	**7.5**	2.9	124	**2.0**	1.8	123	**2.2**	9.0	**352.17; **
Ca²⁺	302	**1.8**	6.2	112	**7.7**	2.5	113	**3.1**	1.4	113	**3.2**	9.3	**313.74; **
Mg²⁺	301	**0.7**	5.8	113	**1.6**	0.5	113	**0.9**	0.5	114	**1.6**	9.4	**125.64; **
Na⁺	302	**0.9**	5.8	112	**3.8**	1.1	113	**2.7**	0.9	114	**3.7**	9.2	**40.45; **
K⁺	301	**1.2**	5.8	112	**1.6**	0.4	113	**1.0**	0.3	114	**1.5**	9.4	**188.97; **
NH₄⁺	297	**1.4**	5.9	112	**0.0**	0.1	123	**0.0**	0.1	124	**0.08**	9.1	**1.47; n.s.**
Cl⁻	298	**1.7**	6.2	122	**1.6**	1.0	123	**1.0**	0.7	122	**1.6**	9.1	**37.12; **
SO₄²⁻	288	**7.6**	8.2	121	**13.7**	6.8	121	**11.4**	6.3	121	**12.4**	11.2	**3.95; ***

N – number of cases; SD – standard deviation; ** – statistical significance at $p < 0.001$; * – statistical significance at $p < 0.05$; n.s. – not significant; degrees of freedom = 2 in each cases. One-way ANOVA was computed for streams only.

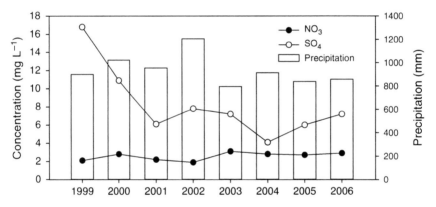

Fig. 5.2 Annual rainfall amount (mm) and main pollutant (sulphate (SO_4^{2-}) and nitrate (NO_3^-)) concentrations in bulk precipitation (mg l⁻¹) from Pasečná station during 1999–2006

5.3.4.1 Conductivity, Alkalinity and Sulphate

All three catchments show a steady decline in annual mean conductivity, alkalinity and SO_4^{2-} concentrations since 1999, which broadly reflect the overall decrease of the same parameters in bulk precipitation. During the 7-year period, the highest changes in annual mean conductivity, alkalinity and SO_4^{2-} concentration occurred in drained pasture (Mlýnský stream) corresponding to a decline of 19 μS cm⁻¹, 0.14 mEq l⁻¹ and 9.1 mg SO_4^{2-} l⁻¹ (Horský: 2.8 μS cm⁻¹, 0.06 mEq l⁻¹, 8.9 mg SO_4^{2-} l⁻¹; Bukový: 1.8 μS cm⁻¹, 0.09 mEq l⁻¹ and 11.6 mg SO_4^{2-} l⁻¹). Changes in conductivity result from changes of base cation concentrations rather than from SO_4^{2-} fluctuations.

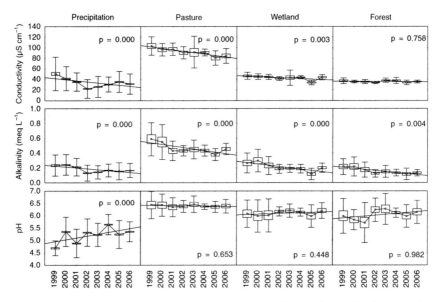

Fig. 5.3 Time series of conductivity and chemical characteristics (alkalinity, pH) of precipitation and stream waters (discharge point from the catchment) during the period 1999–2006 (The trends were approximated by simple linear regression models – significant at p-level = 0.05)

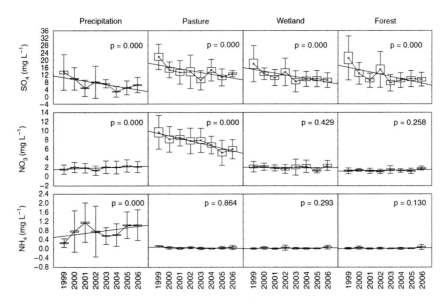

Fig. 5.4 Time series of chemical characteristics (SO_4^{2-}, NO_3^-, NH_4^+) of precipitation and stream waters (discharge point from the catchment) during the period 1999–2006 (The trends were approximated by simple linear regression models – significant at p-level = 0.05)

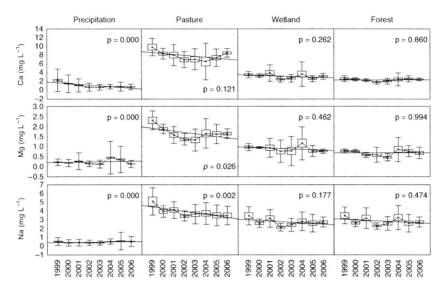

Fig. 5.5 Time series of chemical characteristics (Ca^{2+}, Mg^{2+}, Na^+) of precipitation and stream waters (discharge point from the catchment) during the period 1999–2006 (The trends were approximated by simple linear regression models – significant at p-level = 0.05)

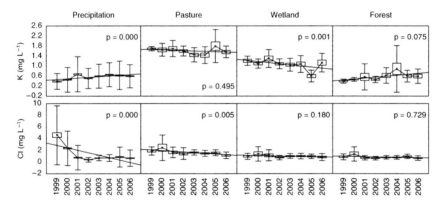

Fig. 5.6 Time series of chemical characteristics (K^+, Cl^-) of precipitation and stream waters (discharge point from the catchment) during the period 1999–2006 (The trends were approximated by simple linear regression models – significant at p-level = 0.05)

5.3.4.2 pH, Nitrate and Ammonium

Since 1999, the pH in bulk precipitation has increased by about 0.6 units. Nitrate concentration in bulk precipitation has moderately increased from 2.1 to 2.9 mg NO_3^- l^{-1} though NO_3^- concentration in stream water markedly decreased (from 9.7

to 6.0 mg NO_3^- l^{-1}). Nitrate concentrations in stream water from the drained pasture are evidently higher than those in stream water from wetland and forest also in 2006 (Mlýnský 6.0, Horský 2.4 and Bukový 1.8 mg NO_3^- l^{-1}). Annual NH_4^+ concentration in bulk precipitation showed a moderate increase with a large fluctuation, but annual mean concentrations in water of all catchments were remarkably constant over time.

5.3.4.3 Calcium and Chlorides

Concentrations have decreased both in bulk precipitation and in stream water from the drained pasture. On the other hand, the annual mean concentration of Ca^{2+} and chlorides are remarkably constant over time both in the wetland stream (Ca^{2+}: 2.5–3.9 mg l^{-1}, Cl$^-$: 0.7–1.4 mg l^{-1}) and in the forest stream (Ca^{2+}: 1.9–2.7 mg l^{-1}, Cl$^-$: 0.6–1.3 mg l^{-1}).

5.3.4.4 Magnesium, Sodium and Potassium

Annual mean Mg, Na and K concentrations in bulk precipitation are relatively low and constant over time. During the monitoring period, higher Mg, Na and K concentrations were measured in stream water from the drained Mlýnský catchment in comparison with Horský and Bukový.

5.4 Discussion

The three small catchments were selected as experimental sites for long-term monitoring in 1997. Since 1999, relationships between trends of matter concentrations in precipitation and in discharged water have been studied. The greatest changes have been recorded in the drained pasture (Mlýnský stream), where concentrations of almost all the monitored parameters decreased during the 7-year period. To a smaller extent, the concentrations decreased also in wetlands (Horský stream) and in forest (Bukový stream).

Annual mean concentrations of base cations (Ca^{2+}, Mg^{2+}, Na^+, K^+) in bulk precipitation were relatively low and constant over time. In discharged water, markedly higher concentrations of all base cations were found over time in Mlýnský catchment in comparison with Horský and Bukový catchments, although the storage (pool) of these ions in soil of the Mlýnský catchment is the lowest (Procházka et al., 2001).

The total annual transport of Ca^{2+} and Mg^{2+} from the Mlýnský stream catchment is five to six times higher than that from the Horský and Bukový catchments. The high transport from the drained Mlýnský catchment is emphasised by an increased

Table 5.3 Average annual matter budget (kg ha^{-1} year^{-1}) in precipitation water and in the water discharged from individual catchments

2000–2004	NO$_3^-$	NO$_2^-$	NH$_4^+$	Cl$^-$	SO$_4^{2-}$	Ca^{2+}	Mg^{2+}	Na$^+$	K$^+$
Precipitation	17.9	0.5	8.0	11.6	72.9	10.9	1.8	3.8	5.5
Mlýnský	76.1	0.2	0.5	16.8	119.1	72.0	15.3	34.5	14.9
Horský	12.6	0.1	0.2	7.0	76.4	21.4	5.5	17.1	7.6
Bukový	11.5	0.0	0.1	5.5	57.3	14.0	3.4	13.7	2.6

water discharge associated with lower evapotranspiration. The Mlýnský stream transports about 72 kg Ca from each hectare annually, and about the same amount of K, Na and Mg together (Table 5.3). This represents the equivalent of 500 kg of inorganic fertiliser per hectare that would have to be applied every year – in order to compensate for the losses of base cations and thus neutralise the protons released through mineralisation (Procházka et al., 2006).

Since the 1990s a decrease of acid deposits in mountains of Central Europe has been observed (Oulehle et al., 2006). In the Šumava Mountains (Bohemian Forest), Kopáček et al., (1998, 2001) and Hruška et al. (2005) described the slowdown of acidification of local lakes due to reduction of SO$_4^{2-}$ deposits.

The occurrence of base-rich alkaline groundwater has been confirmed at Beddgelert Forest of north Wales (Reynolds et al., 2004) in common with upland acid-sensitive sites elsewhere in Wales (Neal et al., 2001, 2004) and in Scotland (Soulsby et al., 1998). Chemical composition of discharged waters from the three studied catchments of Šumava mountains is controlled by processes in soil. Considering granite as a main geological substrate in the catchments studied, no accumulation of Ca in groundwater and therefore no subsequent effect of groundwater on Ca content in discharged water is expected (Stevens et al., 1989).

Nitrate concentration in bulk precipitation in the study area has moderately increased, though NO$_3^-$ concentration in stream water markedly decreased. The marked decline of NO$_3^-$ concentration in the drained pasture stream is dominated by the effect of non-fertilising after change of political situation in 1989. Nevertheless, NO$_3^-$ concentrations in stream water from the drained pasture are evidently higher than in stream water from the wetland and forested catchments. Our findings of annual deposition of 18 kg NO$_3^-$ per hectare (Procházka et al., 2006) agree well with annual matter depositions in areas of GEOMON catchments monitoring network described by Fottová (2003). For the Liz catchment of the Šumava mountains, the annual depositions of NO$_3^-$ during 7 years of monitoring (1994–2000) were between 12 and 40 kg ha^{-1} year^{-1} at the average of 21 kg a^{-1} year^{-1}. However, the Mlýnský catchment discharged four times more NO$_3^-$ in comparison with its atmospheric deposition and six times more than was discharged from the Horský and Bukový catchments. We suppose that the considerably higher losses of ions from the Mlýnský catchment resulted from drainage and subsequent mineralisation of soil (Procházka et al., 2006).

The lower mean content of organic matter in the soil of the Mlýnský stream catchment supports the expectation that drained soils are subject to a faster mineralisation of organic matter (Procházka *et al.*, 2001). During the decomposition of organic matter, acids (NO_3^-, SO_4^{2-}) and CO_2 are formed and protons are released. The proton H^+ then exchanges for Ca and Mg, important base cations readily soluble in the soil water, and, if not required immediately by growing vegetation, are transported away with the water outflow from the catchment. This process of 'soil acidification', i.e. removing base cations that can neutralise acid rain, might be more important than the impact of acid rain itself (Thimonier *et al.*, 2000). Rees and Ribbens (1995) showed the exacerbated problem of long-term acid deposition and, more recently, large-scale afforestation in the region of Galloway (Scotland, UK) in relation to soil and water acidification.

Deposition of SO_4^{2-} in precipitation and stream SO_4^{2-} concentration trends show a delayed response in SO_4^{2-} output in catchments to decreased deposition. Sulphate is still the dominant acid in precipitation in Šumava Mountains, but its concentration relative to NO_3^- has markedly declined. Prechtel *et al.* (2001) suggest that differences in response to declining S deposition are linked to the size of soil S pools. Kopáček *et al.* (2001) presented supporting evidence from detailed S budgets for alpine and forest catchments, which indicated that, under current deposition scenarios, the forest catchments with substantial soil cover would take many years to leach the accumulated soil S and ultimately reach equilibrium conditions.

The effect of climatic variability and long-term changes in pollutant deposition has patterns of variation of water quality. The contrasting hydrochemistry of the three catchments further demonstrates differences of land use together with the complexity and heterogeneity of water balances and groundwater systems in upland catchments. Catchments with wetland and forest appear to be more resistant to recovery in terms of conductivity and alkalinity in comparison with the relatively strong signal fast response from the drained pasture. High NO_3^- and bicarbonate concentrations in the water discharged from the pasture result most probably from fast mineralisation (oxidation) of soil organic matter. As a consequence, loss of base cations from soil as well as soil and water acidification is expected to continue in the drained acidified Mlýnský catchment.

5.5 Conclusions

Changes in the chemistry of bulk precipitation and stream water between 1999 and 2006 are described for small drained, wetland and forest catchments located within the Bohemian Forest in south-west of the Czech Republic.

Over the 7-year monitoring period, the annual mean pH of bulk precipitation increased from 4.8 to 5.4 whilst the annual mean SO_4^{2-} concentration decreased from 16.8 mg SO_4^{2-} l^{-1} in 1999 to 7.2 mg SO_4^{2-} l^{-1} in 2006. The decrease in atmospheric deposition is reflected by decreased annual mean SO_4^{2-} concentrations in all three streams.

Nitrate concentration in bulk precipitation in the study area has moderately increased, though in stream water from pasture it markedly decreased. Nevertheless,

NO_3^- concentrations in stream water from the drained pasture (7.5 mg NO_3^- l^{-1}) are evidently higher than those in stream waters from wetland (2.0 mg NO_3^- l^{-1}) and forest (2.2 mg NO_3^- l^{-1}).

Markedly higher concentrations of all base cations in stream water are over time from the drained pasture (Mlýnský catchment) as compared with the wetland (Horský) and the forest (Bukový). The annual mean concentrations of base cations in bulk precipitation are relatively low and constant over time. As a consequence, higher loss of base cations from soil of drained pasture can be expected to continue in the drained and acidified Mlýnský catchment. Synergic effect of mineralisation (accompanied by acidification of the soil), together with acid depositions emphasises acidification of the soil in the drained pasture.

With regard to the latter, the long-term data sets at the three catchments in Šumava Mountains and other similar sites are of immense value in providing relevant pilot data for model and guideline of predictions and quantitative evidence of sustainability management for mountain and submontane areas. The presented results support the concept of landscape sustainability formulated by Ripl (2003) in terms of chemical efficiency: drained soil shows long-term higher matter losses and acidification in comparison with wetlands and forest.

Acknowledgement The study has been supported by MSM project no. 6007665806, Projects NPVII 2B06023 and Z60870520.

References

Bossard, M., Feranec, J., & Otahel, J. (2000). CORINE land cover technical quide – Addendum 2000. *Technical report no.40, May 2000, European Environmental Agency.*

Fottová, D. (2003). Trends in sulphur and nitrogen deposition fluxes in the GEOMON network, Czech Republic, between 1994 and 2000. *Water, Air, & Soil Pollution, 150,* 73–89.

Hakrová, P. (2003). Studium podmínek pro podporu druhové diverzity travních porostů. *PhD thesis. University of South Bohemia, Faculty of Agriculture. České Bud jovice,* 132 p. (in Czech).

Hakrová, P., & Procházka, J. (eds.) (1998). Project Ministry of Youth, Sports and Education VS 96 072, Running report, II. Expert part. *University of South Bohemia, Faculty of Agriculture, České Budějovice* (in Czech).

Hruška, J., Hofmeister, J., Oulehle, F., Kopáček, J., Vrba, J., Metelka, V., Tesař, M., Šír, M., Máca, P., & Beudert, B. (2005). Biogeochemické cykly ekologicky významných prvků v měnících se přírodních podmínkách lesních ekosystémů NP Šumava. *Závěrečná zpráva projektu VaV/1D/1/29/04,* 69 p. (in Czech).

Kopáček, J., Hejzlar, J., Stuchlík, E., Fott, J., & Veselý, J. (1998). Reversibility of acidification of mountain lakes after reduction in nitrogen and sulphur emissions in central Europe. *Limnology and Oceanography, 43,* 357–361.

Kopáček, J., Veselý, J., & Stuchlík, E. (2001). Sulphur and nitrogen fluxes and budgets in Bohemian Forest and Tatra Mountains during the industrial revolution (1850–2000). *Hydrology and Earth System Sciences, 5*(3), 391–405.

Neal, C., Reynolds, B., Neal, M., Pugh, B., Hill, B., & Wickham, H. (2001). Long-term changes in the water quality if rainfall, cloud water and stream water for moorland, forested and clear-felled catchments at Plynlimon, mid-Wales. *Hydrology and Earth System Sciences, 5,* 459–476.

Neal, C., Reynolds, B., Neal, M., Wickham, H., Hill, L., & Pugh, B. (2004). The impact of conifer harvesting on stream water quality: The Afon Hafren, mid-Wales. *Hydrology and Earth System Sciences, 8,* 460–484.

Oulehle, F., Hofmeister, J., Cudlín, P., & Hruška, J. (2006). The effect of reduced atmospheric deposition on soil and soil solution chemistry at a site subjected to long-term acidification, Načetín, Czech Republic. *Science of Total Environment, 370,* 532–544.

Prechtel, A., Alewell, C., Armbruster, M., Bittersohl, J., Cullen, J.M., Evans, C.D., Helliwell, R., Kopáček, J., Marchetto, A., Matzner, E., Meesenburg, H., Moldan, F., Moritz, K., Veselý, J., & Wright, R.F. (2001). Response of sulphur dynamics in European catchments to decreasing sulphate deposition. *Hydrology and Earth System Sciences, 5,* 311–325.

Procházka, J., Hakrová, P., Pokorný, J., Pecharová, E., Hezina, T., Šíma, M., & Pechar, L. (2001): Effect of different management practices on vegetation development, losses of soluble matter and solar energy dissipation in three small sub-mountain catchments. In J. Vymazal (Ed.), *Transformations of Nutrients in Natural and Constructed Wetlands* (pp. 143–175). Leiden, The Netherlands: Backhuys.

Procházka, J. (2004). Hodnocení koloběhu vody, látek a disipace sluneční energie s různým způsobem hospodaření na příkladu vybraných dílčích povodí. *PhD thesis. University of South Bohemia, Faculty of Agriculture. České Budějovice,* 210 p. (in Czech).

Procházka, J., Včelák, V., Wotavová, K., Štíchová, J., & Pechar, L. (2006). Holistic concept of landscape assessment: Case study of three small catchments in the Šumava Mountains. *Ekológia (Bratislava), 25*(Supplement 3/2006), 5–17.

Rees, R.M., & Ribbens, J.C.H. (1995). Relationships between afforestation, water chemistry and fish stock in an upland catchment in south-west Scotland. *Water, Air, & Soil Pollution, 85,* 303–330.

Reynolds, B., Stevens, P.A., Brittain, S.A., Norris, D.A., Hughes, S., & Woods, C. (2004). Long-term changes in precipitation and stream water chemistry in small forest and moorland catchments at Bedgelert Forest, north Wales. *Hydrology and Earth System Sciences, 8,* 436–448.

Ripl, W. (1995). Management of water cycle and energy flow for ecosystem control: The energy-transport-reaction (ETR) model. *Ecological Modelling, 78,* 61–76.

Ripl, W. (2003). Water: The bloodstream of the biosphere. *Philosophical Transaction of the Royal Society London B, 358,* 1921–1934.

Stevens, P.A., Hornung, M., & Hughes, S. (1989). Solute concentration, fluxes and major nutrient cycles in mature Sitka spruce plantation in Beddgelert Forest, North Wales. *Forest Ecological Management, 27,* 1–20.

Soulsby, C., Chen, M., Ferrier, M., Helliwell, R.C., Jenkins, A., & Harriman, R. (1998). Hydrochemistry of shallow groundwater in an upland Scottish catchment. *Hydrological Processes, 12,* 1111–1118.

Thimonier, A., Dupouey, J.L., & le Tacon, F. (2000). Recent losses of base cations from soil of *Fagus sylvatica* L. stands in northeastern France. *Ambio, 29,* 314–321.

Chapter 6
Dynamics of Litterfall and Decomposition in Peatland Forests: Towards Reliable Carbon Balance Estimation?

Raija Laiho[1]($\boxtimes$), Kari Minkkinen[1], Jani Anttila[1], Petra Vávřová[1,2], and Timo Penttilä[2]

Abstract The vast carbon (C) stores in peat soils may be seriously affected by different land-uses, or changes in the prevailing climatic patterns. Land use in peatlands usually includes artificial drainage. Any land-use mediated changes in C emissions from peatlands need to be estimated for greenhouse gas (GHG) reporting. This is not an easy task, since all factors that affect the dynamics of litter inputs and decomposition of organic matter are affected, and the changes may be different in different peatland types, or under different climates. This chapter describes an approach for estimating the C balance of peatland forests on site level that combines measured and modelled information on litterfall and decomposition. Further, we outline the most critical data needs. According to our estimates, the soils of boreal peatland forests may act as either net sinks or net sources of C depending on the ratio of litter inputs to decomposition outputs. The dynamics of below-ground litters, especially, and moss litters are most poorly known.

Keywords Carbon balance, decomposition, litter, peatland forests

6.1 Introduction

Peatlands have been drained to improve forest growth in many parts of northern Europe/Eurasia, and to a lesser extent in other parts of the world, e.g. UK and North America. Altogether, the area of drained peatland forests has been estimated as 15 million hectares (Joosten & Clarke, 2002). Following drainage, many ecosystem characteristics and functions change as compared to the undrained situation. All factors that affect the dynamics of inputs and decomposition of organic matter are

[1] Peatland Ecology Group, University of Helsinki, Department of Forest Ecology, Helsinki, Finland

[2] Finnish Forest Research Institute, Vantaa Research Unit, Vantaa, Finland

($\boxtimes$) Corresponding author: e-mail: raija.laiho@helsinki.fi

J. Vymazal (ed.) *Wastewater Treatment, Plant Dynamics and Management in Constructed and Natural Wetlands,*
© Springer Science + Business Media B.V. 2008

affected: soil moisture and aeration (Boggie, 1977; Silins & Rothwell, 1999), soil temperature and acidity (Minkkinen *et al.*, 1999), soil decomposer community (Jaatinen *et al.*, 2007), and last but by far not least, vegetation composition (Laine *et al.*, 1995) and the amount and quality of litterfall (Laiho *et al.*, 2003). These factors further have largely unexplored interactions (Laiho, 2006). The balance between inputs and decomposition of organic matter in turn translates into the carbon (C) balance of the site. Previous research has shown that a drained boreal or temperate peatland forest may act either as a sink or a source of C depending on the case (Minkkinen & Laine, 1998; Minkkinen *et al.*, 1999; Hargreaves *et al.*, 2003), but the factors controlling this variation remain insufficiently known.

EU countries have committed to reporting their greenhouse gas (GHG) emissions annually to the European Commission and to the Secretariat of the United Nations Framework Convention on Climate Change (UNFCCC). Due to the vast C store in peat, any land use on peatland may be of great significance and its effects on the C stock should be accurately known for reliable GHG reporting. For instance, during the development of the reporting system for Finnish forests, it has become obvious that peatland forests will determine the C balance of this sector. If they lose C from their soils, it will by far override any sink function of upland forests, which largely depends on changes in their tree stand C pools only. Also in countries with smaller peatland areas, their contribution to GHG exchange may be significant (Byrne *et al.*, 2004).

Recent research that has mainly been based on gas-exchange studies has given us valuable insights into the net sink/source function of peatlands (e.g. Alm *et al.*, 2007a). The eddy covariance method can be used to estimate the whole ecosystem balance, while the chamber methods allow the analysis of spatial variation and the factors in effect (Alm *et al.*, 2007b). Still, these methods provide only limited possibilities for analyzing the components of the C cycle in more detail, and thus, it remains difficult to predict the changes in C fluxes and C balance under changing environmental conditions. The aim of this chapter is to describe an approach of estimating C balance on site level by combining measured and modelled information on litterfall and decomposition. In longer term, our goal is to develop a peatland ecosystem model running these C fluxes and capable of responding to changes in, e.g. water level and nutrient regime. Such a model is needed for reliable predictions of changes in C balance under changing environmental conditions, and, e.g. comparing the effects of optional land-use or management measures. Data needs for model development will be outlined.

6.2   Litter Inputs

Peatlands are generally perceived as systems where either *Sphagnum* mosses or sedges play the most significant roles in C cycling and sequestration; however, the tree layer tends to dominate biomass composition whenever present (Reinikainen *et al.*, 1984) and this dominance becomes even more evident following drainage (Laiho *et al.*, 2003). Only when the nutrient regime of the site is too poor to sustain forest growth will the tree component remain minor in drained sites (Vasander, 1982). This does not necessarily mean that ground vegetation would be an insignificant component in the C balance of peatland forests, however.

The only litterfall data sequence from peatland forests reported so far is that of Laiho *et al.* (2003; Fig. 6.1) who studied oligotrophic pine fens, a peatland type commonly drained for forestry (Keltikangas *et al.*, 1986). Their data were in part based on direct measurements, but relied on indirect estimates concerning some components such as moss and root litters. However, they may be used as an indicator of which vegetation components should be studied in more detail. The importance of arboreal (tree and shrub) litter in peatland forest is prominent (Fig. 6.1). In more nutrient-rich sites this would be largely tree-litter, while on nutrient-poor sites with often a prominent shrub layer, shrub litter increases in importance. Yet, mosses seem to be able to retain a relatively high biomass, with presumably high litter inputs, even following drainage. The pattern suggested by the estimates of Laiho *et al.* (2003) is confirmed by further measurements in a range of peatland types; moss production was always highest in the drained peatland forest (Fig. 6.2), while the recession suggested to take place soon following drainage (Fig. 6.1) depended on peatland type. Also the changes taking place in species composition following drainage depended on peatland type (J. Anttila and R. Laiho, 2007; see Laine *et al.*, 1995). More data on moss production in different conditions are still needed.

For the tree layer, biomass may be predicted based on tree diameter distribution (e.g. Laiho *et al.*, 2003), and litterfall may be estimated based on turnover rates of different biomass components (Muukkonen & Lehtonen, 2004; Muukkonen, 2005; Lehtonen *et al.*, 2004). Peatland stands may differ from (managed) mineral soil stands structurally (Sarkkola *et al.*, 2005), and therefore, models based on non-peatland stand data should be tested with peatland data. Such test data from

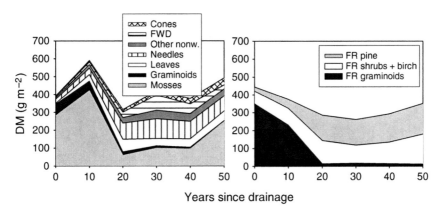

Fig. 6.1 Annual above-ground (*left*) and belowground (*right*) litterfall by litter type in an age sequence of pristine and drained pine-dominated oligotrophic peatland forests in southern Finland. Year zero = pristine situation (Data from Laiho *et al.*, 2003); FWD = fine woody debris; Other non-w. = unidentified non-woody tree litter (male flowers, bark fragments, etc.); FR = fine root litter; estimated for roots with a diameter <2 mm only

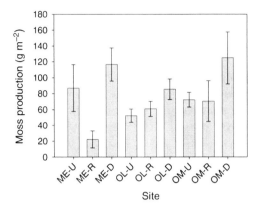

Fig. 6.2 Moss litter production, estimated as annual dry mass production, in some peatland sites of varying nutrient regime and drainage status, in southern Finland. Bars represent averages of 6–12 plots, and error bars represent standard errors of mean (J. Anttila and R. Laiho, 2007); OL = oligotrophic; ME = mesotrophic; OM = ombrotrophic; DR = drained for 45 years; RD = drained for 5 years; UD = undrained

peatland stands are still scarce, and more measurements are needed. Two approaches may be used for estimating litterfall from ground vegetation. First, either an average ground vegetation biomass is predicted for different site types at different stages of succession, or the biomass is predicted based on inventoried species coverages (Muukkonen *et al.*, 2006). In both cases, litter production may then be estimated based on biomass by species or species groups.

The dynamics of root litter inputs into the soil and their decomposition are still poorly known in peatland forests, as in most forest ecosystems, due to methodological difficulties (e.g. Wallén, 1992). Root biomass estimates have been presented for different types of sites (e.g. Håland & Brække, 1989; Laiho & Finér, 1996) but there is little information on their turnover rates. Finér and Laine (1998, 2000) applied two methods for estimating fine root production and turnover in drained peatland sites, but their results were somewhat obscured by the high spatial and temporal variation. This is probably the most critical gap in our knowledge of the C fluxes in peatland forests.

6.3   Decomposition

Different litter types may have greatly differing decomposition rates; that is why it is important to consider the changes in vegetation during the peatland forest succession (see review by Laiho, 2006). Generally speaking, moss, especially *Sphagnum*, litters tend to decompose slowly. Arboreal litters decompose slower than graminoid and herbaceous litters, and broadleaf litters generally decompose faster than conifer litters. Even though litter types differ in decomposability, the environmental conditions prevailing on the site may also have a significant effect on mass loss rates (Figs. 6.3–6.5).

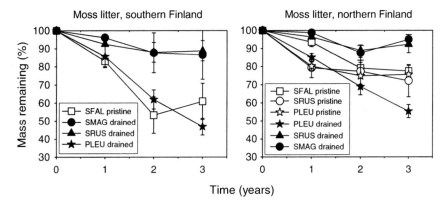

Fig. 6.3 Decomposition of moss litter incubated in the (moss) litter layer in pristine and drained oligotrophic–mesotrophic peatland forest sites in southern and northern Finland (R. Laiho and T. Penttilä, 2007); SFAL = *Sphagnum fallax;* SMAG = *S. magellanicum;* SRUS = *S. russowii;* PLEU = *Pleurozium schreberi*

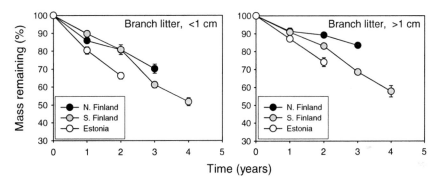

Fig. 6.4 Decomposition of branch litter of two diameter classes incubated in the litter layer in drained mesotrophic peatland forests on a climatic gradient from northern Finland to Estonia (P. Vávřová, T. Penttilä and R. Laiho, 2007)

The essential question is how much of the litter will remain undecomposed in the long run, or whether all is eventually decomposed (Latter *et al.*, 1998; Prescott, 2005). In that respect, some *Sphagnum* mosses growing in relatively dry sites seem more 'promising' than, e.g. forest (feather) mosses and branch litter (Figs. 6.3 and 6.4). Needle litter is known to remain partially undecomposed in forest sites (Berg *et al.*, 1996), and the proportion seems to be rather similar in drained peatland forests (Domisch *et al.*, 2006).

Several models have been developed to simulate decomposition in peatlands (e.g. Frolking *et al.*, 2001; Bauer, 2004); however, these models are, as a rule, not yet capable of predictions for rapidly evolving plant communities under changing hydrological conditions. The Yasso model (Liski *et al.*, 2005) is currently used in

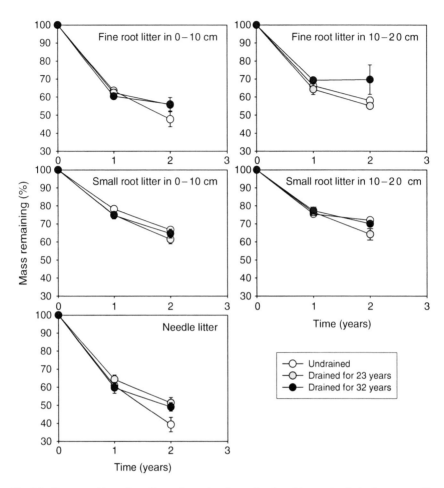

Fig. 6.5 Decomposition of root litter of two size classes incubated in two depths in the peat profile, and of needle litter incubated in the litter layer, in a pristine and two drained oligotrophic peatland forests in southern Finland (Data from Laiho *et al.*, 2004); Fine roots = diameter <2 mm; Small roots = diameter 2–5 mm

estimating the decomposition dynamics of above-ground litter in the GHG reporting for Finnish forests (Statistics Finland, 2006), both on upland and peatland sites. Testing in peatland conditions has, however, indicated poor sensitivity to variations in site nutrition and moisture conditions (T. Penttilä *et al.*, 2008).

6.4   Examples of Carbon Balance Estimations

We estimated the soil C balance of two peatland forest sites, which represent dwarf-shrub type (**DsT**; initially ombrotrophic, Vatkg in the Finnish system; see Westman & Laiho, 2003) and *Vaccinium vitis-idaea* type (**VT**, initially oligotrophic, Ptkg), located in Vesijako Research Forest, Padasjoki, southern Finland. The sites were drained for forestry already in 1915 and currently support Scots pine (*Pinus sylvestris*) dominated stands with stand stem volumes of 164 m³ ha⁻¹ and 185 m³ ha⁻¹, respectively. The same sites were intensively studied for their Heterotrophic soil respiration (SR) fluxes during 2001–2004 using the chamber method (Minkkinen *et al*, 2007; Alm *et al*., 2007b). Their results are currently used as best estimates of SR for all forestry-drained peatlands in the GHG reporting for Finnish forests (Statistics Finland, 2006). Here we estimate, as follows, the C fluxes to soil from aboveground (AG) and below-ground (BG) litter components to obtain the soil C balance, i.e. the difference of C inputs and SR outputs, for the corresponding time period.

The net C flux to soil from AG litter was estimated as the difference of total C input in AG litter and the C output as carbon dioxide (CO_2) flux from AG litter decomposition to atmosphere. We measured AG litter fall from the tree stands with litter traps resulting in annual means of 323 gm⁻² and 434 gm⁻² dry mass for the DsT and the VT site, respectively. AG litter from the ground vegetation, consisting of mire shrubs (*Ledum palustre, Betula nana*), forest shrubs (*V. myrtillus, V. vitis-idaea, Empetrum nigrum*), bryophytes (*Hylocomium splendens, Pleurozium schreberi, Dicranum polysetum, Sphagnum angustifolium, S. capillifolium, S. magellanicum, S. russowii*), and the few herbs and grasses (mainly *Rubus chamaemorus, Eriophorum vaginatum*), was estimated using an average of the values from Laiho *et al*. (2003): 250, 12, and 4 g m⁻² dry mass of annual litter for bryophytes, shrub leaves, and herbs and grasses, respectively. The same values were used for both sites.

Minkkinen *et al*. (2007) used three treatments for detecting the following sources of organic matter (OM) decomposition contributing to the CO_2 flux from soil to atmosphere:

1. Heterotrophic decomposition of the peat soil (= SR); live roots and new litter were eliminated
2. SR + live roots (new litter eliminated)
3. SR + live roots + new litter

We estimated the decomposition of newly fallen AG litter as the difference in the average annual CO_2 fluxes of treatments (3) and (2) during 2001–2003. This approach probably underestimated the rates of moss litter decomposition to an unknown extent, as new moss litter did not deposit to the measurement plots during the monitoring period.

The annual estimates of BG litterfall from roots of trees and ground vegetation were considered inputs to soil organic matter as such, as the respective decomposition of roots was accounted for in the SR fluxes (Minkkinen *et al*., 2007). For estimating fine root (<2 mm) litter from the tree stands, we first calculated the stand-level foliar biomass for the Scots pine and Norway spruce trees by applying Marklund's (1988) biomass functions to measured tree diameter at breast height (DBH) and height data, resulting in 258 and 313 g m⁻² of pine foliage, and 140 and 86 g m⁻² of spruce foliage, for the DsT and VT stands, respectively. We then applied the following formula for estimating stand-level, annual inputs of fine root litter to soil, relying on fine root to foliage biomass ratios reported for Scots pine and

Norway spruce on upland sites (Helmisaari *et al.*, 2007) and average pine fine root turnover rates reported for drained peatland sites (Finér & Laine, 1998; see also Laiho *et al.*, 2003):

$$DM_{i\_litter\_fr} = t_{i\_fr} \times rf_i \times DM_{i\_fol} \qquad (1)$$

where:

$DM_{i\_litter\_fr}$ = dry mass of stand-level annual fine root litter from tree species *i*
t_{i_fr} = turnover rate of fine root biomass for tree species *i*
rf_i = ratio of live fine root biomass to foliar biomass for tree species *i*
DM_{i_fol} = stand-level foliar biomass for tree species *i*

using the following parameter values for both sites:

t_{i_fr} = 0.65 for Scots pine and Norway spruce
rf_i = 0.676 for Scots pine, 0.256 for Norway spruce

Root litter from ground vegetation and understorey birch was estimated using an average of the values reported by Laiho *et al.* (2003).

The SR fluxes also included CO_2 evolving from decaying coarse roots of trees removed in commercial harvests during some decades before the monitoring period. Thus, an estimate of coarse root litter needed to be added as the corresponding input to soil OM. This estimate was obtained by simulating the earlier stand development with the MOTTI stand simulator (Hynynen *et al.*, 2005). The commercial thinnings performed in the stands were taken into account following the procedures of the true stand management histories as well as possible. For the trees removed in thinnings, we predicted coarse root litter biomass with Marklund's (1988) biomass functions and added 11% to the resultant values to obtain coarse root biomasses down to root diameter 2 mm (Petersson & Ståhl, 2006). The stand-level dry mass estimates of coarse root litter, presumed to have very low rates of decomposition, were then aggregated over the whole simulation period and transformed to average annual values of 37 and 83 g m^{-2} for DsT and VT, respectively.

The total soil C balance obtained from the input components presented above and the SR output estimates of −240 (DsT) and −266 g (VT) C m^{-2} year^{-1} (Minkkinen *et al.*, 2007) resulted in a net sink of 126 g C m^{-2} year^{-1} (DsT) and a net source of −58 g C m^{-2} year^{-1} (VT) between soil and atmosphere during the monitoring period (Fig. 6.6). These estimates do not account for the C sink due to the increasing tree stand biomass. The reliability of the SR estimates has been discussed earlier (Minkkinen *et al.*, 2007; Alm *et al.*, 2007a). Regarding C inputs to soil, the estimates presented here for different litter fractions include uncertainty to an unknown extent. While the measured magnitudes of AG tree litter may be considered relatively reliable, the AG litter component of an almost similar order of magnitude, i.e. litter from mosses, had to be based on indirect estimation. Differences in the estimated decomposition rates of AG litter and the resultant net input to the soil OM between the two site types are worth noting. While all AG litter appears to be rapidly decomposed in the oligotrophic (VT) site, the ombrotrophic

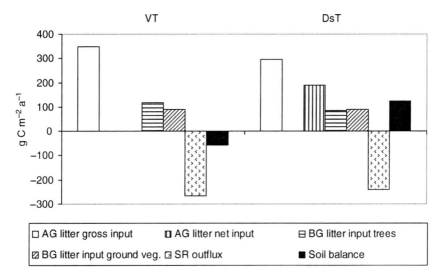

Fig. 6.6 An example of C balance estimation. Litter inputs, soil respiration, and soil C balance in two sites in a drained peatland in southern Finland, 2001–2003. DsT = dwarf-shrub type drained peatland forest (Vatkg in Finnish), VT = *Vaccinium vitis-idaea*-type drained peatland forest (Ptkg)

(DsT) site obviously accumulates new soil OM from the newly fallen litter. This difference has also become apparent from currently ongoing decomposition experiments where mesh bags were installed on the soil surface on both sites: a considerable amount of new OM has deposited on the bags at the DsT site and much less if any at the VT site. Below-ground litter from roots comprised solely the estimated C input to the soil on the VT site, and ca. half of the input on the DsT site (Fig. 6.6). Here we suspect the root litter from the ground vegetation (shrubs) to be still underestimated for the DsT site as the shrub root biomass estimates for both sites were derived from VT sites where shrubs generally are less abundant.

6.5 Future Research and Development Needs

For more reliable estimation of the C balance of forested organic soils we need to improve our level of knowledge regarding at least two crucial components of C inputs to soils, i.e. the production and turnover rates of (1) moss biomass, and (2) root biomass, and the responses of these to environmental factors, both climatic and nutritional. For the most problematic root component, an indirect method may be developed. From the findings of, e.g. Vanninen and Mäkelä (1999) and Helmisaari *et al.* (2007) on upland conifer stands, and from the data of Laiho and Finér (1996), it appears that models for estimating the variation between different sites in the ratio of fine root to foliage biomass of boreal tree stands could be based on the C:N or some other measure of nitrogen (N) availability in the rooting zone. This may also apply to the integrated fine root biomass of trees and ground vegetation, as it would seem that, in given radiation conditions, the level of total fine root biomass tends to remain fairly stable even after hydrological disturbance (Laiho *et al.*, 2003) although variation in the biomass ratio between trees and other vegetation may occur among sites of different N availability (Helmisaari *et al.*, 2007). Further, models for estimating the turnover rates of fine root biomass, possibly also related to soil N or water availability, also need to be developed.

Some potentially important C fluxes have not been considered so far; these include fluxes associated with water (inputs to soil as C in canopy throughfall and outputs as leaching with run-off water), and C input from mycorrhizal external hyphae. The latter process was recently highlighted by Godbold *et al.* (2006).

Acknowledgements This work was supported by the Academy of Finland (203585, 205090), and the Ministry of Agriculture and Forestry via the research programmes Pools and fluxes of carbon in Finnish forests and their socio-economic implications, and Greenhouse impact of the use of peat and peatlands in Finland.

References

Alm, J., Shurpali, N.J., Minkkinen, K., Aro, L., Hytönen, J., Laurila, T., Lohila, A., Maljanen, M., Martikainen, P.J., Mäkiranta, P., Penttilä, T., Saarnio, S., Silvan, N., Tuittila, E.S., & Laine, J. (2007a). Emission factors and their uncertainty for the exchange of CO_2, CH_4 and N_2O in Finnish managed peatlands. *Boreal Environment Research, 12*, 191–209.

Alm, J., Shurpali, N.J., Tuittila, E.S., Laurila, T., Maljanen, M., Saarnio, S., & Minkkinen, K. (2007b). Methods for determining emission factors for the use of peat and peatlands – flux measurements and modelling. *Boreal Environment Research, 12*, 85–100.

Boggie, R. (1977). Water-table depth and oxygen content of deep peat in relation to root growth of *Pinus contorta*. *Plant and Soil, 48*, 447–454.

Byrne, K.A., Chojnicki, B., Christensen, T.R., Drösler, M., Freibauer, A., Friborg, T., Frolking, S., Lindroth, A., Mailhammer, J., Malmer, N., Selin, P., Turunen, J., Valentini, R., & Zetterberg, L. (2004). *EU peatlands: Current carbon stocks and trace gas fluxes*. Carboeurope GHG, Report SS4. 58 p. http://gaia.agraria.unitus.it/ceuroghg/ReportSS4.pdf

Joosten, H., & Clarke, D. (2002). *Wise use of mires and peatlands*. Saarijärvi, Finland: International Mire Conservation Group: Greifswald, Germany; and International Peat Society: Jyväskylä, Finland.

Bauer, I.E. (2004). Modelling effects of litter quality and environment on peat accumulation over different time-scales. *Journal of Ecology, 92*, 661–674.

Berg, B., Ekbohm, G., Johansson, M.B., McClaugherty, C.A., Rutigliano, F., & Virzo De Santo, A. (1996). Maximum decomposition limits of forest litter types: a synthesis. *Canadian Journal of Botany, 74*, 659–672.

Domisch, T., Finér, L., Laine, J., & Laiho, R. (2006). Decomposition and nitrogen dynamics of litter in peat soils from two climatic regions under different temperature regimes. *European Journal of Soil Biology, 42*, 74–81.

Finér, L., & Laine, J. (1998). Root dynamics at drained peatland sites of different fertility in southern Finland. *Plant and Soil, 201*, 27–36.

Finér, L., & Laine, J. (2000). The ingrowth bag method in measuring root production on peatland sites. *Scandinavian Journal of Forest Research, 15*, 75–80.

Frolking, S., Roulet, N.T., Moore, T.R., Richard, P.J.H., Lavoie, M., & Muller, S.D. (2001). Modeling northern peatland decomposition and peat accumulation. *Ecosystems, 4,* 479–498.

Godbold, D.L., Hoosbeek, M.R., Lukac, M., Cotrufo, M.F., Janssens, I.A., Ceulemans, R., Polle, A., Velthorst, E.J., Scarascia-Mugnozza, G., de Angelis, P., Miglietta, F., & Peressotti, A. (2006). Mycorrhizal hyphal turnover as a dominant process for carbon input into soil organic matter. *Plant and Soil, 28,* 15–24.

Håland, B., & Brække, F.H. (1989). Distribution of root biomass in a low-shrub pine bog. *Scandinavian Journal of Forest Research, 4,* 307–316.

Hargreaves, K.J., Milne, R., & Cannell, M.G.R. (2003). Carbon balance of afforested peatland in Scotland. *Forestry, 76,* 299–317.

Helmisaari, H.S., Derome, J., Nöjd, P., & Kukkola, M. (2007). Fine root biomass in relation to site and stand characteristics in Norway spruce and Scots pine stands. *Tree Physiology, 27,* 1493–1504.

Hynynen, J., Ahtikoski, A., Siitonen, J., Sievänen, R., & Liski, J. (2005). Applying the MOTTI simulator to analyse the effect of alternative management schedules on timber and non-timber production. *Forest Ecology and Management, 207,* 5–18.

Jaatinen, K., Fritze, H., Laine, J., & Laiho, R. (2007). Effects of short- and long-term water-level drawdown on the populations and activity of aerobic decomposers in a boreal peatland. *Global Change Biology, 13,* 491–510.

Keltikangas, M., Laine, J., Puttonen, P., & Seppälä, K. (1986). Peatlands drained for forestry during 1930–1978: Results from field surveys of drained areas. *Acta Forestalia Fennica, 193,* 1–94. (In Finnish with English summary.)

Laiho, R. (2006). Decomposition in peatlands: Reconciling seemingly contrasting results on the impacts of lowered water levels. *Soil Biology & Biochemistry, 38,* 2011–2024.

Laiho, R., & Finér, L. (1996). Changes in root biomass after water-level drawdown on pine mires in southern Finland. *Scandinavian Journal of Forest Research, 11,* 251–260.

Laiho, R., Vasander, H., Penttilä, T., & Laine, J. (2003). Dynamics of plant-mediated organic matter and nutrient cycling following water-level drawdown in boreal peatlands. *Global Biogeochemical Cycles, 17*(2), 1053, doi:10.1029/2002GB002015.

Laiho, R., Laine, J., Trettin, C.C., & Finér, L. (2004). Scots pine litter decomposition along drainage succession and soil nutrient gradients in peatland forests, and the effects of inter-annual weather variation. *Soil Biology & Biochemistry, 36,* 1095–1109.

Laine, J., Vasander, H., & Laiho, R. (1995). Long-term effects of water level drawdown on the vegetation of drained pine mires in southern Finland. *Journal of Applied Ecology, 32,* 785–802.

Latter, P.M., Howson, G., Howard, D.M., & Scott, W.A. (1998). Long-term study of litter decomposition on a Pennine peat bog: Which regression? *Oecologia, 113,* 94–103.

Lehtonen, A., Sievänen, R., Mäkelä, A., Mäkipää, R., Korhonen, K.T., & Hokkanen, T. (2004). Potential litterfall of Scots pine branches in southern Finland. *Ecological Modelling, 180,* 305–315.

Liski, J., Palosuo, T., Peltoniemi, M., & Sievänen, R. (2005). Carbon and decomposition model Yasso for forest soils. *Ecological Modelling, 189,* 168–182.

Marklund, L.G. (1988). Biomass functions for pine, spruce and birch in Sweden. *Sveriges lantbruksuniversitetet, institutionen för skogstaxering, Rapport 45.* Umeå. 73 p. (In Swedish with English summary.)

Minkkinen, K., & Laine, J. (1998). Long-term effect of forest drainage on the peat carbon stores of pine mires in Finland. *Canadian Journal of Forest Research, 28,* 1267–1275.

Minkkinen, K., Vasander, H., Jauhiainen, S., Karsisto, M., & Laine, J. (1999). Post-drainage changes in vegetation and carbon balance in Lakkasuo mire, Central Finland. *Plant and Soil, 207,* 107–120.

Minkkinen, K., Laine, J., Shurpali, N.J., Mäkiranta, P., Alm, J., & Penttilä, T. (2007). Heterotrophic soil respiration in forestry-drained peatlands. *Boreal Environment Research, 12,* 115–126.

Muukkonen, P. (2005). Needle biomass turnover rates of Scots pine (*Pinus sylvestris* L.) derived from the needle-shed dynamics. *Trees - Structure and Function, 19,* 273–279.

Muukkonen, P., & Lehtonen, A. (2004). Needle and branch biomass turnover rates of Norway spruce (*Picea abies*). *Canadian Journal of Forest Research, 34,* 2517–2527.

Muukkonen, P., Mäkipää, R., Laiho, R., Minkkinen, K., Vasander, H., & Finér, L. (2006). Relationship between biomass and percentage cover in understorey vegetation of boreal coniferous forests. *Silva Fennica, 40*, 231–245.

Petersson, H., & Ståhl, G. (2006). Functions for below-ground biomass of *Pinus sylvestris*, *Picea abies*, *Betula pendula* and *Betula pubescens* in Sweden. *Scandinavian Journal of Forest Research, 21*, 84–93.

Prescott, C.E. (2005). Do rates of litter decomposition tell us anything we really need to know? *Forest Ecology and Management, 220*, 66–74.

Reinikainen, A., Vasander, H., & Lindholm, T. (1984). Plant biomass and primary production of southern boreal mire-ecosystems in Finland. *Proceedings of the 7th International Peat Congress* (pp. 1–20). Dublin: The Irish National Peat Committee/International Peat Society.

Sarkkola, S., Hökkä, H., Laiho, R., Päivänen, J., & Penttilä, T. (2005). Stand structural dynamics on drained peatlands dominated by Scots pine. *Forest Ecology and Management, 206*, 135–152.

Silins, U., & Rothwell, R.L. (1999). Spatial patterns of aerobic limit depth and oxygen diffusion rate at two peatlands drained for forestry in Alberta. *Canadian Journal of Forest Research, 29*, 53–61.

Statistics Finland. (2006). *Greenhouse gas emissions in Finland 1990–2004. National Inventory Report to the UNFCCC.* December 2006. Statistics Finland, Helsinki.

Vanninen, P., & Mäkelä A. (1999). Fine root biomass of Scots pine stands differing in age and soil fertility in southern Finland. *Tree Physiology, 19*, 823–830.

Vasander, H. (1982). Plant biomass and production in virgin, drained and fertilized sites in a raised bog in southern Finland. *Annales Botanici Fennici, 19*, 103–125.

Wallén, B. (1992). Methods for studying below-ground production in mire ecosystems. *Suo, 43*, 155–162.

Westman, C.J., & Laiho, R. (2003). Nutrient dynamics of peatland forests after water-level drawdown. *Biogeochemistry, 63*, 269–298.

Chapter 7
Near Infrared Reflectance Spectroscopy for Characterization of Plant Litter Quality: Towards a Simpler Way of Predicting Carbon Turnover in Peatlands?

Petra Vávřová[1,2] (✉), Bo Stenberg[3], Marjut Karsisto[2], Veikko Kitunen[2], Tarja Tapanila[2], and Raija Laiho[1]

Abstract The ability of near infrared reflectance spectroscopy (NIRS) for the rapid determination of several chemical properties of plant litters was investigated. The chemical properties included fractions that potentially affect decomposition dynamics in peatlands: total carbon, total nitrogen, extractable (soluble) substances, holocellulose (sum of cellulose and hemicelluloses), sulfuric acid (H_2SO_4) insoluble lignin (Klason lignin), copper oxide (CuO) oxidation products of lignin: vanillin (V1), vanillic acid (V2), acetovanillone (V3), 4-hydroxybenzaldehyde (P1), 4-hydroxyacetophenone (P2), 4-hydroxybenzoic acid (P3), and ferulic acid (C2), as well as carbon (C) to nitrogen (N), and N to lignin ratios. To maximize variability, the samples included litters of nine plant species, representing five groups of plant litter: graminoid, deciduous foliage, conifer foliage, wood, and moss. For each quality parameter we compared (1) the model calibrated for all the various litter types (combined dataset) and (2) the model calibrated only for pine branch litter (branch dataset). Relationships were found between the chemical properties and near infrared (NIR) spectra using partial least squares (PLS) regression. Using both the combined and the branch datasets, very good NIR calibrations were possible for total C and N, ash content, nonpolar (dichlormethane) extractives (NPE), ethanol extractives (EE), and total extractives (TE) (sum of all soluble substances), holocellulose, C:N, and N:lignin ratios based on r^2 that varied from 0.80 to 0.99. Using only the combined dataset very good calibrations were also possible for acetone extractives (AE), water extractives (WE), lignin, P2, P3, C2, V1, V2, and V3, with r^2 from 0.81 to 0.97. The models combining all the different litter types performed

[1] Peatland Ecology Group, Department of Forest Ecology, University of Helsinki, Finland

[2] Finnish Forest Research Institute, Vantaa Research Unit, Finland

[3] Department of Soil Sciences, SLU, Skara, Sweden

(✉) Corresponding author: e-mail: petra.vavrova@helsinki.fi

J. Vymazal (ed.) *Wastewater Treatment, Plant Dynamics and Management in Constructed and Natural Wetlands*,
© Springer Science + Business Media B.V. 2008

better than the models constructed for the pine branch litter only, based on r^2 and residual prediction deviation (RPD), the ratio of the Y-variable standard deviation to the root mean square error of prediction (RMSEP). However, RMSEP was generally smaller when a branch litter property was predicted by the branch model compared to the prediction using the combined model, indicating the potential of improving the NIR calibrations by constructing specific models for different groups of litter. NIRS proved to be an accurate and fast method for the determination of plant litter quality that might be highly relevant for decomposition and C dynamics in peatlands.

Keywords Decomposition, litter quality, NIRS, peatlands

7.1 Introduction

Litter decomposition is one of the key processes in ecosystem functioning. In peatlands, there is a relatively small imbalance between primary production and decomposition; only 2–16% of the net primary production of a peatland ecosystem gets deposited as peat (Päivänen & Vasander, 1994). Still, it has resulted in peatlands being a significant sink of atmospheric carbon (C), representing about 30% of the global soil C pools with their estimated reservoir of 455 Pg (10^{15} g) (Gorham, 1991).

The rate of litter decomposition and mineralization is regulated by the combination of several factors. The quality of plant litter is one of the most important ones (Laiho, 2006). Initial quality of litter has been used as a decay predictor in many studies. The characteristics used include nitrogen (N) concentration (e.g., Berg & Staaf, 1980; Aber & Melillo, 1980; Tian *et al.*, 1992), lignin concentration (Berendse *et al.*, 1987; Meentemeyer, 1978; Taylor *et al.*, 1991; Murphy *et al.*, 1998), the lignin:N ratio (Edmonds, 1987; Blair, 1988; Stohlgren, 1988; Taylor *et al.*, 1989), the C:N ratio (Taylor *et al.*, 1989), and concentration of water-soluble compounds (Berg & Ekbohm, 1991). Yet, the conventional chemical analyses involved in characterization of litter quality are costly, time-consuming, and therefore unlikely to be used routinely for the large sample sets used in complex ecological studies.

Near infrared reflectance spectroscopy (NIRS, 780–2,500 nm) is an established analytical technique which offers the potential for the accurate, quick, and inexpensive characterization of various organic materials. NIRS has been mainly applied in agricultural science, e.g., assessing forage quality (Norris *et al.*, 1976; Winch & Major, 1981; Duncan *et al.*, 1987; Shenk & Westerhaus, 1991; Garnsworthy *et al.*, 2000). In ecological studies, NIRS has been successfully used for the determination of organic compounds in soils (e.g., Dalal & Henry, 1986; Confalonieri *et al.*, 2001; Ludwig *et al.*, 2002; Martin *et al.*, 2002; Coûteaux *et al.*, 2003), peat (Beining *et al.*, 2001), or to characterize the composition of forest foliage and forest litters (Card *et al.*, 1988; Wessman *et al.*, 1988; McLellan *et al.*, 1991a, b; Joffre *et al.*,

1992; Lacaze & Joffre, 1994; Martin & Aber, 1994; Bolster *et al.*, 1996). NIRS has also been used as a direct decay predictor (Gillon *et al.*, 1993, 1999b; Bouchard *et al.*, 2003; Bruun *et al.*, 2005). Litter decomposability was actually predicted more accurately by NIRS than by the initial litter composition measured by standard chemical methods (Gillon *et al.*, 1999b).

Organic samples absorb near infrared (NIR) radiation mainly by C–H, N–H, and O–H bonds. The nature and number of bonds present in a sample, and thus the amount of radiation that might be absorbed, is determined by the chemical composition of the sample. Therefore, the NIR spectrum reflected back from a sample contains information about its chemical composition. Direct interpretation of NIR spectra is difficult because although the spectral characteristics of different compounds are unique, they are also broad and thus frequently overlap (Curran, 1989). However, NIRS can be used as an indirect method that estimates the chemical composition of samples by comparing their spectra with spectra of samples whose chemical composition is already known. During a calibration process, a multivariate statistical model is built to describe the relationship between NIR information and concentration of a known chemical component found in the samples. This model is then used to predict composition of samples with unknown composition. By selecting a small subset of samples for both conventional analyses and NIR calibration, the majority of samples only need to be analyzed for their NIR spectra – thus allowing the number of samples studied to be significantly increased without increasing the cost.

NIR spectroscopy has many advantages: it is a fast, inexpensive, and nondestructive technique that requires relatively small amounts of sample with little or no preparation. NIRS does not use or produce chemicals, and we can determine numerous parameters from one spectral measurement. However, NIRS is not a stand-alone method; its accuracy depends on the accuracy of the reference method used for the calibration and the accuracy of the NIR analyses. The robustness of a calibration also depends on the strength of the spectral features of the dependent parameter and if the features are direct or indirect.

The purpose of this study was to test the ability of NIRS to characterize the chemical composition of a variety of plant litters typical of different peatland sites, by developing NIR calibrations for N and several C fractions potentially affecting decomposition dynamics in peatlands.

7.2 Material and Methods

7.2.1 Litter Material

The sample set was selected with the aim of providing strong calibrations for the chemical parameters tested by maximizing the variability among sample compositions and spectral features. Litters of nine plant species, representing five groups of plant litter: graminoid, deciduous foliage, conifer foliage, wood, and moss, were collected at eight peatland sites. Six of the sites were situated in Finland, one in Estonia, and one in the Czech Republic (Table 7.1). The sites varied in their nutrient regime, drainage, and/or climatic conditions, and therefore the materials were assumed to represent large natural variation in the plant litter quality.

After collection, the litters were air dried at room temperature to about 6–8% moisture content and kept frozen prior to further analyses in order to minimize any changes in the chemical quality during storage.

The litters were grouped into three datasets: a_1, a_2, and b (Table 7.1). Combined dataset a_1 included all the litter types studied for calibration of total C, total N, and ash content. Combined dataset a_2 included *Betula nana* leaves, *Carex rostrata* leaves, *Eriophorum vaginatum* leaves, *Pinus sylvestris* needles, and *P. sylvestris* branches for calibration of all the quality parameters except total C, total N, and ash content. Branch dataset (b) included pine branch litters for calibration of all the quality parameters.

This grouping enabled us to compare (1) broad-range (using combined datasets) and (2) narrow-range (using branch dataset) calibrations. For narrow-range calibrations we selected the branch litter as it was the litter type for which we had the biggest number of samples analyzed (Table 7.1).

Table 7.1 Litter types studied and number of samples per each litter type used for the calibrations

Litter type	N	MES			OLI			OMB			Site	Dataset*
		P	RD	D	P	RD	D	P	RD	D		
Betula nana leaves	9	1	1	1	1	1	1	1	1	1	5	$a_{1,2}$
Carex lasiocarpa leaves	4	1	1		1	1					5	a_1
C. rostrata leaves	4	1	1		1	1					5	$a_{1,2}$
Eriophorum vagina-tum leaves	9	1	1	1	1	1	1	1	1	1	5	$a_{1,2}$
Pinus sylvestris needles	9	1	1	1	1	1	1	1	1	1	5	$a_{1,2}$
P. sylvestris branches < 1 cm	12			12							2, 4, 7	$a_{1,2,}$ b
P. sylvestris branches > 1 cm	13			13							2, 4, 7	$a_{1,2,}$ b
Pleurozium schreberii	3			1			1			1	5	a_1
Sphagnum fallax	6	6									1, 2, 3, 5, 6, 8	a_1
S. fuscum	3							1	1	1	5	a_1
S. papillosum	6	6									1, 2, 3, 5, 6, 8	a_1

MES = mesotrophic fen; OLI = oligotrophic fen; OMB = ombrotrophic bog; P = pristine; RD = recently drained; D = drained; 1 = Chlum u Třeboně, Czech Republic; 2 = Väätsa, Estonia; 3 = Parainen, southern Finland; 4 = Vesijako, southern Finland; 5 = Orivesi, southern Finland; 6 = Rovaniemi, northern Finland; 7 = Kivalo, northern Finland; 8 = Pallasjärvi, northern Finland. Numbers represent number of samples taken at each site, N = total number of samples per each litter type
*datasets used for calibrations of different chemical properties: a_1 = combined dataset including all litter types studied, a_2 = combined dataset including *B. nana* leaves, *C. rostrata* leaves, *E. vaginatum* leaves, *P. sylvestris* needles and *P. sylvestris* branch litters, b = branch dataset including *P. sylvestris* branch litters only

7.2.2 Chemical Analyses

The chemical composition of the litters was analyzed according to Karsisto *et al.* (2003). The analyses were selected to divide the litter into organic chemical fractions that are supposed to behave differently during decomposition (Berg & Matzner, 1997), as the further interest of our research group will focus on predicting decomposition rates of these litters. These fractions were extractable (soluble) substances, holocellulose (sum of cellulose and hemicelluloses), H_2SO_4-insoluble lignin (Klason lignin), and copper oxide (CuO) oxidation products of lignin: vanillin (V1), vanillic acid (V2), acetovanillone (V3), 4-hydroxybenzaldehyde (P1), 4-hydroxyacetophenone (P2), 4-hydroxybenzoic acid (P3), and ferulic acid (C2).

The amounts of extractable substances were determined gravimetrically by sonicating milled samples with a solvent in a sonicator water bath and weighing the samples after filtration and drying. Mass loss during each extraction was considered to be the content of extractable compounds. Dichlormethane was used to remove nonpolar extractives (NPE; basics in, e.g., Ryan *et al.*, 1990; TAPPI standards) and acetone, ethanol, and hot water to remove polar extractives (Ryan *et al.*, 1990).

Extractive-free samples were then hydrolyzed in H_2SO_4. Primary hydrolysis of each 0.3 g sample was performed with 1.00 ml 72% H_2SO_4 for 1 h in a sonicator bath at 30°C. Hydrolysates were then diluted to 4% H_2SO_4 by ultrapure water and a secondary hydrolysis was performed for 1 h at 120°C and pressure 1.2 bar. The acid-insoluble residue was defined as Klason lignin after filtration and drying. To get information about the structure of lignin molecules, modified degradative alkaline CuO oxidation method was used (Hedges & Ertel, 1982). Holocellulose was determined from extractive-free samples using the sodium chlorite method (Quaramby & Allen, 1989).

Total content of elements C and N was measured using the Vario MAX C/N (Elementar Analysensysteme, Hanau, Germany) combustion analyzer. Dry mass content was determined by drying samples at 105°C overnight and the ash content by combustion at 550°C for 4 h.

The minimum, maximum, mean values, and standard deviations of the measured chemical properties of the litters are shown in Table 7.2, with coefficients of correlation (r) in Table 7.3. Although most of the quality parameters were not normally distributed, no transformations were done. Two samples were removed from modeling total N content, C:N and N:lignin ratios because of their extremely high reference values which might have been caused by an error in the reference measurement.

7.2.3 Spectral Measurements

NIR spectra were obtained with a FieldSpec Pro FR spectroradiometer (ASDI Colorado). Dried and milled samples were presented to the instrument on a petri dish. Spectra consisting of 20 averaged scans were collected at five different sections of the dish and averaged for each sample. Thus, a NIR spectrum of a sample

Table 7.2 Selected descriptive statistics for the reference data (measured chemical properties) of the plant litters studied

Y	N	Dataset	Mean	Min	Max	SD	Skewness
total C (%)	78	a$_1$	50.53	43.71	56.61	4.03	−0.27
total C (%)	25	b	53.00	50.92	55.15	1.19	−0.05
total N (%)	76	a$_1$	0.56	0.16	1.18	0.23	0.20
total N (%)	25	b	0.33	0.16	0.60	0.12	0.38
ash (%)	56	a$_2$	1.33	0.16	4.80	0.99	1.21
ash (%)	25	b	0.50	0.16	0.93	0.23	0.40
NPE (%)	56	a$_2$	6.82	1.55	15.84	4.12	0.66
NPE (%)	25	b	5.46	1.55	10.03	2.41	0.49
AE (%)	56	a$_2$	2.37	0.11	7.79	2.04	0.86
AE (%)	25	b	0.55	0.11	1.02	0.29	0.21
EE (%)	56	a$_2$	4.09	0.43	11.09	2.67	0.53
EE (%)	25	b	1.82	0.43	4.09	1.14	0.37
WE (%)	56	a$_2$	5.02	1.05	11.00	3.35	0.42
WE (%)	25	b	1.90	1.05	2.95	0.55	−0.14
TE (%)	56	a$_2$	18.29	5.08	37.49	10.41	0.54
TE (%)	25	b	9.73	5.08	15.06	3.03	0.17
lignin (%)	56	a$_2$	34.75	20.79	46.08	8.35	−0.23
lignin (%)	25	b	43.02	41.10	46.08	1.20	0.40
holl (%)	56	a$_2$	53.66	23.96	74.16	14.21	−0.74
holl (%)	25	b	59.80	45.99	74.16	7.07	0.25
P1 (mg g⁻¹)	57	a$_2$	1.40	0.22	4.13	0.87	0.88
P1 (mg g⁻¹)	25	b	2.04	0.77	4.13	0.83	0.64
P2 (mg g⁻¹)	57	a$_2$	0.38	0.11	1.02	0.20	1.31
P2 (mg g⁻¹)	25	b	0.30	0.11	0.61	0.13	0.72
P3 (mg g⁻¹)	57	a$_2$	0.58	0.17	1.40	0.34	1.07
P3 (mg g⁻¹)	25	b	0.33	0.17	0.56	0.11	0.76
C2 (mg g⁻¹)	57	a$_2$	4.25	0.36	15.78	5.16	1.24
C2 (mg g⁻¹)	25	b	1.69	0.86	3.53	0.76	1.06
V1 (mg g⁻¹)	57	a$_2$	21.48	3.28	60.14	18.36	0.67
V1 (mg g⁻¹)	25	b	39.19	16.57	60.14	12.69	−0.07
V2 (mg g⁻¹)	57	a$_2$	2.81	0.29	7.94	2.38	0.65
V2 (mg g⁻¹)	25	b	5.09	1.89	7.94	1.68	−0.33
V3 (mg g⁻¹)	57	a$_2$	4.92	0.00	11.00	3.22	0.33
V3 (mg g⁻¹)	25	b	8.05	4.61	11.00	1.85	−0.36
C:N	76	a$_1$	115.58	45.68	311.83	67.22	1.15
C:N	25	b	183.31	91.92	311.83	64.17	0.36
N:lig	76	a$_1$	0.016	0.004	0.039	0.009	0.69
N:lig	25	b	0.008	0.004	0.013	0.003	0.29

Y = chemical parameter, N = number of samples, NPE = nonpolar (dichlormethane) extractives, AE = acetone extractives, EE = ethanol extractives, WE = hot water extractives, TE = total extractives (sum of NPE, AE, EE, and WE), lignin = Klason lignin, holl = holocellulose (sum of cellulose and holocelluloses), P1 = 4-hydroxybenzaldehyde, P2 = 4-hydroxyacetophenone, P3 = 4-hydroxybenzoic acid, C2 = ferulic acid, V1 = vanillin, V2 = vanillic acid, V3 = acetovanillone, C:N = carbon-to-nitrogen ratio, N:lig = nitrogen-to-lignin ratio.

* See Table 7.1 for an explanation of the meaning of a$_1$, a$_2$ and b

Table 7.3 Coefficients of correlations (*r*) between the litter chemical properties of the (a) combined and (b) branch dataset. Coefficients in bold equal $r^2 > 0.5$

	C	N	ash	NPE	AE	EE	WE	TE	lignin	holl	P1	P2	P3	C2	V1	V2	V3	C:N	N:lig
a																			
C	1																		
N	−0.28	1																	
ash	−0.47	**0.74**	1																
NPE	**0.81**	0.07	0.05	1															
AE	−0.11	**0.71**	0.64	0.23	1														
EE	0.20	0.50	0.49	0.49	**0.85**	1													
WE	0.18	0.60	0.65	0.57	**0.87**	**0.93**	1												
TE	0.42	0.48	0.48	**0.76**	**0.78**	**0.92**	**0.96**	1											
lignin	0.12	−0.44	−0.69	−0.36	−0.57	−0.59	−0.70	−0.63	1										
holl	−0.59	−0.38	−0.17	−0.70	−0.64	**−0.76**	**−0.75**	**−0.84**	0.14	1									
P1	−0.02	−0.62	−0.46	−0.26	**−0.71**	−0.59	−0.64	−0.60	0.45	0.52	1								
P2	0.21	0.34	0.21	0.27	0.61	0.68	0.62	0.60	−0.09	−0.69	−0.14	1							
P3	0.27	0.12	0.38	0.60	0.31	0.51	0.59	0.62	**−0.84**	−0.22	−0.17	0.15	1						
C2	**−0.79**	0.42	0.58	−0.49	0.21	−0.06	0.01	−0.17	−0.51	0.50	−0.25	−0.38	0.13	1					
V1	0.19	**−0.71**	**−0.74**	−0.23	**−0.74**	−0.59	**−0.72**	−0.61	**0.76**	0.35	**0.82**	−0.15	−0.46	−0.55	1				
V2	0.25	−0.64	**−0.72**	−0.19	**−0.73**	−0.60	−0.70	−0.59	**0.78**	0.28	**0.72**	−0.18	−0.45	−0.54	**0.89**	1			
V3	0.30	−0.65	**−0.77**	−0.15	−0.66	−0.54	−0.64	−0.53	**0.83**	0.17	**0.74**	−0.03	−0.51	−0.66	**0.95**	**0.91**	1		
C:N	0.24	**−0.89**	−0.68	−0.11	−0.63	−0.42	−0.52	−0.44	0.45	0.35	0.69	−0.11	−0.13	−0.50	**0.77**	0.66	0.70	1	
N:lig	−0.33	**0.92**	**0.86**	0.13	0.69	0.54	0.67	0.54	**−0.72**	−0.24	−0.56	0.24	0.41	0.56	**−0.79**	**−0.76**	**−0.79**	**−0.82**	1
b																			
C	1																		
N	0.40	1																	
ash	**0.80**	**0.71**	1																
NPE	**0.79**	**0.86**	**0.72**	1															
AE	0.46	0.46	0.34	0.41	1														

(continued)

Table 7.3 (continued)

	C	N	ash	NPE	AE	EE	WE	TE	lignin	holl	P1	P2	P3	C2	V1	V2	V3	C:N	N:lig
EE	0.00	−0.14	−0.15	−0.12	0.34	1													
WE	**0.72**	0.49	0.23	0.37	0.52	0.32	1												
TE	**0.80**	**0.77**	0.59	**0.86**	0.64	0.37	0.65	1											
lignin	0.22	0.40	0.35	0.21	0.05	−0.09	0.15	0.16	1										
holl	**−0.87**	**−0.85**	−0.63	**−0.83**	−0.54	−0.20	**−0.74**	**−0.92**	−0.36	1									
P1	−0.48	−0.52	−0.43	−0.45	−0.16	0.19	−0.31	−0.36	−0.03	0.37	1								
P2	−0.43	−0.53	−0.44	−0.45	−0.12	0.30	−0.25	−0.30	−0.06	0.32	**0.98**	1							
P3	−0.24	−0.41	−0.36	−0.39	−0.17	0.21	−0.05	−0.25	0.06	0.18	**0.78**	**0.82**	1						
C2	0.46	0.52	0.68	0.65	0.27	−0.31	0.18	0.46	−0.05	−0.45	−0.66	−0.63	−0.53	1					
V1	−0.55	−0.50	−0.51	−0.55	−0.05	0.43	−0.29	−0.33	0.02	0.44	**0.76**	**0.72**	0.45	**−0.85**	1				
V2	−0.17	−0.14	−0.20	−0.33	−0.02	0.21	−0.07	−0.20	0.43	0.11	0.46	0.46	0.61	−0.59	0.52	1			
V3	−0.37	−0.36	−0.46	−0.48	−0.10	0.18	−0.15	−0.35	0.22	0.30	**0.81**	**0.75**	0.64	**−0.83**	**0.88**	0.67	1		
C:N	**−0.75**	**−0.96**	**−0.71**	**−0.81**	−0.42	0.13	−0.46	**−0.71**	−0.34	**0.82**	0.54	0.53	0.41	−0.53	0.57	0.20	0.45	1	
N:lig	**0.80**	**0.99**	**0.71**	**0.88**	0.48	−0.13	0.49	**0.78**	0.32	**−0.85**	−0.53	−0.54	−0.43	0.55	−0.52	−0.19	−0.39	**−0.96**	1

C = total C content, N = total N content, NPE = nonpolar (dichlormethane) extractives, AE = acetone extractives, EE = ethanol extractives, WE = hot water extractives, TE = total extractives (sum of NPE, AE, EE and WE), lignin = Klason lignin, holl = holocellulose (sum of cellulose and holocelluloses), P1 = 4-hydroxybenzaldehyde, P2 = 4-hydroxyacetophenone, P3 = 4-hydroxybenzoic acid, C2 = ferulic acid, V1 = vanillin, V2 = vanillic acid, V3 = acetovanillone, C:N = carbon-to-nitrogen ratio, N:lig = nitrogen-to-lignin ratio

consisted of 100 averaged scans. Each spectrum consisted of reflectance measurements between 350 and 2,500 nm, i.e., reflectance in both visible and NIR part was measured. The instrument registers bands at 1.4–2 nm intervals, which are interpolated to give spectra with 1 nm increment. Reflectance (R) was transformed to absorbance (A) using A = log (1/R). Spectra were also reduced by half by averaging adjacent bands, and the 350–400 nm part of the spectra was removed as it contained a lot of noise.

7.2.4 Multivariate Data Analyses

Data analyses were performed by partial least squares (PLS) regression (Esbensen, 2002; Martens & Naes, 1991) using The Unscrambler, version 9.2, software package (Camo Process AS; Oslo, Norway). PLS regression reduces large numbers of correlated NIRS data into a limited number of orthogonal components (loading vectors), which are used in a multivariate regression with the chemical properties of the litters as the dependent variables. The models use spectral absorbances at all wavelengths as the X-matrix, and a chemical property as the Y-variable. We used the PLS-1 algorithm (Martens & Naes, 1991) that handles only one Y-variable at a time, so that a model is specifically built only for that single variable. As the entire spectrum is used by PLS, each wavelength contributes to the model. Thus, problems known to be associated with stepwise selection of individual wavelengths are avoided (Grossmann et al., 1996).

Spectral data preprocessing strongly influences the prediction performance of calibration models (e.g., Gillon et al., 2004), so we investigated a number of preprocessing options. The aim of preprocessing is to enhance the spectral features that carry the information of interest and to suppress or eliminate superfluous features.

Initial models were constructed using spectra with no transformation. Then we tested two different smoothing methods: moving average and Savitzky–Golay smoothing (Savitzky & Golay, 1964), with different segment sizes (numbers of wavelengths) taken into account during the smoothing process. Next, we tested transformation by first and second derivative, again with different segment sizes over which the derivative was taken and smoothed using Savitzky–Golay algorithm. Last, we used standard normal variate (SNV), detrending (DT), (Barnes et al., 1989) and SNV–DT combined with Savitzky–Golay smoothing. The smaller the segment size taken into account during the smoothing process or the derivation, the higher is the sensitivity, and therefore also the noise. On the other hand, a large segment size may result in loss of potentially relevant information.

Models were built on the transformed data (1) for the entire spectrum measured in both the visible (400–780 nm) and the NIR region (780–2,500) and (2) for the NIR region only (780–2,500 nm). For each quality parameter, we compared (1) the model calibrated for various litter types (combined dataset, a_1 and a_2) and (2) model calibrated for pine branch litters only (branch dataset, b).

The prediction abilities of the models were tested using full cross-validation which systematically removes one sample from the dataset, establishes a model with the remaining samples, and uses that model to predict the value of the *Y*-variable for the removed sample. This process continues with all samples successively being removed from the dataset until all the samples are predicted (Martens & Naes, 1991). Therefore, the predictions are independent as they are based on regression models that do not include the predicted data.

Full cross-validation also facilitates estimation of the optimal number of terms in the calibrations. All the residuals from predicting the extracted samples are pooled to provide a standard error of cross-validation (SECV). The minimum SECV determines the number of terms to be used. This is very important in order to avoid overfitting; if too many terms are used in the model, information which is not related to the variable of interest is also included. The prediction ability of such a model is poor even though the samples used for calibration fit the calculated regression line perfectly.

The r^2 values, root mean square error of calibration (RMSEC), root mean square error of prediction (RMSEP, in this study calculated as the root mean square error of cross-validation), and the ratio of the *Y*-variable standard deviation to RMSEP (RPD) were used for estimating the accuracy of the models. RMSEC tells how well the reference values are fit by the calculated regression line in the calibration model, while RMSEP shows the accuracy of prediction on independent (extracted) samples. RMSEC/RMSEP is expressed in the same units as the dependent variable so it indicates the error in the same units as our *Y*.

RMSEC/RMSEP and RPD are calculated as follows:

$$RMSEC/RMSEP = \sqrt{\frac{1}{I} \sum_{i=1}^{I} (\hat{y}_i - y_i)^2}$$

$$RPD = \frac{SD}{RMSEP}$$

where

$\hat{y}_i$ = predicted value of the *i*th observation
y_i = measured value of the *i*th observation
I = number of observations in the calibration set
SD = standard deviation of the reference variable (*Y*)

An RPD of 3 is considered satisfactory for use in certain applications (Williams & Sobering, 1996). If large errors can be expected in the reference method, an RPD of 2 is also acceptable (Chang *et al.*, 2001). RPDs of 1 or less indicate that calibration performance is no better than a guess (Stenberg *et al.*, 2004).

A good model should have (1) high correlation coefficient, (2) low RMSEC, low RMSEP, and a small difference between RMSEC and RMSEP, (3) high RPD, and (4) a relatively low number of factors in order to avoid inclusion of signal noise in the modeling.

7.3   Results and Discussion

Models based only on the NIR part (780–2,500 nm) of the spectra performed better in predicting litter quality parameters, compared to the models using the entire spectrum in both visible and NIR part (400–2,500 nm). This is inconsistent with earlier studies where the visible region also provided useful predictive information for litter properties, mass loss, and foliage moisture content, and predictions were more accurate when the full spectrum was taken into account rather than just the NIR or visible region (Gillon *et al.*, 1993, 2004). Our samples were visually examined prior to the collection of NIR spectra. A range of colors was evident between and within different litter types. We assume that excluding the visible region prevents the masking of information within the NIR region. If a relationship was found between the color of samples and their chemical composition during the calibration process (this relationship would probably not have any logical justification), the predictions for independent samples might be inaccurate due to possibly different color of the predicted samples.

Initial models constructed on spectra with no transformation did not yield satisfactory calibrations, and the prediction ability of the models was improved by mathematical treatments of the spectra prior to calibrations. Examples of NIR spectra without transformation, transformed by first derivative, second derivative, and SNV–DT are shown in Fig. 7.1a–d.

NIR spectra transformed by the second derivative combined with Savitzky–Golay smoothing over 4 + 4 data points showed the best performance in predicting most of the quality parameters (Table 7.4). The first derivative combined with Savitzky–Golay smoothing over 4 + 4 data points was the best spectral pretreatment method for predicting the content of V2 (both combined and branch dataset) and lignin (branch dataset). Standard normal variate and de-trended (SNV--DT) spectra showed the best performance in predicting N and N to lignin ratio in both the combined and branch dataset, C and holocellulose in the combined dataset, and ash content in the branch dataset. Our results are in accordance with earlier studies where first or second derivative and SNV–DT spectra, gaps between 4 and 20 points, and smooth between 2 and 10 gained the best models for predicting various properties of organic soils or plant materials (Beining *et al.*, 2001; Chodak *et al.*, 2002; Coûteaux *et al.*, 2003; Stenberg *et al.*, 2004). There is no best spectral pretreatment that could be generally used for predicting properties of all types of samples, as the optimal pretreatment method is very instrument and sample-presentation dependent. Usually, trial and error is the only way to optimize the spectral pretreatment by searching the lowest prediction error (Chodak *et al.*, 2002; Coûteaux *et al.*, 2003).

The predictive power of NIR spectroscopy varied for the different litter properties as well as the datasets. The ability of NIRS to predict values of litter properties was grouped into three categories based on RPD values (Table 7.4), following Chang *et al.* (2001). Category A (RPD > 2.0) included properties with measured vs. predicted r^2 values between 0.75 and 1.00. These well-predicted properties included total N, total C, ash content, NPE, ethanol extractives (EE), total extractives (TE),

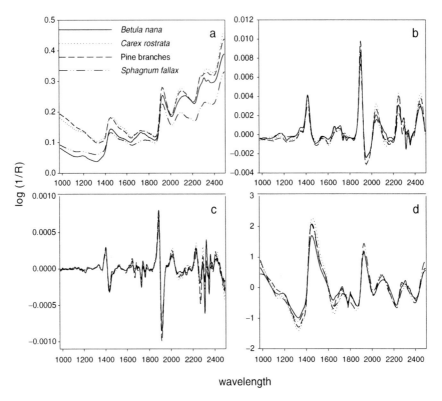

Fig. 7.1 NIR spectra of litter samples of *Betula nana*, *Carex rostrata*, *Pinus sylvestris* and *Sphagnum fallax* without transformation (a), transformed by 1st derivative (b), 2nd derivative (c), and SNV–DT (d)

holocellulose, C:N, and N:lignin ratio for both combined and branch dataset. For the combined dataset, all the other chemical properties except for P1 fit into the A category, while predictions of these properties using the branch dataset were less accurate according to r^2 and RPD results. Category B (RPD = 1.4 ~ 2.0) included litter properties with measured vs. predicted r^2 values between 0.50 and 0.75. These were P1 for the combined dataset and water extractives (WE) and lignin for the branch dataset. Category C ($r^2 < 0.50$, RPD < 1.4) included only models for branch dataset predicting acetone extractives (AE), P1, P2, P3, V1, V2, V3, and C2.

Prediction of litter properties in category B could probably be improved by using different calibration strategies, but properties in category C may not be reliably predicted using NIRS (Chang *et al.*, 2001).

Total C content is generally well predicted by NIR spectroscopy. The accuracy of our calibrations (RMSEP 0.91 and 0.41, r^2 0.95 and 0.88, RPD 4.43 and 2.90 for combined and branch dataset respectively; Table 7.4, Fig. 7.2) was among the best results obtained in published studies. The values of RMSEP have varied between

Table 7.4 PLS regression results

Y	N	Dataset	Math treatment	PC	r^2	RMSEC	r^2	RMSEP	RPD	Category**
total C	78	a$_1$	SNV–DT	5	0.96	0.79	0.95	0.91	4.43	A
total C	25	b	2, 4, 4	4	0.97	0.21	0.88	0.41	2.90	A
total N	76	a$_1$*	SNV–DT	9	0.92	0.06	0.88	0.08	2.91	A
total N	25	b	SNV–DT	5	0.93	0.03	0.83	0.05	2.40	A
ash	56	a$_2$	2, 4, 4	7	0.95	0.22	0.88	0.35	2.83	A
ash	25	b	SNV–DT	5	0.93	0.06	0.85	0.09	2.56	A
NPE	56	a$_2$	2, 4, 4	4	0.96	0.86	0.94	0.97	4.25	A
NPE	25	b	2, 4, 4	4	0.96	0.48	0.89	0.78	3.09	A
AE	56	a$_2$	2, 4, 4	5	0.94	0.48	0.88	0.70	2.91	A
AE	25	b	2, 4, 4	1	0.22	0.25	0.07	0.28	1.04	C
EE	56	a$_2$	2, 4, 4	5	0.94	0.64	0.86	0.99	2.70	A
EE	25	b	2, 4, 4	5	0.97	0.18	0.86	0.42	2.71	A
WE	56	a$_2$	2, 4, 4	6	0.98	0.45	0.96	0.69	4.86	A
WE	25	b	2, 4, 4	4	0.87	0.19	0.58	0.35	1.57	B
TE	56	a$_2$	2, 4, 4	4	0.96	2.11	0.94	2.58	4.03	A
TE	25	b	2, 4, 4	4	0.94	0.71	0.84	1.17	2.59	A
lignin	56	a$_2$	2, 4, 4	6	0.99	0.85	0.98	1.29	6.47	A
lignin	25	b	1, 4, 4	6	0.83	0.49	0.54	0.82	1.46	B
holl	56	a$_2$	SNV-DT	3	0.95	3.07	0.94	3.45	4.12	A
holl	25	b	2, 4, 4	4	0.94	1.67	0.83	2.85	2.48	A
P1	56	a$_2$	2, 4, 4	2	0.66	0.50	0.62	0.53	1.64	B
P1	25	b	2, 4, 4	5	0.91	0.24	0.37	0.71	1.17	C
P2	56	a$_2$	2, 4, 4	7	0.92	0.06	0.75	0.10	2.03	A
P2	25	b	2, 4, 4	5	0.93	0.03	0.43	0.10	1.30	C
P3	56	a$_2$	2, 4, 4	5	0.91	0.10	0.87	0.12	2.84	A
P3	25	b	2, 4, 4	1	0.18	0.10	0.11	0.10	1.10	C
C2	56	a$_2$	2, 4, 4	3	0.97	0.91	0.96	1.06	4.87	A
C2	25	b	2, 4, 4	1	0.57	0.49	0.48	0.54	1.40	C
V1	56	a$_2$	2, 4, 4	3	0.85	6.93	0.84	7.37	2.49	A
V1	25	b	2, 4, 4	3	0.82	5.32	0.36	10.30	1.23	C
V2	56	a$_2$	1, 4, 4	4	0.81	1.02	0.77	1.13	2.10	A
V2	25	b	1, 4, 4	1	0.13	1.54	0.00	1.72	0.98	C
V3	56	a$_2$	2, 4, 4	3	0.88	1.12	0.86	1.21	2.66	A
V3	25	b	2, 4, 4	1	0.31	1.50	0.21	1.62	1.14	C
C:N	76	a$_1$*	2, 4, 4	6	0.94	16.52	0.89	22.26	3.02	A
C:N	25	b	2, 4, 4	3	0.90	20.15	0.81	27.71	2.32	A
N:lig	76	a$_1$*	SNV-DT	7	0.92	0.0026	0.88	0.0032	2.85	A
N:lig	25	b	SNV-DT	5	0.92	0.0008	0.80	0.0012	2.28	A

PC, number of terms used in the calibration model; RMSEC, root mean square error of calibration; RMSEP, root mean square error of prediction; RPD, residual prediction deviation (the ratio of the Y-variable standard deviation to RMSEP); Math treatment, indicates the mathematical transformation of the spectral data: the first number is the order of the derivative function, the second is the segment length in data points over which the derivative was taken, and the third is the segment length over which the function was smoothed, using Savitzky – Golay algorithm; SNV-DT, standard normal variate detrended.

**Categories of the prediction ability of NIRS for litter chemical properties. A: RPD > 2.0; B: RPD = 1.4 ~ 2.0; C: RPD < 1.4.

See Table 7.1 and 7.2 for an explanation of the meaning of a$_1$, a$_2$, b, NPE, AE, EE, WE, TE, P1, P2, P3, C2, V1, V2, V3, C:N, N:lig

1.07 and 2.29 for plant litters or biomass (Jofre *et al.*, 1992; Gillon *et al.*, 1999a; Bouchard *et al.*, 2003) while RMSEP was 3.01 for organic soils (Coûteaux *et al.*, 2003) and have varied between 0.45 and 0.79 for mineral soils or mineral soil–chemical–organic material mixtures (Chang *et al.*, 2001; Chang & Laird, 2002; McCarty *et al.*, 2002; Coûteaux *et al.*, 2003). The r^2 values in these studies have varied between 0.74 and 0.99, RPD between 2.8 and 5.1.

For total N content, the accuracy of our calibrations (Table 7.4, Fig. 7.2) was within the range of results obtained in other studies where the values of RMSEP have varied between 0.04 and 0.5 for plant litters or biomass (McLellan *et al.*, 1991a; Gillon *et al.*, 1999a; Bouchard *et al.*, 2003; Coûteaux *et al.*, 2003; Shepherd *et al.*, 2003; Stenberg *et al.*, 2004), and between 0.03 and 0.08 for soils (Chang *et al.*, 2001; Chang & Laird, 2002; Coûteaux *et al.*, 2003) and the values of r^2 have varied between 0.85 and 0.99, RPD between 2.52 and 6.80.

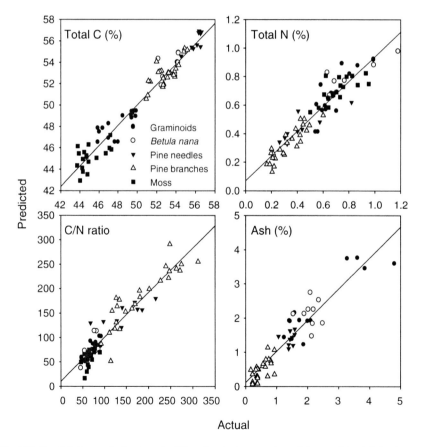

Fig. 7.2 Predicted vs. actual values for litter chemical parameters of the combined dataset

The accuracy of our calibration for the C:N ratio was comparable with the calibrations obtained for mineral and organic soils (Chang & Laird, 2002; Chodak *et al.*, 2002), and much better than predictions for C:N ratio of biomass and litters obtained by Stenberg *et al.* (2004), where the C:N ratio of various plant materials could not be calibrated to satisfaction over the whole range.

Ash content was predicted by NIRS with a good accuracy (Table 7.4, Fig. 7.2), which is in accordance with other studies on plant biomass. Similar predictions of the ash content were obtained for forage (Cozzolino *et al.*, 2000; Windham *et al.*, 1991) and plant litter (Jofre *et al.*, 1992). The main components forming the bulk of biomass ashes are Ca, Si, Al, Mg, K, Na, S, and Cl (FORCE Technology, 2007). Mineral elements cannot be detected directly with NIR spectroscopy, but could be detected indirectly because their presence in organic complexes affects H bonds (Shenk, 1992) or the concentrations of organic constituents (Watson *et al.*, 1976). Several studies have shown the ability of NIRS to predict levels of minerals in soils, probably due to the relationship between elements and soil organic matter (Chang & Laird, 2001) and between physical properties and chemical parameters of soil (Cozzolino & Moron, 2003).

Holocellulose content was predicted by NIRS with a good accuracy (Table 7.4, Fig. 7.4), which was better than the results reported for mixtures of biomass, litter, and organic soil: RMSEP 3.50, r^2 0.83, RPD 2.30 (Ono *et al.*, 2003).

Lignin content prediction for the combined dataset (Table 7.4, Fig. 7.4) was one of the best results obtained, in comparison with other studies. The values of RMSEP have varied between 1.3 and 6.1, r^2 between 0.71 and 0.99 for biomass or plant litters (McLellan *et al.*, 1991a; Brinkmann *et al.*, 2002; Kelley *et al.*, 2004), and have varied between 3.3 and 5, r^2 between 0.70 and 0.98 for calibrations using both soil and biomass or litter samples (Ono *et al.*, 2003; Terhoeven-Urselmans *et al.*, 2006) or for organic residues in soils (Shepherd *et al.*, 2005). The RPD values in these studies have varied between 2.1 and 8.96.

Models predicting lignin content for the branch dataset only did not perform so well (Table 7.4, Fig. 7.3) according to r^2 and RPD results. The values were very similar to the ones obtained for loblolly pine (*P. taeda*) wood with the reported values 0.52 and 1.49 for r^2 and RPD respectively (Yeh *et al.*, 2005). Relatively low variation within the reference data (mean 43.02 and 29.8 (% of dry mass), SD 1.2 and 0.9 for *P. sylvestris* and *P. taeda* respectively) is probably responsible for such unsatisfactory correlations. However, RMSEP of lignin content of branch samples was much smaller when using the branch model compared to the prediction of lignin content of branch samples using the combined model (Table 7.5).

Models concerning extractives were all very good for the combined dataset with r^2 from 0.86 to 0.96, RMSEP from 0.69 to 2.58, and RPD from 2.70 to 4.86 as compared to literature data with r^2 0.82, RMSEP from 1.50 to 3.40, and RPD from 1.87 to 2.10 (Schimleck & Yazaki, 2003; Ono *et al.*, 2003). For the branch dataset, models predicting NPE, EE, and TE were very good while the ones predicting AE and WE did not perform so well according to r^2 and RPD results (Table 7.4, Fig. 7.3) probably because of the low variation within the reference data (Table 7.2). However, RMSEP of AE and WE content of branch samples

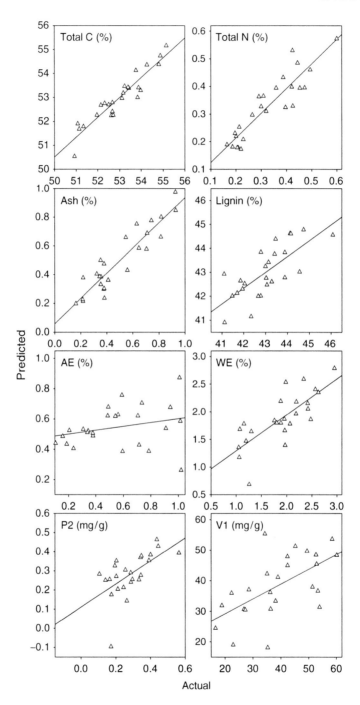

Fig. 7.3 Predicted vs. actual values for litter chemical parameters of the branch dataset

Table 7.5 Prediction ability of (a) combined and (b) branch models to predict chemical properties of branch litters. Values in bold indicate decrease in RMSEP for more than 30% when branch model was used

Y	(a) branch properties predicted by the combined models			(b) branch properties predicted by the branch models		
	RMSEP	RPD	Category*	RMSEP	RPD	Category*
total C	0.78	1.53	B	0.41	2.9	A
total N	0.05	2.26	A	0.05	2.4	A
Ash	0.22	1.03	C	0.09	2.56	A
NPE	1.16	2.09	A	0.78	3.09	A
AE	0.38	0.76	C	0.28	1.04	C
EE	0.66	1.73	B	0.42	2.71	A
WE	0.59	0.94	C	0.35	1.57	B
TE	2.13	1.42	B	1.17	2.59	A
lignin	1.38	0.87	C	0.82	1.46	B
holl	4.34	1.63	B	2.85	2.48	A
P1	0.75	1.11	C	0.71	1.17	C
P2	0.09	1.39	C	0.1	1.3	C
P3	0.13	0.86	C	0.1	1.1	C
C2	0.85	0.89	C	0.54	1.4	C
V1	10.76	1.18	C	10.3	1.23	C
V2	1.65	1.02	C	1.72	0.98	C
V3	1.60	1.16	C	1.62	1.14	C
C:N	28.38	2.26	A	27.71	2.32	A
N:lig	0.0024	1.27	C	0.0012	2.28	A

RMSEP, root mean square error of prediction; RPD, residual prediction deviation (the ratio of the Y variable standard deviation to RMSEP)
* Categories of the prediction ability of NIRS for litter chemical properties. A: RPD > 2.0; B: RPD = 1.4 ~ 2.0; C: RPD < 1.4.
See Table 7.2 for an explanation of the meaning of NPE, AE, EE, WE, TE, P1, P2, P3, C2, V1, V2, V3, C:N, N:lig

was much smaller when using the branch models compared to the prediction of AE and WE content of branch samples using the combined models (Table 7.5).

The above-mentioned results show that very good NIR calibrations were possible for total C and N, ash content, NPE, EE, TE, holocellulose, C:N and N:lignin ratio using both combined and branch dataset and AE, WE, lignin, P2, P3, C2, V1, V2, and V3 using combined dataset (Table 7.4 and Figs. 7.2–7.5). Acceptable calibrations were obtained for P1 using combined dataset, and WE and lignin using branch dataset, taking into account the possible inaccuracy of the reference methods.

Generally, models combining all the different litter types performed better than the models constructed for the pine branch litters only, according to r^2 and RPD results. The reason might be a too low variation within the branch dataset to yield significant correlations between NIR spectra and the chemical properties (Table 7.2) as well as the relatively low number of samples in this dataset. However, RMSEP was generally smaller when the branch litter properties were predicted by the branch model compared to the predictions using the combined model (Table 7.5), the difference was more than 30% for most of the parameters. This indicates the

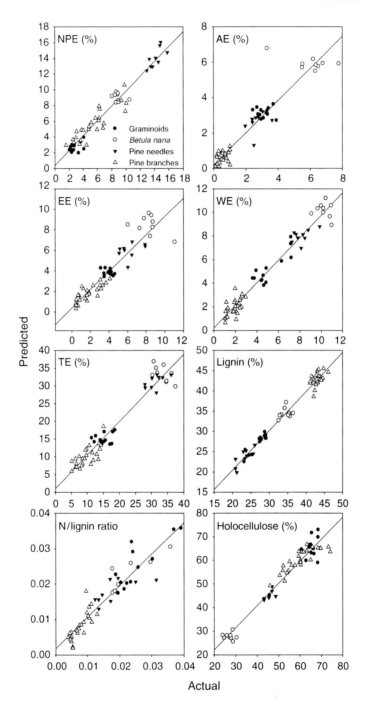

Fig. 7.4 Predicted vs. actual values for litter chemical parameters of the combined dataset

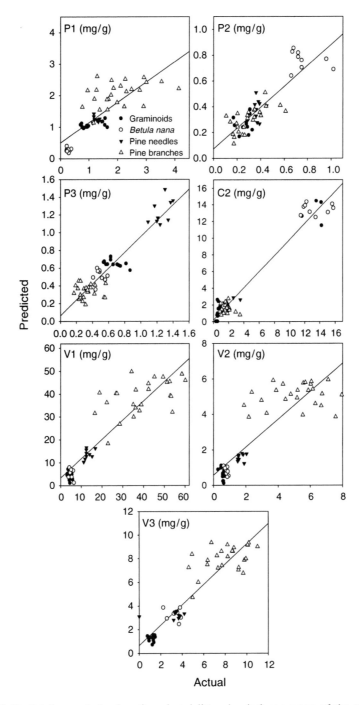

Fig. 7.5 Predicted vs. actual values for selected litter chemical parameters of the combined dataset

potential of improving the NIR calibrations by constructing specific models for different groups of litter.

The predictive ability of NIRS will be further tested on an extended sample set, and specific models will be built for different litter types. Analyzing a much larger sample set will enable us to test the NIR calibrations by independent validation on separate sample sets and the models will be used for routine NIRS predictions of selected litter properties. Our further interest focuses on direct predicting of litter decomposability by NIRS.

Acknowledgments This study was funded by the Academy of Finland projects 203585 and 205090. We also thank the European Science Foundation for supporting this study by a short-term exchange grant to Petra Vávřová, and Alison Gillette for linguistic revision.

References

Aber, J.D., & Melillo, J.M. (1980). Litter decomposition: Measuring relative contribution of organic matter and nitrogen to forest soils. *Canadian Journal of Botany, 58*, 416–421.

Barnes, R.J., Dhanoa, M.S., & Lister, S.J. (1989). Standard normal variate transformation and detrending of near-infrared diffuse reflectance spectra. *Applied Spectroscopy, 43*, 772–777.

Beining, B., Holden, N.M., Ward, S.M., & Farrell, E.P. (2001). The distribution and prediction of gross heat value in Irish industrial peat. *International Peat Journal, 10*, 49–55.

Berendse, F., Berg, B., & Bosatta, E. (1987). The effect of lignin and nitrogen on the decomposition of litter in nutrient-poor ecosystems: A theoretical approach. *Canadian Journal of Botany, 65*, 1116–1120.

Berg, B., & Ekbohm, G. (1991). Litter mass-loss rates and decomposition patterns in some needle and leaf litter types. Long-term decomposition in a Scots pine forest. VII. *Canadian Journal of Botany, 69*, 1449–1456.

Berg, B., & Matzner, E. (1997). The effect of N deposition on the mineralization of C from plant litter and humus. *Environmental Reviews, 5*, 1–25.

Berg, B., & Staaf, H. (1980). Decomposition rate and chemical changes of Scots pine needle litter. II. *Ecological Bulletin, 32*, 373–390.

Blair, J.M. (1988). Nitrogen, sulfur and phosphorus dynamics in decomposing deciduous leaf litter in the southern Appalachians. *Soil Biology and Biochemistry, 20*, 693–701.

Bolster, K.L., Martin, M.E., & Aber, J.D. (1996). Determination of carbon fraction and nitrogen concentration in tree foliage by near infrared reflectance: A comparison of statistical methods. *Canadian Journal of Forest Research, 26*, 590–600.

Bouchard, V., Gillon, D., Joffre, R., & Lefeuvre, J.C. (2003). Actual litter decomposition rates in salt marshes measured using near-infrared reflectance spectroscopy. *Journal of Experimental Marine Biology and Ecology, 290*, 149–163.

Brinkmann, K., Blaschke, L., & Polle, A. (2002). Comparison of different methods for lignin determination as a basis for calibration of Near-infrared reflectance spectroscopy and implications of lignoproteins. *Journal of Chemical Ecology, 28*, 2483–2501.

Bruun, S., Stenberg, B., Breland, T.A., Gudmundsson, J., Henriksen, T.M., Jensen, L.S., Korsæth, A., Luxhøi, J., Pálmason, F., Pedersen, A., & Salo, T. (2005). Empirical predictions of plant material C and N mineralization patterns from near infrared spectroscopy, stepwise chemical digestion and C/N ratios. *Soil Biology and Biochemistry, 37*, 2283–2296.

Card, D.H., Peterson, D.J., Matson, P.A., & Aber, J.D. (1988). Prediction of leaf chemistry by the use of visible and near-infrared reflectance spectroscopy. *Remote Sensing of Environment, 26*, 123–147.

Chang, C.W., & Laird, D.A. (2002). Near-infrared reflectance spectroscopic analysis of soil C and N. *Soil Science, 167*, 110–116.

Chang, C.W., Laird, D.A., Mausbach, M.J., & Hurburgh, C.R. (2001). Near infrared reflectance spectroscopy-principal components regression analyses of soil properties. *Soil Science Society of America Journal, 65*, 480–490.

Chodak, M., Ludvik, B., Khanna, P., & Beese, F. (2002). Use of near infrared spectroscopy to determine biological and chemical characteristics of organic layers under spruce and beech stands. *Journal of Plant Nutrition and Soil Science, 165*, 27–33.

Confalonieri, M., Fornaiser, F., Ursino, A., Boccardi, F., Pintus, B., & Odoardi, M. (2001). The potential of near infrared reflectance spectroscopy as a tool for the chemical characterization of agricultural soils. *Journal of Near Infrared Spectroscopy, 9*, 123–131.

Coûteaux, M.M., Rovira, P., & Berg, B. (2003). Near infrared reflectance spectroscopy for determination of organic matter fractions including microbial biomass in coniferous forest soils. *Soil Biology and Biochemistry, 35*, 1587–1600.

Cozzolino, D., Fassio, A., & Giminez, A. (2000). The use of near-infrared reflectance spectroscopy (NIRS) to predict the composition of whole maize plants. *Journal of the Science of Food and Agriculture, 81*, 142–146.

Cozzolino, D., & Moron, N. (2003). The potential of near-infrared reflectance spectroscopy to analyse soil chemical and physical characteristics. *Journal of Agricultural Science, 140*, 65–71.

Curran, P.J. (1989). Remote sensing of foliar chemistry. *Remote sensing of Environment, 30*, 271–278.

Dalal, R.C., & Henry, R.J. (1986). Simultaneous determination of moisture, organic carbon, and total nitrogen by near infrared reflectance spectrophotometry. *Soil Science Society of America Journal, 50*, 120–123.

Duncan, P., Clanton, D., & Clark, D. (1987). Near infrared reflectance spectroscopy for quality analysis of sandhills meadow forage. *Journal of Animal Science, 64*, 1743–1750.

Edmonds, R.L. (1987). Decomposition rates and nutrient dynamics in small-diameter woody litter in four forest ecosystems in Washington, USA. *Canadian Journal of Forest Research, 17*, 499–509.

Esbensen, K.H. (2002). *Multivariate data analysis: In practice.* Oslo: CAMO Process AS.

FORCE Technology (2007). Biolex – a database of solid biomass fuels. Retrieved April 14, 2007 from http://biolex.dk-teknik.dk

Garnsworthy, P.C., Wiseman, J., & Fegeros, K. (2000). Prediction of chemical, nutritive and agronomic characteristics of wheat by near infrared spectroscopy. *Journal of Agricultural Science, 135*, 409–417.

Gillon, D., Jofre, R., & Dardenne, P. (1993). Predicting the stage of decay of decomposing leaves by near infrared reflectance spectroscopy. *Canadian Journal of Forest Research, 23*, 2552–2559.

Gillon, D., Houssard, C., & Jofre, R. (1999a). Using near-infrared reflectance spectroscopy to predict carbon, nitrogen and phosphorus content in heterogenus plant material. *Oecologia, 118*, 173–182.

Gillon, D., Joffre, R., & Ibrahima, A. (1999b). Can litter decomposability be predicted by near infrared reflectance spectroscopy? *Ecology, 80*, 175–186.

Gillon, D., Dauriac, F., Deshayes, M., Valette, J.C. and Moro, C. (2004). Estimation of foliage moisture content using near infrared reflectance spectroscopy. *Agricultural and Forest Meteorology, 124*, 51–62.

Gorham, E. (1991). Northern peatlands: Role in the carbon cycle and probable responses to climatic warming. *Ecological Applications, 1*, 182–195.

Grossmann, Y.L., Ustin, S.L., Jacquemoud, S., Sanderson, E.W., Schmuck, G., & Verdebout, J. (1996). Critique of stepwise multiple linear regression for the extraction of leaf biochemistry information from leaf reflectance data. *Remote Sensing of Environment, 56*, 182–193.

Hedges, J.I., & Ertel, J.R. (1982). Characterization of lignin by capillary chromatography of cupric oxide oxidation products. *Analytical Chemistry, 54*, 174–178.

Jofre, R., Gillon, D., Dardenne, P., Agneessens, R., & Biston, R. (1992). The use of near-infrared spectroscopy in litter decomposition studies. *Annales des Sciences Forestières, 49,* 481–488.

Karsisto, M., Savitski, M., Kitunen, V., Penttilä, T., Laine, J., & Laiho, R. (2003). Quantification of organic fractions in litter and peat organic matter. In J.O. Honkanen & P.S. Koponen (Eds.), *Current perspectives in environmental science and technology* (pp. 135–137). Joensuu, Finland: Finnish Society for Environmental Sciences.

Kelley, S.S., Rowell, R.M., Davis, M., Jurich, C.K., Ibach, R. (2004). Rapid analysis of the chemical composition of agricultural fibers using near infrared spectroscopy and pyrolysis molecular beam mass spectrometry. *Biomass and Bioenergy, 27,* 77–88.

Lacaze, B., & Joffre, R. (1994). Extracting biochemical information from visible and near infrared reflectance spectroscopy of fresh and dried leaves. *Journal of Plant Physiology, 144,* 277–281.

Laiho, R. (2006). Decomposition in peatlands: Reconciling seemingly contrasting results on the impacts of lowered water levels. *Soil Biology & Biochemistry, 38,* 2011–2024.

Ludwig, B., Khanna, P.K., Bauhus, J., & Hopmans, P. (2002). Near infrared spectroscopy of forest soils to determine chemical and biological properties related to soil sustainability. *Forest Ecology and Management, 171,* 121–132.

Martens, H., Naes, T. (1991). *Multivariate Calibration.* (Chichester: John Wiley & Sons)

Martin, M.E., & Aber, J.D. (1994). Analysis of forest foliage. III: Determining nitrogen, lignin and cellulose in fresh leaves using near infrared reflectance data. *Journal of Near Infrared Spectroscopy, 2,* 25–32.

Martin, P.D., Malley, D.F., Manning, G., & Fuller, L. (2002). Determination of soil organic carbon and nitrogen at the field level using near infrared spectroscopy. *Canadian Journal of Soil Scence, 82,* 413–422.

McCarty, G.W., Reeves III, J.B., Reeves, V.B., Follet, R.F., Kimble, J.M. (2002). Mid-infrared and near-infrared diffuse reflectance spectroscopy for soil carbon measurement. *Soil Science Society of America Journal, 66,* 640–646.

McLellan, T.M., Aber, J.D., & Martin, M.E. (1991*a*). Determination of nitrogen, lignin and cellulose content of decomposing leaf material by near infrared reflectance spectroscopy. *Canadian Journal of Forest Research, 21,* 1684–1688.

McLellan, T.M., Martin, M.E., Aber, J.D., Melillo, J.M., Nadelhoffer, K.J., & Dewey, B. (1991*b*). Comparison of wet chemistry and near infrared reflectance measurements of carbon-fraction chemistry and nitrogen concentration of forest foliage. *Canadian Journal of Forest Research, 21,* 1689–1693.

Meentemeyer, V. (1978). Macroclimate and lignin control of liter decomposition rates. *Ecology, 59,* 465–472.

Murphy, K.L., Klopatek, J.M., & Klopatek, C.C. (1998). The effects of litter quality and climate on decomposition along an elevational gradient. *Ecological Applications, 8,* 1061–1071.

Norris, K.H., Barnes, R.F., Moore, J.E., & Shenk, J.S. (1976). Predicting forage quality by near infrared reflectance spectroscopy. *Journal of Animal Science, 43,* 889–897.

Ono, K., Hiraide, M., & Amari, M. (2003). Determination of lignin, holocellulose, and organic solvent extractives in fresh leaf, litterfall, and organic material on forest floor using near-infrared reflectance spectroscopy. *Journal of Forest Research, 8,* 1341–6979.

Päivänen, J., & Vasander, H. (1994). Carbon balance in mire ecosystems. *World Resource Review, 6,* 102–111.

Quaramby, C., & Allen, S.E. (1989). Organic constituents. In S.E. Allen (Ed.), *Chemical Analysis of Ecological Materials* (pp. 160–200). Oxford: Blackwell.

Ryan, M.G., Melillo, J.M., & Ricca, A. (1990). A comparison of methods for determining proximate carbon fractions of forest litter. *Canadian Journal of Forest Research, 20,* 166–171.

Savitzky, A., & Golay, M.J.E. (1964). Smoothing and differentiation of data by simplified least squares procedures. *Analytical Chemistry, 36,* 1627–1639.

Schimleck, L.R., & Yazaki, Y. (2003) Analysis of *Pinus radiata* D. Don Bark by Near Infrared Spectroscopy. *Holzforschung, 57,* 520–526.

Shenk, J.S. (1992). NIRS analysis of natural agricultural products. In K.I. Hildrum, T. Isaaksson, T. Naes, & A. Tandberg (Eds.), *Near infra-red spectroscopy. Bridging the gap between data analysis and NIR applications* (pp. 235–240). London: Ellis Horwood.

Shenk, J.S., & Westerhaus, M.O. (1991). Population definition, sample selection, and calibration procedure for near infrared reflectance spectroscopy. *Crop Science, 31,* 469–474.

Shepherd, K.D., Vanlauwe, B., Gachengo, C.N., Palm, C.A. (2005). Decomposition and mineralization of organic residues predicted using near infrared spectroscopy. *Plant and Soil, 277,* 315–333.

Stenberg, B., Jensen, L.S., Nordkvist, E., Breland, T.A., Pedersen, A., Gudmundsson, J., Bruun, S., Salo, T., Palmason, F., Henriksen, T.M. & Korsaeth, A. (2004). Near infrared reflectance spectroscopy for quantification of crop residue, green manure and catch crop C and N fractions governing decomposition dynamics in soil. *Journal of Near Infrared Spectroscopy, 12,* 331–346.

Stohlgren, T.J. (1988). Litter dynamics in two Sierran mixed conifer forests, II: Nutrient release in decomposing leaf litter. *Canadian Journal of Forest Research, 18,* 1136–1144.

Taylor, B., Parkinson, D., & Parsons, W.J.F. (1989). Nitrogen and lignin content as predictors of litter decay rates: a microcosm test. *Ecology, 70,* 97–104.

Taylor, B.R., Prescott, C.E., Parsons, W.J.F., &Parkinson, D. (1991). Substrate control of litter decomposition in four Rocky Mountain coniferous forests. *Canadian Journal of Boany, 69,* 2242–2250.

Tian, G., Kang, B.T., & Brussard, L. (1992). Biological effects of plant residues with contrasting chemical compositions under humid tropical conditions – decomposition and nutrient release. *Soil Biology and Biochemistry, 24,* 1051–1060.

Terhoeven-Urselmans, T., Michel, K., Helfrich, M., Flessa, H., & Ludwig, B. (2006). Near-infrared spectroscopy can predict the composition of organic matter in soil and litter. *Journal of Plant Nutrition and Soil Science, 169,* 168–174.

Watson, C.A., Etchevers, G. & Shuey, W.C. (1976). Relationship between ash and protein contents of flour mill streams determined with the InfraAnalyzer and standard approved methods. *Cereal Chemistry, 53,* 803–804.

Wessman, C.A., Aber, J.D., Peterson, D.L., & Melillo, J.M. (1988). Foliar analysis using near-infrared reflectance spectroscopy. *Canadian Journal of Forest Research, 18,* 6–11.

Williams, P., & Sobering, D.C. (1996). How do we do it: A brief summary of the methods we use in developing near infrared calibrations. In A.M.C. Davies & P. Williams (Eds.), *Near infrared spectroscopy: The future waves* (pp. 185–188). Chichester, UK: NIR Publications.

Winch, J.E., & Major, H. (1981). Predicting nitrogen and digestibility of forages using near infrared reflectance photometry. *Canadian Journal of Plant Science, 61,* 45–51.

Windham, W.R., Hill, N.S., & Stuedemann, J.A. (1991). Ash in forage, esophageal, and faecal samples analyzed using near-infrared reflectance spectroscopy. *Crop Science, 31,* 1345–1349.

Yeh, T.F., Yamada, T., Capanema, E., Chang, H.M., Chiang, V., & Kadla, J.F. (2005). Rapid Screening of Wood Chemical Component Variations Using Transmittance Near-Infrared Spectroscopy. *Journal of Agriculture and Food Chemistry, 53,* 3328–3332.

Chapter 8
Leachate Treatment in Newly Built Peat Filters: A Pilot-Scale Study

Pille Kängsepp[1,2], Margit Kõiv[3], Mait Kriipsalu[4], and Ülo Mander[3](✉)

Abstract The purpose of this short-term pilot-scale ($1\,m^3$) experiment was to focus on the efficiency of leachate treatment during the initial period of a newly built peat filter. The initial start-up period of a filter is dynamic and differs for different types of peat and leachate. Reduction of biochemical oxygen demand (BOD), chemical oxygen demand (COD), and nutrient concentrations from leachate of different age and composition was targeted in experimental filters of three types of peat. Well-mineralised fen peat with undisturbed structure showed a significantly better reduction when treating methanogenic leachate with respect to the concentrations of COD, total nitrogen (N_{tot}), and ammonium-nitrogen (NH_4-N), which were reduced by up to 36, 62, and 99%, respectively, compared to the poorly mineralised milled *Sphagnum* peat (with no reduction of COD, 52% reduction of N_{tot}, and 67% of NH_4-N). Good results in the reduction of BOD and total phosphorus (P_{tot}) (50%) in well-mineralised *Sphagnum* peat were achieved when treating acidogenic leachate. Generally, even if a considerable amount of pollutants was removed in newly built peat filters (milligrams per metre per day) during the first days in operation, the target values were still above the Estonian limit for wastewater discharge into the environment. The best results were achieved for removal of P_{tot}, which was reduced below the target values.

Keywords After-treatment, biochemical oxygen demand (BOD), chemical oxygen demand (COD), nitrogen, phosphorus

[1] Institute of Molecular and Cell Biology, University of Tartu, Riia 23, Tartu, 51010, Estonia

[2] School of Pure and Applied Natural Sciences, Kalmar University, Kalmar 39182, Sweden

[3] Institute of Geography, University of Tartu, Vanemuise 46, Tartu, 51014, Estonia

[4] Institute of Forestry & Rural Engineering, Estonian University of Life Sciences, Kreutzwaldi 64, Tartu, 51014, Estonia

(✉) Corresponding author: e-mail: ulo.mander@ut.ee

J. Vymazal (ed.) *Wastewater Treatment, Plant Dynamics and Management in Constructed and Natural Wetlands,*
© Springer Science + Business Media B.V. 2008

8.1   Introduction

The composition of landfill leachate varies greatly, depending on waste quantity and composition, the decomposition rate and age of the waste, and landfilling technology. High concentrations have been reported for common water-quality parameters, such as chemical oxygen demand (COD) up to 60,000 mg l^{-1} (Tchobanoglous *et al.*, 1993), for total phosphorus (P_{tot}) up to 100 mg l^{-1} (Tchobanoglous *et al.*, 1993), and for ammonium-nitrogen (NH_4-N) up to 1,200 mg l^{-1} (McBean & Rovers, 1999). The spread of pollution is not acceptable, and leachate must be collected and treated. To promote a sustainable closed-loop treatment strategy, leachate must be treated close to the site of its creation. This creates a need to develop on-site leachate treatment systems.

Peat filters have been widely used in the treatment of wastewaters of various origins and quality (Geerts *et al.*, 2001) as well as leachate (Kadlec, 2003; Kinsley *et al.*, 2004). Peat has a unique combination of physical, chemical, and biological properties, making it attractive for use in trickling filters, bio-filters, and other conventional treatment systems. Peat is also an abundant material in Estonia, where peatland covers 22% of total land area, reaching 2.4 billion tonnes (Orru, 1992). The efficiency of peat filters in the treatment of leachate differs from site to site. This may be due to the properties of peat, the way how the peat was obtained, stored, and placed in the filter-bed, as well as the properties of leachate and its loading rates. It has not been thoroughly studied whether the peat filters start to function in purifying leachate shortly after their construction. The main objective of this study was to determine the treatment efficiency of various peat types in newly built filters for the reduction of biological oxygen demand (BOD_7), COD, and nutrient concentrations during the first weeks. Leachate from two landfills of different ages was studied, and three types of peat were included. This research was part of a long-term leachate treatment project, where leachate treatment efficiency as well as the properties of peat and combinations of filter materials were studied.

8.2   Materials and Methods

8.2.1   Description of Sites

Leachate was collected from Aardlapalu and Väätsa municipal solid waste landfills in Estonia. Aardlapalu landfill has been in operation since 1971. It receives about 50,000 t of mixed municipal waste per year. The waste is compacted but not properly covered. The landfill has no bottom lining. Approximately 500 m³ day⁻¹ of leachate is collected by drainage collectors around the outer perimeter and stored in a large sedimentation pond. The leachate is relatively diluted. During the period of the study, the mean composition of leachate from Aardlapalu landfill for COD, BOD_7, total nitrogen (N_{tot}), total phosphorus (P_{tot}), and pH were 720, 29, 260, 3.7 mg l^{-1}, and 7.3, respectively (Centre of Ecological Engineering, 2001).

Väätsa landfill is the first sanitary landfill in Estonia that meets the requirements of the EU landfill directive (European Commission, 1999/31/EC) and the requirements of Government of Estonia for landfills (RTL 2004, 56, 938). The first stage of the landfill (1 ha) was in service from 2000 to 2005, and about 60,000 t of mixed waste have been deposited. The landfill has a standard bottom lining. Approximately 10–20 m³ of leachate is collected daily and treated in a biological treatment system, as described by Sooäär (2003). During the period of the study, the mean composition of leachate from Väätsa landfill for COD, BOD_7, N_{tot}, P_{tot}, and pH in raw leachate was 4,133, 2,193, 401, 1.7 mg l⁻¹, and 7.4, respectively (Sooäär, 2003).

8.2.2 Experimental Peat Filters

Experiments were performed in custom-designed vertical flow peat filters (F1–F3) with a volume of 1 m³, rectangular shape (1 × 1 × 1 m), and a permeable floor (50 × 50 mm metal net covered with a 1 × 1 mm plastic net). Three different types of peat were selected for investigation in this study, and the chemical composition of the peat has been shown in Table 8.1.

Filter F1 was filled with well-mineralised fen peat obtained by pushing the metal frame of the experimental filter body into the peat bog, and excavating it without disturbing the natural structure of the peat. The peat was classified as minerotrophic mire, containing a mix of well-decomposed *Phragmites* and *Carex* peat. The hydraulic conductivity of natural peatland was measured on-site with a conventional 500 cm² double-ring infiltrometer according to Hillel (1998). The hydraulic conductivity of the layer from 5 to 50 cm below the surface was 8×10^{-6} m s⁻¹, and the layer from 50 to 100 cm was 8×10^{-5} m s⁻¹. Filter F2 was filled with milled fluffy poorly mineralised *Sphagnum* peat mined from Umbusi raised peat bog. This highmoor deposit is classified as swamp peat and is exploited industrially for bedding and heating purposes. Filter F3 was filled with fluffy well-mineralised *Sphagnum* peat from the lower deposits of a depleted Lokuta industrial peatland. The peat was sieved through a 26 mm sieve to remove stones and roots.

Table 8.1 Composition of peat in three filters before experiment (Filter 1 includes determinations of four different layers, from the surface of peatland)

	F1				F2	F3
	0–25 cm	25–50 cm	50–75 cm	75–100 cm		
pH_{KCl}	6.67	6.09	5.90	5.72	3.5	6.7
N (%)	0.525	2.637	2.838	3.000	1.0	0.5
P (mg kg⁻¹ DM)	123	71	41	64	9	23
K (mg kg⁻¹ DM)	47	72	32	38	98	93
Ca (mg kg⁻¹ DM)	5,123	16,424	20,985	20,541	5,440	8,294
Mg (mg kg⁻¹ DM)	721	2,024	2,418	3,054	975	1,416
Organic matter (% DM)	11	47	50	85	85	37

8.2.3   Hydraulic Loading

Filters F1–F3 received pre-treated leachate. Leachate from Aardlapalu was pre-treated by extended aeration during 11 days in a $3\,m^3$ Roth Micro-Step fine-bubble compact wastewater treatment unit. After aeration, the leachate was loaded on both F1 and F2 for $4\,h$ each day over a period of 4 days. The flow rates to F1 and F2 (90.3 and $2\,l\,min^{-1}$ respectively) had to be different due to the different type of peat, its density, compaction, and moisture content. During the test period, $0.5\,m^3$ of leachate was percolated through F1, and $1.5\,m^3$ through F2.

In Väätsa, leachate was biologically pre-treated in a compact two-phase activated sludge leachate treatment plant, which is accomplished with an aerobic–anoxic sedimentation pond after the plant. Leachate from the outflow of the sedimentation pond was loaded on F3 by timer-adjusted pumping during 22 days, providing 24 pumping events per day, each of which was 15 min long ($2.9\,m^3\,day^{-1}$). During the experiments, the total loading of leachate on filter F3 was about $104\,m^3$.

8.2.4   Sampling and Analyses

Water samples from F1 and F2 were taken daily. Water samples from F3 were taken on days 2, 4, 6, 8, 15, and 22. BOD_7, COD, N_{tot}, NH_4-N, nitrate-nitrogen (NO_3-N), nitrite-nitrogen (NO_2-N), P_{tot}, and pH were determined in the certified Laboratory of the Tartu Wastewater Treatment Plant using standard methods. The content of organic matter, nitrogen (N), phosphorus (P), potassium (K), calcium (Ca), magnesium (Mg), and pH_{KCl} in peat was determined at the Laboratory of Plant Biochemistry of the Estonian University of Life Sciences using standard methods. Reduction in concentration (%), mass loading ($g\,m^{-2}\,day^{-1}$), and load removed ($g\,m^{-2}\,day^{-1}$) were calculated according to methods described by Kadlec (1999). Statistical analysis was carried out using *Statistica 7.0* (StatSoft Inc.). The non-parametric Kruskal–Wallis test for the multiple comparisons of mean ranks between the water-quality parameters was used.

8.3   Results and Discussion

The limit values for BOD_7, COD, N_{tot}, and P_{tot} in leachate treatment in Estonia are 25, 125, 75, and $2.0\,mg\,l^{-1}$, respectively, and removal efficiency should be higher than 90, 75, 75, and 80%, respectively (RT I 2001, 69, 424; RTL 2004, 56, 938).

8.3.1   Pre-treatment of Leachate

Aeration of the Aardlapalu methanogenic leachate over 11 days led to 70% and 14% reduction of BOD_7 and COD respectively. NH_4-N was not significantly reduced by aeration. The slight increase in the concentration of NO_2-N and NO_3-N indicated that some nitrification occurred during aeration, and some loss of ammonia by air stripping was observed. During the period of aeration, an increase in pH from 7.5 to 9.0 was measured.

The overall biological treatment efficiency at Väätsa landfill is high (Sooäär, 2003). Reduction of COD, BOD_7, and N_{tot} was 98, 89, and 71%, respectively. These results show that biological pre-treatment is successful for the treatment of high-strength leachate from landfills in the acidogenic phase, where concentrations of fatty acids are high. Increase in the pH (from 7.4 to 8.7) and P_{tot} (from 1.7 to 3.6 mg l^{-1}) was observed. The average values of BOD_7 (52 mg l^{-1}), COD (451 mg l^{-1}), N_{tot} (117 mg l^{-1}), and P_{tot} (3.6 mg l^{-1}) in the effluent from the bio-treatment (Sooäär, 2003), however, still exceeded limit values for wastewater discharge (RT I 2001, 69, 424), thus requiring additional polishing.

8.3.2   Treatment Efficiency in Newly Built Peat Filters

The concentrations of COD, BOD_7, N_{tot}, NH_4-N, NO_3-N, NO_2-N, P_{tot}, and pH was measured in the influent and effluent from the peat filters, as compared in Fig. 8.1.

The average value of pH in leachate decreased in all filters. The greatest decrease was in F2 (from 9.0 to 3.5). The change in pH may be related to the organic acid components, which were flushed out from the peat, as well as the loss of alkalinity (HCO_3^-) during the denitrification process, which has been described by Patterson *et al.* (2001).

In F3 the average reduction of BOD was 50%, which, however, is still above the limit values. In filters F1 and F2, an increase in BOD was observed. A reduction in COD was achieved in F1 (36%) and F3 (10%). The slight increase in BOD in the effluent from F1 and F2 and limited reduction of COD in the F2 could be due to the washout of suspended solids from the peat, which has also been observed by Kløve (2001) and Nieminen (2003). This increase might also be due to the bacterial breakdown of large molecules of organic matter to smaller ones, or their conversion into other types of organic matter. In these cases, further investigations via an analytical screening of leachate composition might be useful to identify what kind of compounds have caused the higher BOD and COD values in the effluent.

There were significant differences in influent concentrations of N_{tot} and NH_4-N to the peat filters due to the different efficiency during the pre-treatment phase. The highest reduction of N_{tot} from leachate was achieved in F1 (62%), followed by F2 (57%), and F3 (16%). Although some of N_{tot} was removed in all filters (Fig. 8.1), the limit values were not met. Maximum removal of NH_4-N was achieved by F1 (99%), followed by F2 (67%), and F3 (28%). Reduction of NH_4-N in peat might be by sorption or by nitrification. The sorption mechanisms of NH_4-N and conditions are described in greater detail by McNevin *et al.* (1999). Moreover, the adsorption test developed by Heavey (2003) confirms that the cation-exchange sites in peat only provide temporary storage prior to a subsequent nitrification process. The

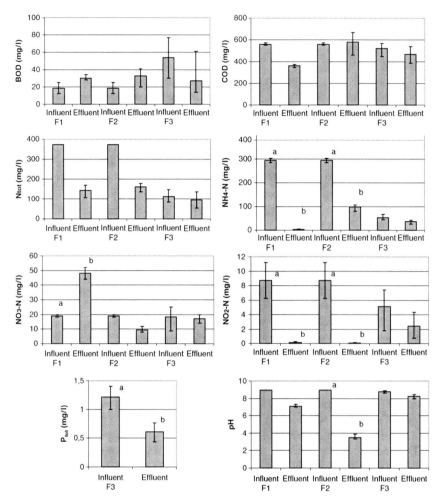

Fig. 8.1 Average values of N_{tot}, NH_4-N, NO_3-N, BOD_7, COD, pH, and P_{tot}. Significantly different ($p < 0.05$) influent and effluent values according to Kruskal–Wallis test are indicated by a and b

nitrification occurred in the well-mineralised fen peat in F1. Almost 300 mg l^{-1} of NH_4-N disappeared, compared to only around 30 mg l^{-1} of NO_3-N produced, which indicates the losses via denitrification. In filters F2 and F3, no significant nitrification took place during the first days, which is not surprising concerning the time span. However, after 15 days some nitrification of NH_4-N in F3 (Fig. 8.2) was noticed.

Although NO_3-N in the influent reduced in time, there was an increase in the effluent (Fig. 8.2). This partly can be explained by a decrease in NH_4-N, which might depend on transformation to NO_3-N, but increased evaporation may also

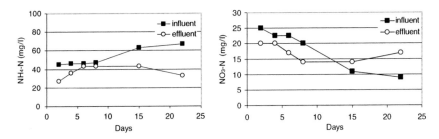

Fig. 8.2 Temporal variability in concentration of NH_4-N and NO_3-N in the influent and the effluent of filter F3

account for some of these changes. The value of NO_2-N has decreased by ca 2 mg l^{-1}, most probably by oxidation to NO_3-N, which should also contribute to the increase in NO_3-N in the effluent. The temperature of leachate during the summer period in Väätsa was $18.0 \pm 1.6°C$. Thus, there should be favourable conditions for an increasing growth in nitrifying bacteria. The growth of nitrifying bacteria takes time, and there might even be several other reasons why the nitrification process might be delayed. Mæhlum (1998) indicated that nitrification may be limited due to lack of oxygen. The amount of oxygen in the peat was not measured directly, but batch loading of leachate during the experiment was selected in order to support the natural inflow of oxygen through the top and bottom of the filter body.

The removal of P_{tot} in F3 was successful (50%) and effluent value was below the limits. The reduction of P_{tot} could have been caused by the sorption, sedimentation, and combination of complex compounds, as also demonstrated by Mann (1990), Kadlec and Knight (1996), and Richardson *et al.* (1996). Some quantity of P may be bound to the bio-film (Mann, 1990; Richardson *et al.*, 1996). Phosphorus transforms easily from organic to inorganic forms and forms chemical complexes with organic and inorganic ligands, which can be adsorbed into the soil or precipitated. In aerobic wetland conditions, phosphorus appears in dissolved complexes together with Ca and Mg ions in alkaline conditions, and with iron (Fe) or aluminium (Al) ions in soil with acidic to neutral pH (Mæhlum, 1998).

8.3.3 Mass Loading and Mass Load Removal

Every mass unit (milligram) of pollutants removed by area of the filter (square metre) per day results in a reduction of environmental load and better protection of surface water and groundwater. The loading of pollutants and the removal of pollutants from leachate in peat as grams per square metres per day were calculated, and are shown in Table 8.2.

Table 8.2 Mass loading and average load removed in peat filters F1–F3 (g m^{-2} day^{-1}). Negative values indicate an increased value in the filter (leakage into effluent)

Parameter	Mass loading on filter (g m^{-2} day^{-1})			Removed in filter (g m^{-2} day^{-1})		
	F1	F2	F3	F1	F2	F3
BOD$_7$	1	10	155	−1	−10	78
COD	40	270	1,510	14	−15	158
N$_{tot}$	25	180	325	17	103	51
P$_{tot}$	n.a.	n.a.	4	n.a.	n.a.	2
NH$_4$-N	20	140	150	20	90	43

F3 received the highest mass load of pollutants per 1 m^2 and per day, also resulting in the largest mass removal of COD and BOD. These results show that even if the mass loading and flow rate are relatively high, the newly built peat filter can withstand the 'stress situation' and remove a high amount of pollutants from the leachate. Unfortunately, excess loading rate will certainly reduce the potential lifetime of the peat. It was noted that mass removal of N$_{tot}$ and NH$_4$-N in the leachate were highest in F2. The lowest amount of pollutants was obtained in F1, where flow rate was lowest. However, the removal efficiency (%) might be improved if the flow rate is smaller due to longer retention time in the peat. Since the flow rate and total mass loading on filter F1 and F2 were lower than on F3, these filters were expected to give higher treatment efficiency (%) and a longer lifetime. However, since the maximum hydraulic loading of leachate was quite low in F1 (well-mineralised fen peat), huge areas are needed for full-scale application, if natural peatlands are to be used for treatment purposes. This, in combination with local legal requirements, could be limiting factors for its use as filter material.

It is likely that the purification mechanism in peat varies over time. In the beginning, the physicochemical removal of pollutants dominates, and fixed bio-film around peat particles has not developed. The time needed to develop a bio-film might be different for various types of peat and leachate. However, it seems that differences in purification mechanisms do not have a serious impact on the removal of pollutants. The physicochemical processes in a good peat filter are sufficient to achieve adequate trapping of compounds already at the very beginning. This indicates that during the long construction time of a filter, before it becomes operational, no specific preparation of peat (e.g. wetting peat with clean water as is often used in order to facilitate the homogeneous settlement of particles and avoid channelling flow) is needed. However, to achieve sufficient removal of as many pollutants of interest as possible, time is needed for the establishment of micro-organisms and also in order to obtain purification through biological processes.

8.4 Conclusions

All three types of peat are suitable for the removal of N$_{tot}$ and NH$_4$-N pollutants from leachate in the start-up period. The well-mineralised undisturbed fen peat (mixture of *Phragmites* and *Carex* peat) demonstrated significantly better purification of COD, N_{tot}, and NH_4-N (concentration reduction in percentage) of old landfill leachate than milled poorly mineralised *Sphagnum* peat. Fluffy well-mineralised *Sphagnum* peat can be used in the treatment of leachate from the acidic phase. The P_{tot} reduction from acidogenic leachate was good in well-mineralised *Sphagnum* peat, and it was sufficient to fulfil requirements for P_{tot} concentration in the effluent discharge to the environment.

These short-term measurements have shown that in newly built peat filters, purification of leachate can be obtained, probably due to physicochemical retention. Even if a considerable amount of pollutants were removed in newly built peat filters (milligram per square metres per day), concentration values in the effluent were still above the Estonian limit values for wastewater discharge to the environment, except for P_{tot}. However, every mass unit of pollutants removed by the filter per day in the start-up period results in a reduction of environmental load and better protection of surface water and groundwater. It can be expected that the peat filter will show better removal efficiencies in a long-term perspective when steady-state conditions are created and bio-film has had time to form. This encourages us to recommend fen peat filters for the secondary and tertiary treatment of landfill leachate, and even semi-natural fen areas for on-site leachate treatment.

Acknowledgements The Estonian Ministry of Education and Science is acknowledged for Funding (Project No. 0182534s03). Colleagues from the Estonian University of Life Sciences – Tõnu Salu, Toomas Timmusk, and Marek Joost, as well as staff from Väätsa landfill – are acknowledged for their field assistance.

References

Centre of Ecological Engineering. (2001). *Evaluation of construction of wetland system in Aardlapalu landfill* (Tartu Aardlapalu prügila märgalapuhasti rajamise eelduste hindamine. Ökoloogiliste Technoloogiate Keskus, in Estonian). Tartu, Estonia: Centre of Ecological Engineering.

European Community. (1999). 1999/31/EC. European Union Council Directive of 26 April 1999 on the landfill of waste. *Official Journal of the European Communities, L 182*, 16.07.1999, pp. 1–19.

Geerts, S.M., McCarthy, B., Axler, R., Henneck, J., Christopherson, S.H., Crosby, J., & Guite, M. (2001). *Performance of peat filters in the treatment of domestic wastewater in Minnesota*. Paper presented at the 9th National Symposium on Individual and Small Community Sewage Systems, American Society of Agricultural Engineers, St. Joseph, Michigan, USA.

Heavey, M. (2003). Low-cost treatment of landfill leachate using peat. *Waste Management, 23*, 447–454.

Hillel, D. (1998). *Environmental Soil Physics*. San Diego: Academic.

Kadlec, R.H. (1999). Constructed wetlands for treating landfill leachate. In G. Mulamoottil, A. McBean, & F. Rovers (Eds.), *Constructed wetlands for the treatment of landfill leachates.* (pp. 17–31). Boca Raton, FL: CRC/Lewis.

Kadlec, R.H. (2003). Integrated natural systems for landfill leachate treatment. In J. Vymazal (Ed.), *Wetlands – nutrients, metals and mass cycling* (pp. 1–34). Leiden, the Netherlands: Backhuys.

Kadlec, R.H., & Knight, R.C. (1996). *Treatment wetlands*. Boca Raton, FL: CRC/Lewis.

Kinsley, C.B., Crolla, A.M., Kuyucak, N., Zimmer, M., & Laflèche, A. (2004). A pilot peat filter and constructed wetland system treating landfill leachate. In Liénard, A., Burnett, H. (Eds.)

Proceedings of the 9th international conference on wetland systems for water pollution control (pp. 635–643). Avignon, France: ASTEE.

Kløve, B. (2001). Characteristics of nitrogen and phosphorus loads in peat mining wastewater. *Water Research, 35,* 2353–2362.

Mæhlum, T. (1998). Cold-climate constructed wetlands: Aerobic pre-treatment and horizontal subsurface flow systems for domestic sewage and landfill leachate purification. Dissertation, Agricultural University of Norway, Ås, Norway.

Mann, R.A. (1990). Phosphorus removal by constructed wetlands: Substratum adsorption. In P.F. Cooper & B.C. Findlater (Eds.), *Use of constructed wetlands in water pollution control.* (pp. 97–105). Oxford: Pergamon.

McBean, E.A., & Rovers, F. (1999). Landfill leachate characteristics as inputs for the design of wetlands used as treatment systems. In G. Mulamoottil, A. McBean, & F. Rovers (Eds.), *Constructed wetlands for the treatment of landfill leachates* (pp. 1–16). Boca Raton, FL: CRC/Lewis.

McNevin, D., Barford, J., & Hage, J. (1999). Adsorption and biological degradation of ammonium and sulphide on peat. *Water Research, 33,* 1449–1459.

Nieminen, M. (2003). Effects of clear-cutting and site preparation on water quality form a drained Scots pine mire in southern Finland. *Boreal Environment Research, 8,* 53–59.

Orru, M. (1992). *Estonian peat resources (Eesti turbavarud,* in Estonian). Tallinn: Geological Survey of Estonia.

Patterson, A.A., Davey, K., & Farnan, N. (2001). Peat bed filters for on-site treatment of septic tank effluent. In R.A. Patterson & M.J. Jones (Eds.), *Advancing on-site wastewater systems* (pp. 315–322). Armidale, Australia: Lanfax Labs Armidale).

Richardson, C.J., Qian, S.S., & Craft, C.B. (1996). Predictive models for phosphorus retention in wetlands. In J. Vymazal (Ed.), *Nutrient cycling and retention in wetlands and their use for wastewater treatment* (pp. 125–150). Prague: Ecology and Use of Wetlands.

RT I 2001, 69, 424. (2001). Requirements for wastewater discharged into water bodies or into soil. (Heitvee veekogusse või pinnasesse juhtimise kord, in Estonian). Regulation No. 269 of 31 July 2001 of the Government of the Republic of Estonia. (RTI 2001, 69, 424; 2003, 83, 565; 2006, 10, 67). Tallinn, Estonia *State Gazette.*

RTL 2004, 56, 938. (2004). Requirement for the establishment, operation and closure of the land-fills. (Prügila rajamise, kasutamise ja sulgemise nõuded, in Estonian). Regulation No. 38 of 29. April 2004 of the Government of the Republic of Estonia (RTL 2004, 56, 938; 108, 1720; 2006, 91, 1685). Tallinn, Estonia. *State Gazette.*

Sooäär, M. (2003). *Activated sludge treatment of leachate from Väätsa landfill, Estonia.* Paper presented at the 4th International Conference on the Establishment of Co-operation between Companies/Institutions in the Nordic Countries and the Countries in the Baltic Sea Region, Eco-tech 03, Kalmar, Sweden.

Tchobanoglous, G., Theisen, H., & Vigil, S. (1993). Integrated solid waste management: *Engineering principles and management issues.* New York: McGraw-Hill.

Chapter 9
Monthly Evapotranspiration Coefficients of Large Reed Bed Habitats in the United Kingdom

Katy E. Read[1]($\boxtimes$), Peter D. Hedges[2], and Phil M. Fermor[1]

Abstract The UK National Biodiversity Action Plan highlighted reed beds as a habitat of conservation concern and provided targets for the management, restoration and creation of reed beds throughout the UK, ideally in blocks greater than 20 ha in size. Previous water-use rate research by Aston University targeted small and fringe reed bed habitats. Extensive literature reviews carried out in 2000 identified a lack of information with respect to the water-use rates of larger reed beds. A research programme was developed utilising the lysimeter technique for determining evapotranspiration (ET) rates in large reed beds at three research sites. At each site, meteorological monitoring equipment and lysimeters were installed within the reed beds. Monitoring was carried out monthly throughout 2001 and 2002. Phenological data was collected from fixed quadrats within the reed beds and from the lysimeters during each site visit. This data was utilised to determine which of the lysimeters were 'successful', i.e. representative of the reed bed into which they where installed. From the 'successful' lysimeters, monthly crop coefficients (Kc) were derived using reference crop evapotranspiration (ETo) data provided by the UK Meteorological Office (MORECS Grass data). The results are compared with relevant published data and show that the findings are comparable with other studies.

Keywords Evapotranspiration, large-scale wetland creation, lysimeters, phytometers, reed bed, water budgets

[1] Middlemarch Environmental Ltd, Triumph House, Birmingham Road, Allesley, Coventry CV5 9AZ, United Kingdom

[2] Aston University, Aston Triangle, Birmingham, B4 7ET, United Kingdom

($\boxtimes$) Corresponding author: e-mail: katy.read@middlemarch-environmental.com

J. Vymazal (ed.) *Wastewater Treatment, Plant Dynamics and Management in Constructed and Natural Wetlands,*
© Springer Science + Business Media B.V. 2008

9.1   Introduction

To ameliorate for the habitat loss and degradation experienced in the UK, the UK National Biodiversity Action Plan (UK BAP) set targets for the appropriate management, restoration and creation of a range of wetland and terrestrial habitats. To meet the targets, new wetlands will have to be created throughout the UK in areas with suitable hydrology, geology, sediments, topography, ecology and meteorology, all of which must be considered in the design of the new wetlands.

One of the UK BAP habitats with a specific Habitat Action Plan (HAP) are reed beds (UKBG, 1995), which are defined as wetland dominated by stands of common reed *Phragmites australis*, wherein the water table is at or above ground level for most of the year. In the UK, the most extensive reed beds are found in river flood plains and low-lying coastal flood plains, although they also occur along natural lake margins, within estuaries and within artificial sites such as old gravel pits (Hawke & Jóse, 1996). There are approximately 5,000 ha of reed bed in the UK, and although this area is comprised of over 900 sites, only 20 are greater than 20 ha in size. The greatest concentration of UK reed beds is found in East Anglia, with 36% of the UK's total resource occurring in Norfolk and Suffolk (Gilbert *et al.*, 1996).

Reed beds are amongst the most important habitats for birds in the UK and as they are nationally scarce, many of these species are also limited in their distribution (Hawke & Jóse, 1996). The reed bed HAP (UKBG, 1995) provided targets for reed bed restoration and creation in the UK and recommends that reed bed creation should ideally take place in blocks of at least 20 ha. The target is for the creation of 1,200 ha of new reed bed on land of low nature conservation interest by 2010.

To ensure that newly created reed bed habitats are hydrologically sustainable, a water budget must be completed as part of the design process. To successfully carry out detailed water budgets, information with respect to the water use (measured as evapotranspiration [ET]) of the target habitats must be available.

In response to an identified paucity of information, this project was initiated to enable the determination of ET rates for large reed bed habitats, utilising the methodology developed by Fermor (1997). The literature shows that numerous authors have used lysimeters (sometimes known as phytometers) to estimate ET rates from wetland habitats (e.g. Gel'bukh, 1964; Bernatowicz *et al.*, 1976; Gavenčiak, 1972; Gilman & Newson, 1983; Fermor *et al.*, 2001).

9.2   Study Sites

Three reed bed study sites (see Fig. 9.1) were identified through discussions with English Nature, the Royal Society for the Protection of Birds (RSPB) and Warwickshire Wildlife Trust. Lysimeters and meteorological equipment were established at each site between February and April 2000.

Fig. 9.1 Location of reed bed study sites

9.2.1 *Aqualate Mere*

Aqualate is the largest of the natural meres in the north-west Midlands, located at UK national grid reference SJ 770 205. The site is privately owned but managed by English Nature (now Natural England) and is designated as a Site of Special Scientific Interest (SSSI) and National Nature Reserve (NNR). The mere itself is surrounded by reed bed (totalling 4 ha), which has been classified as National Vegetation Classification S4 (see Rodwell, 1991, for classification). The reed bed is well developed and reaches up to 40 m in width (Coleshaw & Walker, 2001). Data from the nearby weather station (at Moreton) shows an annual long-term average (1961–1990) rainfall of 698 mm spread evenly throughout the year, with no significant reduction during the summer months. The annual average potential evapotranspiration (PET) is 640.8 mm and exceeds rainfall between April and August. Average monthly temperatures range between 4°C in the winter months and 16°C in the summer months.

9.2.2   Brandon Marsh

Brandon Marsh SSSI is an 87 ha wetland reserve located 3 miles east of Coventry at UK national grid reference SP 386 761. The site is leased to and managed by Warwickshire Wildlife Trust. The reserve developed as a result of mining subsidence and sand and gravel abstraction, and comprises open water, reed bed, marsh, grassland and woodland. Goose Pool (an old settling pool) was chosen as a suitable reed bed site for use in this study, having developed over the last 15 years as a result of succession. Average annual rainfall for nearby Finham is 658 mm and is distributed evenly throughout the year. Average annual PET totals 611.1 mm, with PET generally exceeding rainfall between April and September. The average temperature falls to a minimum of 4.1°C in winter and reaches a maximum of 17.2°C in the summer.

9.2.3   Leighton Moss

Leighton Moss SSSI is located at Warton, Lancashire at UK national grid reference SD 483 746, and is owned and managed by the Royal Society for the protection of Birds (RSPB). The site is located close to Morecambe Bay and would historically have been tidal. However, the creation of flood banks and dykes, and cessation of pumping in 1917, started the natural succession which has allowed the open water and reed bed habitats, that form most of the Moss today, to develop. Lysimeters were installed within one of the large areas of reed bed at the reserve. Average yearly rainfall at nearby Beetham Hall was 1,186 mm with the pattern of monthly rainfall showing a significant reduction in rainfall between April and August. Annual PET totals 573.9 mm, and only exceeds rainfall between May and July. Average minimum temperatures of 2.0°C were recorded during winter, increasing to 19.5°C during summer.

9.3   Experimental Design

9.3.1   Lysimeter Installation

To render the ET of a wetland more accessible for measurement, Gilman (1993) suggested building physical barriers to prevent lateral flows (groundwater and surface water), thus effectively removing these components from the water budget. This can be achieved using a lysimeter, which contains an elementary area of the target wetland in a watertight tank. Lysimeter studies are based on a simplified water balance equation, which allows the calculation of ET using Equation 9.1.

$$ET = R - \Delta s \qquad (9.1)$$

where: ET is the evapotranspiration in millimetres/day; R is the rainfall in millimetres/day; and Δs is the change in the volume of water storage within the lysimeter per day.

Between 10 and 12 lysimeters were installed within the reed beds at each site. The lysimeters used in this study were 0.55 m in diameter, 1 m deep circular rigid plastic tanks and were installed by hand. At each selected location a metal ring was sunk into the soil beneath the reeds, water was removed from inside the metal ring, and the enclosed reed turves cut and carefully removed. Material was then excavated from inside of the ring, the lysimeter sunk into the resulting hole and the ring removed. The previously excavated sediment was then reinstated into the lysimeter and the reed turves were replaced, thereby ensuring that the reed bed inside the lysimeter was representative of that outside (see Fig. 9.2). The lip of the lysimeter was set approximately 100 mm above the reed bed top water level.

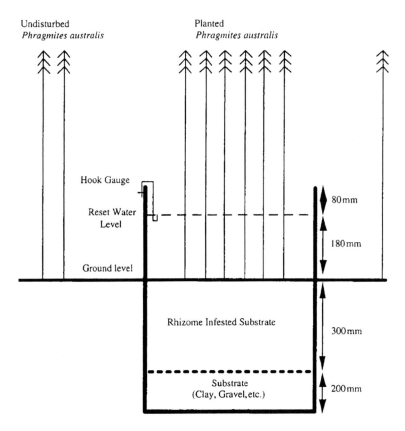

Fig. 9.2 Reed bed lysimeter

An adjustable hook gauge was installed on each lysimeter to provide a reference water level, and was set between 80 and 120 mm below the rim, depending on the site and time of year. In addition to the lysimeters, at each site meteorological monitoring equipment was installed which included a US Class A evaporation pan and splayed-base rain gauge.

9.3.2   Monitoring Regime

During each monthly monitoring visit, the following data were recorded: (1) the rainfall; (2) the volume of water required to return the water levels within the evaporation pan and each lysimeter to datum; and (3) the phenological characteristics of the reeds, both within the lysimeters and within four 0.50 × 0.50 m fixed-location quadrats within the surrounding reed bed (March–September only).

9.4   Determination of Successful Lysimeters

ASCE (1996) stated that reasonable values for ET can only be obtained using lysimetry when plant density, height and leaf area of the vegetation inside the lysimeter are close to that of the surrounding vegetation. However, D.J.G. Gowing (2000, Researcher and Lecturer, Department of Biological Sciences, Open University, Milton Keynes, personal communication) stated that leaf area was not a suitable measure for reed bed comparison, and therefore stem density and maximum reed height were used for comparison. The phenological data was therefore used to determine which of the lysimeters were deemed to be 'successful' and mirrored the surrounding reed bed.

To determine which of the lysimeters at each site were 'successful', an estimation of the 'Standing Crop' from the quadrants and lysimeters was determined using Equation 9.2.

$$
\begin{array}{lll}
\text{Mean Standing Crop} = & \text{Mean Crop} \times & \text{Mean Crop Density} \\
(\text{m stems}^{-1}\ \text{m}^{-2}) & \text{Height (m)} & (\text{stems m}^{-2})
\end{array}
\tag{9.2}
$$

Monthly standing crop data from the quadrants was averaged and the 95% confidence limit determined. Those lysimeters with a standing crop within the above limits of the quadrat data between June and August were classed as being 'successful' and used in the subsequent calculation of ET rates.

Using this method, it was determined that there were no 'successful' lysimeters during the first year of the experiment (2000) at any site due to the fact that the reeds within the lysimeters were not well enough established. Of the lysimeters at Aqualate Mere, four were deemed to be successful in 2001 and six in 2002. At Brandon Marsh although there was one successful lysimeter in 2001 and six in 2002, the reed bed itself had been infested with reedbug – a reed-boring larvae of wainscot moths. This caused dieback in the reeds and therefore they were not considered to be suitable for use in this study, as they did not represent a healthy, well-growing reed bed habitat. At Leighton Moss there were no successful lysimeters in 2001 or 2002. This was attributed to prolonged flooding within the lysimeters, excessive shading of the replanted reeds and saline water ingress when the sea wall was breached during high tides.

9.5 ET(Reed)

ET(Reed) data from successful lysimeters was calculated using Equation 9.1 and is presented in Fig. 9.3.

Figure 9.3 shows the calculated mean monthly ET(Reed), which was low in January and steadily increased to reach a peak of 2.75 mm day^{-1} in June, with July and August values being similar. ET(Reed) then fell steadily between September and December when ET(Reed) reached 0.30 mm day^{-1}.

9.6 Kc(Reed)

The ET rates from a given habitat are dependant on the climatic conditions at the time that the data are collected. For reed bed design, a standardised procedure is required to enable ET(Reed) to be determined. Using the methodology developed by the Food and Agriculture Organisation (Doorenbos & Pruitt, 1977, updated in Allen *et al.*, 1998), ET data was combined with standard Reference Crop

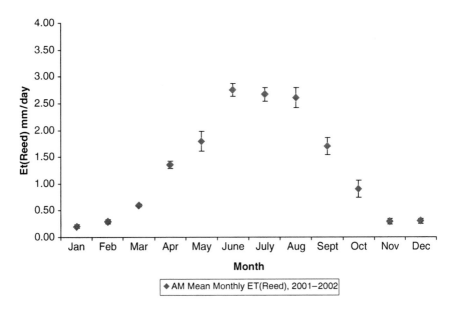

Fig. 9.3 Aqualate Mere mean monthly ET(Reed) including standard error bars, 2001–2002

Evapotranspiration (ETo) data to calculate a crop coefficient (Equation 9.3). The UK crop coefficients obtained can subsequently be used by wetland habitat designers to determine ET rates from target habitats using standard ETo data.

$$Kc(Reed) = ET(Reed)/ETo \qquad (9.3)$$

Where ET(Reed) is the ET from large reed beds in mm day^{-1}; ETo is in mm day^{-1}; and, Kc(Reed) is the crop coefficient for large reed beds.

9.6.1 *Reference Crop Evapotranspiration*

ETo data was obtained from the UK Meteorological Office Rainfall and Evaporation Calculating System (MORECS), which was designed to provide estimates of average ET over $40 \times 40 \text{km}^2$ grid squares throughout the UK (Hough *et al.*, 1996). The MORECS system uses an empirical modification of the Penman–Montieth formula which is applicable to the UK only (Hough *et al.*, 1996).

The MORECS 2.0 ETo data used in this project was based on a 'Grass' surface, which is assumed to have a number of attributes as detailed in Hough *et al.* (1996). A network of stations supplies synoptic data, which is averaged to provide daily readings (where appropriate). The data from each station is then normalised and eventually interpolated to provide values for each $40 \times 40 \text{km}^2$. Hough *et al.* (1996) stated that the errors involved in the use of these interpolation methods are unlikely to be large for temperature and humidity and conclude that acceptable estimates of wind speed are also usually obtained. However, the authors stated that there are greater difficulties with sunshine estimates and concluded that the values for sunshine will be overestimated for squares with large amounts of high ground. In addition, there are problems providing accurate rainfall data as this parameter shows large spatial variation especially in summer. They conclude that the long-term sunshine estimates are likely to be satisfactory, but admit that in the shorter-term, rainfall estimates can be misleading. MORECS data is often used in the UK by wetland designers as the data is readily available.

9.6.2 *Kc(Reed)*

Kc(Reed) data was calculated using the ET(Reed) data presented in section 9.5 and ETo MORECS Grass data provided by the UK Meteorological Office for the appropriate square for Aqualate Mere. Mean Kc(Reed) MORECS Grass including standard error bars is shown in Fig. 9.4.

Figure 9.4 shows that Kc(Reed) MORECS grass increased steadily from 0.07 in January to reach of peak of 0.98 in June. Kc values were maintained throughout July and August and then fell steadily from September to December.

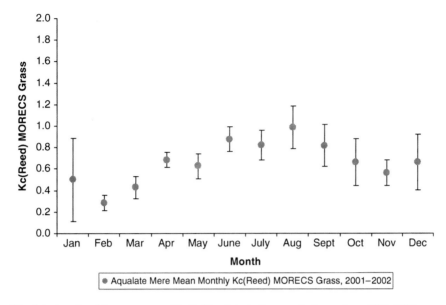

Fig. 9.4 Aqualate Mere mean monthly Kc(Reed) including standard error bars, 2001–2002

9.7 Discussion and Conclusions

On the whole the installation methodology for the lysimeters was successful, apart from excessive shading of the reeds, which is thought to have contributed to establishment problems. However, the water-use rate data from Brandon Marsh was not utilised due to the damage to the reed bed from the reedbug infestation, and neither was the data from Leighton Moss used due to the lack of 'successful' lysimeters. Thus, the ET(Reed) and Kc(Reed) data determined in this study was from the study site at Aqualate Mere only.

A comparison of the Kc(Reed) MORECS Grass data with other Kc(Reed) data from large reed beds is provided in Table 9.1.

The lowest Kc(Reed) values were those presented by Peacock and Hess (2001), which were significantly lower than those from Aqualate Mere. The data from Walton Lake (Fermor et al., 2001) compares favourably with the data from Aqualate Mere with similar Kc(Reed) values throughout the growing season. The Kc(Reed) values presented in this chapter are shown to be comparable with other published data and therefore it can be concluded that the data could be used by wetland designers in the determination of water budgets for large-scale reed bed habitat creation projects in the UK.

Table 9.1 Mean monthly Kc(Reed) from UK large reed beds summary

Reference	ETo source	Jan	Feb	Mar	Apr	May	Jun	Jul	Aug	Sep	Oct	Nov	Dec
Peacock and Hess (2001)	Penman Open Water	–	–	–	–	–	0.58	0.56	–	–	–	–	–
Author – Aqualate Mere	Morecs Grass	0.50	0.29	0.43	0.68	0.63	0.88	0.82	0.99	0.82	0.66	0.56	0.66
Fermor *et al.* (2001) – Walton Lake	Morecs Grass	1.09	0.46	0.48	0.78	0.63	0.77	0.86	0.72	0.75	0.82	0.76	0.97
Gilman *et al.* (1998)*	Penman PET	–	–	–	0.95	0.95	0.95	0.95	1.20	1.07	–	–	–

*Cited by Acreman *et al.* (2002)

Acknowledgements This research could not have been completed without financial support from the Natural Environment Research Council, Aston University, and Middlemarch Environmental Ltd. Thanks are also due to English Nature (now Natural England), the Royal Society for the Protection of Birds and Warwickshire Wildlife Trust.

References

Acreman, M.C., Harding, R.J., Lloyd, C.R. & McNeil, D.D. (2002, April). *Evaporation characteristics of wetlands: Experience from a wet grassland and a reedbed using eddy correlation measurements*. Paper presented to the EGS General Assembly, Nice.

Allen, R.G., Pereira, L.S., Raes, D., & Smith, M. (1998). Crop evapotranspiration: Guidelines for computing crop water requirements. Rome, Italy: FAO Irrigation and Drainage Paper 56.

ASCE. (1996). *Hydrology Handbook*, 2nd edition. ASCE Manuals and Reports on Engineering Practice No. 28. New York: American Society of Civil Engineers.

Bernatowicz, S., Leszczynski, S., & Tyczynska, S. (1976). The influence of transpiration by emergent plants on the water balance in lakes. *Aquatic Botany*, 2, 275–288.

Coleshaw, T., & Walker, G. (2001). *Aqualate Mere management plan: 2001 Revision*. Attingham Park, Shrewsbury, UK: English Nature West Midlands Team.

Doorenbos, J., & Pruitt, W.O. (1977). Crop water requirements. FAO Irrigation and Drainage Paper 24 (reprint). Rome: Food and Agriculture Organisation of the United Nations.

Fermor, P.M. (1997). Establishment of a reedbed within a created surface water fed wetland nature reserve. Ph.D. Thesis, Aston University, Birmingham.

Fermor, P.M., Hedges, P.D., Gilbert, J.C., & Gowing, D.J.G. (2001). Reedbed evapotranspiration rates in England. *Hydrological Process*, 15, 621–631.

Gavenčiak, Š. (1972). Výskum strát vody evaportanpiráciou z vodných rastín (Research on water loss by evapotranspiration from water plants). *Vodohodpod. Čas.*, 20, 16–32.

Gel'bukh, T.M. (1964). Evapotranspiration from overgrowing reservoirs. *Publications of International Association of Hydrological Sciences*, 62, 87.

Gilbert, G., Painter, M., & Smith, K.W. (1996). An inventory of British reedbeds in 1993. *RSPB Conservation Review*, 10, 39–44.

Gilman, K. (1993). Hydrology and freshwater wetland conservation. In *Proceedings of MAFF Conference of River and Coastal Engineers* (pp. 9.2.1–9.2.14). Loughborough, UK: University of Loughborough.

Gilman, K., & Newson, M.D. (1983). *The Anglesey Wetland Study*. Wallingford, UK: Publication of the Institute of Hydrology.

Gilman, K., Hudson, J.A., & Crane, S.B. (1998). Final Report on hydrological evaluation of reed-bed re-creation at Ham Wall, Somerset. LIFE Project 92–1/UK/026.

Hawke, C.J., & Jóse, P.V. (1996). *Reedbed management for commercial and wildlife interests*. Sandy, UK: RSBP.

Hough, M., Palmer, S., Weir, A., Lee, M., & Barrie, I. (1996). The Meteorological Office Rainfall and Evaporation Calculation System: MORECS Version 2.0 (1995). An update to Hydrological Memorandum 45. Bracknell.

Peacock, C.E., & Hess, T.M. (2001). Evapotranspiration from semi-natural reedbeds in Kent, UK. In *Proceedings of the Conference on Changing Wetlands: New Developments in Wetland Science*. University of Sheffield, UK: Society of Wetland Scientists & British Ecological Society.

Rodwell, J.S. (Ed.). (1991). *British Plant Communities*. Cambridge: Cambridge University Press.

UKBG. (1995). Reedbeds: A Coasted Habitat Action Plan. In *Biodiversity. The UK steering group report, Volume 2: Action plans*. Peterborough, UK: English Nature.

Chapter 10
The Hydrological Sustainability of Constructed Wetlands for Wastewater Treatment

Peter D. Hedges[1](⊠), **Phil M. Fermor**[2], and **Jiří Dušek**[3]

Abstract It is argued that there is a need for a hydrological assessment to be undertaken at the design stage of wetlands, which are to be constructed for wastewater treatment, in order to ensure their long-term sustainability. A simple water budget provides a suitable tool for this purpose. The product of an appropriate crop coefficient (Kc), and the reference crop evapotranspiration (ETo) determined from meteorological data, is deemed a suitable procedure for determining the evapotranspiration component of the water budget. The UK and the Czech Republic are taken as case studies, representative of maritime and mid-continent climates, to illustrate a procedure for identifying regions where created wetlands are potentially at most risk from water deficits.

Keywords Crop coefficient, hydrology, municipal wastewater, reed bed, water budgets

10.1 Introduction

Since the early designs for wetlands constructed for the treatment of municipal sewage, there has been considerable research into the ability of these systems to remove nutrients and other pollutants from wastewater. However, little work has been undertaken to investigate their hydrological sustainability. This is an issue which is likely to become of greater significance, as the predicted changes in

---

[1] School of Engineering and Applied Science, Aston University, Aston Triangle, Birmingham, B4 7ET, United Kingdom

[2] Middlemarch Environmental Ltd, Triumph House, Birmingham Road, Allesley, Coventry CV5 9AZ, United Kingdom

[3] University of South Bohemia, Faculty of Biological Sciences, Branišovská 31, 370 05 české Budějovice, Czech Republic

(⊠) Corresponding author: e-mail: p.d.hedges@aston.ac.uk

meteorological patterns due climate change develop, and Europe experiences hotter drier summers with shorter, more intensive wet periods.

It is found that in standard texts on the design of constructed wetlands (e.g. Kadlec & Knight, 1996; Nuttall *et al.*, 1997), although evapotranspiration (ET) is acknowledged as one component of the water balance, it is given little consideration. Quite rightly, the prime criteria used for sizing surface and subsurface systems are pollutant removal rate and the hydraulic parameters of flowrate and gradient. However, the viability of a system will be at risk if there is inadequate inflow to maintain a sensible water budget.

Using long-term meteorological data sets and a simple water budget approach, the susceptibility of constructed wetlands to failure due to macrophyte stress resulting from a lack of water, is explored. The UK is taken as representative of a maritime environment, and the Czech Republic as representative of a mid-continent climate. The procedure adopted is suggested as suitable for identifying those areas of a country where constructed wetlands are most at risk of hydrological failure.

10.2   The Water Budget for Constructed Wetlands

Assuming steady state, the hydrological inputs and outputs for a constructed wetland can be expressed as:

$$O = I + R - ET \tag{10.1}$$

where

O = the outflow from the system
I = the inflow
R = precipitation
ET = evapotranspiration

The potential risks due to hydrological influences can be ascertained by consideration of the likely outflow, but to determine this, quantitative estimates for the remaining three parameters of the water budget are required.

Typical values for inflow rates (I) can be obtained from Kadlec and Knight (1996) and Nuttall *et al.* (1997), who provide reviews of wetland systems constructed for the treatment of municipal wastewater in the USA and the UK respectively. Table 10.1 is a summary of their findings for hydraulic loadings. It should be noted that the higher values for the UK are primarily due to the predominance of tertiary treatment systems.

Long-term data for precipitation (R) are commonly available from meteorological records, but ET rates need to be estimated. One approach to meet the latter requirement is to adjust the evapotranspiration for a reference crop (ETo) by means of an appropriate crop coefficient (Kc) (Doorenbos & Pruitt, 1977; updated in Allen *et al.*, 1998). Taking *Phragmites australis* as the preferred vegetation for

Table 10.1 Summary of municipal wastewater hydraulic loading rates for constructed wetlands

	Surface flow systems		Subsurface flow systems	
	USA[a]	UK[b]	USA[a]	UK[b]
Number of sites reported	31	6	48	20
Max loading (mm day^{-1})	187	451	663	1,619
Min loading (mm day^{-1})	2	22	18	4
Mean loading (mm day^{-1})	33	149	163	208
Median loading (mm day^{-1})	17	69	162	64
Standard deviation (mm day^{-1})	46	168	123	378
Recommended design criteria				
Max (mm day^{-1})	50	–	200	–
Min (mm day^{-1})	7	–	20	–

[a] USA data after Kadelec and Knight (1996)
[b] UK data after Nuttall *et al.* (1997).

constructed wetlands, the potential evapotranspiration, ET(Reed), can be determined from:

$$ET(Reed) = Kc(Reed) \times ETo \qquad (10.2)$$

A variety of values for Kc(Reed) have been published by Fermor (1997) and Fermor *et al.* (2001), but of specific interest to this study are those shown in Table 10.2.

As described elsewhere (Fermor, 1997; Fermor *et al.*, 2001) between 1994 and 1997, lysimeters (see Read *et al.*, Chapter 9, this volume) installed in the root-zone treatment reed beds of Himley Sewage Treatment Works in the West Midlands of England, were employed to measure water-use rates. The ET figures obtained, ET(Reed), when combined with calculated values for a reference crop, ETo, using a rearranged Equation 10.2:

$$Kc(Reed) = ET(Reed)/ETo \qquad (10.3)$$

enabled the monthly crop coefficients for the reeds to be calculated. These values, presented in Table 10.2, are therefore specifically applicable to small reed beds with a constant water supply, which enables the reeds to transpire at their optimum, or potential, rate.

For the purpose of exploring the hydrological sustainability risk, 30-year records of meteorological data were obtained. For the UK these were supplied by the Meteorological Office, and the Czech Republic data were downloaded from the Czech Hydrometeorological Institute's web site (CHMI Climatology Section, 2006).

The Meteorological Office Rainfall and Evaporation Calculating System (MORECS) provides estimates of ET for 40 × 40 km grid squares across the UK (Hough *et al.*, 1996). MORECS employs the modified Penman–Monteith method for calculating ETo for the Reference Crop grass (Monteith, 1965).

The Czech Hydrometeorological Institute publish meteorological data for 22 sites distributed across the country, as shown in Fig. 10.1 and in Table 10.3

Table 10.2 Mean monthly Kc(Reed) for Himley STW

	Jan	Feb	Mar	Apr	May	Jun	July	Aug	Sept	Oct	Nov	Dec
Kc(Reed)	0.67	0.75	0.50	0.96	0.81	1.04	1.19	1.47	2.10	1.50	1.27	0.60
Standard error	–	–	–	0.20	0.09	0.13	0.29	0.62	1.06	0.79	–	–

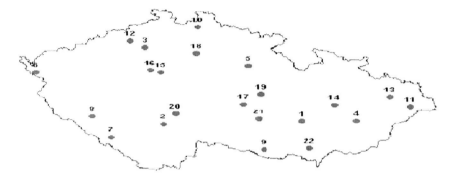

Fig. 10.1 Locations of the Czech Hydrometeorological Institute meteorological stations listed in Table 10.3 (CHMI Climatology Section, 2006)

Table 10.3 Location details for meteorological stations of the Czech Hydrometeorological Institute (CHMI Climatology Section, 2006)

Meteorological station (see Fig. 10.1)		Altitude (m)	Geographical coordinates	
			North latitude	East longitude
1	Brno, Tuřany	241	49°09' 35"	16°41' 44"
2	České Budějovice	388	48°57' 42"	14°28' 05"
3	Doksany	158	50°27' 31"	14°10' 14"
4	Holešov	224	49°19' 07"	17°34' 24"
5	Hradec Králové	278	50°10' 34"	15°50' 19"
6	Cheb	471	50°04' 11"	12°23' 35"
7	Churáňov	1,118	49°04' 06"	13°36' 47"
8	Klatovy	430	49°23' 36"	13°18' 13"
9	Kuchařovice	334	48°53' 00"	16°05' 00"
10	Liberec	398	50°46' 09"	15°01' 30"
11	Lysá hora	1,324	49°32' 46"	18°26' 52"
12	Milešovka	833	50°33' 17"	13°55' 53"
13	Mošnov	251	49°41' 39"	18°07' 12"
14	Olomouc	259	49°34' 10"	17°13' 01"
15	Praha, Karlov	261	50°04' 03"	14°25' 07"
16	Praha, Ruzyně	364	50°06' 03"	14°15' 28"
17	Přibyslav	530	49°34' 58"	15°45' 45"
18	Semčice	234	50°22' 02"	15°00' 16"
19	Svratouch	737	49°44' 06"	16°02' 01"
20	Tábor	461	49°26' 07"	14°39' 42"
21	Velké Meziříčí	452	49°21' 14"	16°00' 31"
22	Velké Pavlovice	196	48°54' 31"	16°49' 28"

(CHMI Climatology Section, 2006). These data include long-term monthly means for the 30-year period from 1961 to 1990 for precipitation, air temperature and sunshine hours. The downloaded meteorological data, together with appropriate adjustment factors relating to location and elevation, were used to calculate long-term mean monthly ETo values for each of the 22 sites, using the Radiation Method (see Doorenbos & Pruitt, 1977; Shaw, 1994). In the Radiation Method:

$$ETo = c\ (W\ Rs) \tag{10.4}$$

where

c = an adjustment factor – a function of mean humidity and daytime wind conditions
W = a weighting factor – a function of temperature and altitude
Rs = solar radiation in equivalent evaporation units (mm day^{-1})

and

$$Rs = (0.25 + 0.50\ n/N)\ Ra \tag{10.5}$$

where

Ra = radiation at the top of the atmosphere
n = actual measured sunshine hours
N = maximum possible sunshine hours

10.3 Assessing the Spatial Distribution of Risk of Hydrological Failure for Surface and Subsurface Wastewater Treatment Wetlands

10.3.1 The United Kingdom

Using 30-year mean rainfall and ETo MORECS values for the period 1961–1990, the distribution of annual available moisture, R-ET(Reed), across the UK was determined as presented in Fig. 10.2 This clearly shows that the East Anglia region to the South West of England is likely to experience an annual net deficit, and hence is where created wetlands are most at risk of hydrological failure.

June was found to be the month with the lowest available moisture, and hence the month when wetlands would be most vulnerable. The R-ET(Reed) map (Fig. 10.3) for June also shows the East Anglia region to be where the risk of failure is greatest.

However, the worst case MORECS square for June R-ET(Reed) provided a deficit of 71 mm. In this case the water budget would balance, were a treatment bed to have an inflow of only 2.4 mm day^{-1}. This is well below the minimum design inflows recommended by Kadlec and Knight (1996) (see Table 10.1).

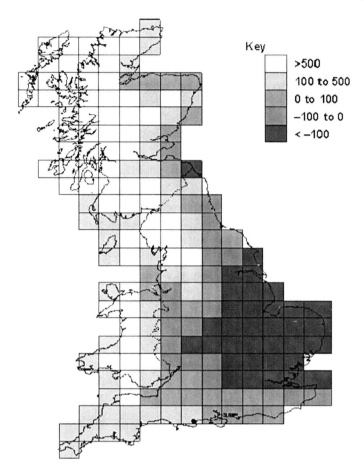

Fig. 10.2 Annual 30-year average R-ET(Reed) in millimetres for MORECS squares covering the UK

10.3.2 The Czech Republic

Unlike the MORECS data, which were produced for $40 \times 40 \text{km}^2$ covering England, Scotland and Wales, only point source data were available for the Czech Republic. To provide a picture for the whole of the country, interpolation between the 22 point sources was achieved using a kriging function within the Idrisi GIS (Eastman, 2001).

Figure 10.4 shows the resulting 30-year average annual available moisture distribution across the Czech Republic. The worst month for the country as a whole proved to be August (Fig. 10.5). In both instances, it is the west-central region and the south-east which show the highest potential risk to created wetlands due to a lack of water.

The meteorological station that would produce the highest risk of wetland hydrological failure was site 9, Kuchařovice, where in August the 30-year mean deficit is −116.3 mm. For a wetland system, this would be satisfied by an inflow of

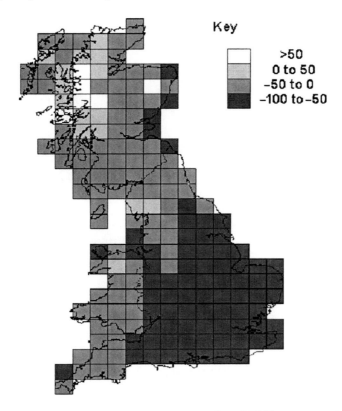

Fig. 10.3 June 30-year average R-ET(Reed) in millimetres for MORECS squares covering the UK

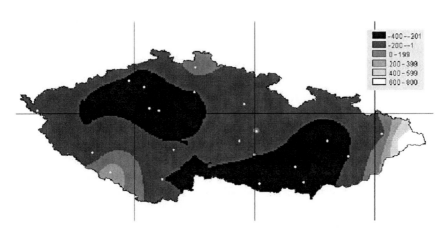

Fig. 10.4 Spatial distribution of annual 30-year average R-ET(Reed) in millimetres for the Czech Republic

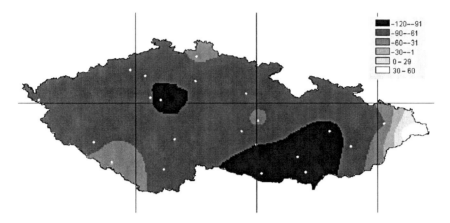

Fig. 10.5 Spatial distribution of August 30-year average R-ET(Reed) in millimetres for the Czech Republic

3.75 mm day^{-1}. As with the worst-case scenario for the UK, this is also well below the recommended minimum inflow shown in Table 10.1.

10.4   Hydrological Sustainability of Constructed Wetlands

The analysis and maps presented earlier (Figs. 10.2–10.5) have been produced using 30-year average meteorological data. Clearly this does not produce a 'worst-case scenario' against which systems should be designed. However, the maps do provide a guide to where the risk of failure is greatest for the countries under consideration.

In the worst-case monthly situations identified from these 30-year mean records, there would be little difficulty in design inflows (I) meeting the water deficit (R-ET(Reed)). However, instances have been reported where water shortfalls have been experienced, and the outflow from reed beds has ceased, albeit for short periods of time.

Figure 10.6 shows detailed records of the inflow and outflow for the twin sub-surface horizontal flow reed beds in Slavošovice, 15 km east of české Budějovice in the Czech Republic. The wetlands have been designed to treat municipal waste-water for 150 person equivalents (PE), with some additional spare capacity to allow for future developments. However, the two beds currently only serve approximately 100 PE. The graph clearly shows that despite additional inputs from precipitation over and above the wastewater inflow, on each of the 4 days reported there has been zero outflow over a period of 5 to 7 h – approximately 26% of the time.

Research studies and design criteria assume that reeds are transpiring at their potential rate, but if this is not the case and the plants are under stress, as would have been the case for the reed beds of Fig. 10.6, what are the implications for

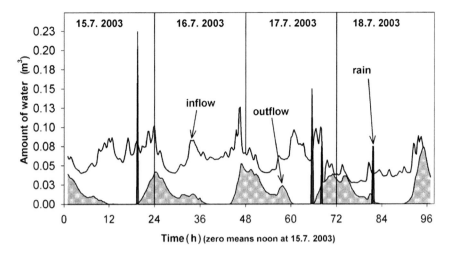

Fig. 10.6 Inflow and outflow over 4 days at the Slavošovice constructed wetlands

vegetation health and treatment efficiency? Vegetation close to the inlet will experience a surfeit of water and continue to transpire at its potential rate. However, as the available moisture reduces with distance from the inlet due to ET loss, plants will progressively become more stressed, and the danger of wilting and potential die-back will increase. Further research is required to answer questions such as how nutrient removal is impacted by vegetation water stress, and what is the sensible throughflow required to maintain treatment efficiency?

Another issue which needs to be addressed relates to systems being designed and constructed to cater for a predicted population equivalent which will occur sometime in the future. As a consequence, in the early days of a system's life, the inflows will be lower than the ultimate design values – as is the case with the Slavošovice reed beds of Fig. 10.6. During this time, the beds will be at their greatest risk of hydrological failure, and if this proves to be an issue it may well be necessary to stage the introduction of the active treatment area to match the growth of the feeder population.

Ideally, for sustainable hydrological design, a threshold return frequency (i.e. once in 10 years) should be adopted, beyond which hydrological failure is unacceptable. Subsequently, the relevant meteorological data should be determined from time series records, and a water balance undertaken to determine whether the design inflow (I) satisfies the shortfall in R-ET(Reed) at this threshold point.

10.5 Conclusion

The risk of hydrological failure of reed beds constructed to treat municipal waste-waters requires greater consideration at the design stage of such systems. To facilitate this, a simple water budget can be employed. As part of the budget calculations, a procedure that employs vegetation-specific crop coefficients, here Kc(Reed), in combination with ETo, is proposed for assessing the ET component.

To assist in decision-making, the water budget approach can also be employed to identify those locations within a country or region most at risk of failure due to inadequate inputs of water into the system.

Finally, research is required to assess the effect of flows through a treatment system that are suboptimal on treatment efficiency. At the design stage, it may be necessary to adopt a return frequency for meteorological conditions, below which hydrological failure is deemed unacceptable.

The hydrological sustainability of constructed wetlands is likely grow in importance with respect to wetland design, as meteorological patterns change due to climate change.

References

Allen, R.G., Pereira, L.S., Raes, D., & Smith, M. (1998). *Crop evapotranspiration: Guidelines for computing crop water requirements*. FAO Irrigation and Drainage Paper 56. Rome: FAO.

CHMI Climatology Section. (2006). Information about climate, Retrieved March 31, 2006, from http://www.chmi.cz/meteo/ok/infklime.html

Doorenbos, J., & Pruitt, W.O. (1977). *Guidelines for predicting crop water requirements*. FAO Irrigation and Drainage Paper 24. Rome: FAO.

Eastman, R.J. (2001). IDRISI for Windows User's Guide Version 2. Worcester, MA: Clark Labs, Clark University.

Fermor, P.M. (1997). Establishment of a reedbed within a created surface water fed wetland nature reserve. Dissertation, Aston University, UK.

Fermor, P.M., Hedges, P.D., Gilbert, J.C., & Gowing, D.J.G. (2001). Reedbed transpiration rates in England. *Hydrological Processes, 18*, 621–631.

Hough, M., Palmer, S., Weir, A., Lee, M., & Barrie, I. (1996). The Meteorological Office Rainfall and Evaporation Calculation System: MORECS Version 2.0. An update to Hydrological Memorandum 45. Bracknell, UK: Meteorological Office.

Kadlec, R.H., & Knight, R.L. (1996). *Treatment wetlands*. New York: CRC Press.

Monteith, J.L. (1965). Radiation and crops. *Experimental Agricultural Review, 1*, 241–251.

Nuttall, P.M., Boon, A.G., & Rowell, M.R. (1997). Review of the design and management of constructed wetlands, Report 180. London: CIRIA, London.

Shaw, E.M. (1994). *Hydrology in practice*, 3rd edn. London: Chapman & Hall.

Chapter 11
Factors Affecting Metal Accumulation, Mobility and Availability in Intertidal Wetlands of the Scheldt Estuary (Belgium)

Gijs Du Laing(✉), Annelies Van de Moortel, Els Lesage, Filip M.G. Tack, and Marc G. Verloo

Abstract We studied factors affecting the accumulation, mobility and availability of metals in intertidal wetlands of the Scheldt estuary (Belgium), both in greenhouse experiments and under field conditions. The surface layer of the intertidal Scheldt sediments was found to be significantly contaminated with trace metals. The metal concentrations can be predicted from clay or organic matter concentrations. They were somewhat higher than predicted at sites within a range of a few kilometres from specific point-sources of metals, whereas they were lower than predicted at sites which are regularly subjected to flooding by high salinity water. In the deeper sediment layers, sulphides seem to play an important role in the metal accumulation. Salts significantly increased the metal mobility in the oxidised sediment layers in the brackish part of the estuary, which was especially observed for cadmium (Cd). The decomposition of stems and leaves of reed plants and willows in the upper sediment layer was found to both increase and decrease the metal mobility. The concentrations in the reed litter itself increased significantly during decomposition under field conditions. The hydrological regime to which metal-polluted sediments were subjected, affected the metal concentrations in the pore water to an important extent. Flooding increased the mobility of iron (Fe), manganese (Mn), nickel (Ni) and chromium (Cr) and decreased the mobility of Cd, copper (Cu) and zinc (Zn) in a calcareous substrate. Fluctuating hydrological conditions resulted in fluctuating metal concentrations in the pore water. Reduction and oxidation of Fe and Mn, decomposition of carbonates and the formation and re-oxidation of sulphides were hereby the main controlling processes.

Keywords Estuary, heavy metals, marsh, sediment, wetland

Laboratory for Analytical Chemistry and Applied Ecochemistry, Department of Applied Analytical and Physical Chemistry, Ghent University, Coupure Links 653, B-9000 Gent, Belgium

(✉) Corresponding author: e-mail: Gijs.DuLaing@UGent.be

J. Vymazal (ed.) *Wastewater Treatment, Plant Dynamics and Management in Constructed and Natural Wetlands*,
© Springer Science + Business Media B.V. 2008

11.1 Introduction

The Scheldt river sources in the north of France, and continues its flow in Belgium through the Walloon Region and the Flemish Region towards the North Sea outlet in the Netherlands. Its river catchment covers a surface of 20,331 km^2 in one of the most populated and industrialised areas of Europe. The downstream Scheldt basin is a typical coastal estuary characterised by a small river discharge, but subjected to a large tidal influence with an amplitude of 4.5 m at the mouth. It is a unique estuary in Europe as the salinity intrusion extends more than 110 km upstream. The variations in tidal level can still be observed at Ghent, 150 km from the river mouth.

During decades, parts of the river Scheldt have been straightened to improve the possibility to navigate. Dikes have been constructed to protect inhabited areas against flooding and to impolder part of the previously flooded river banks, hereby decreasing the area that is subjected to tidal fluctuations of the water level. This gradually decreased the water-holding capacity of the river and its surroundings. In addition, construction of sewers and paved areas increasingly shortcuts rainwater to the rivers and the sea level is rising. All this is causing increased incidences of flooding of inhabited areas. These floodings can occur during heavy rainfall, when sewers and streams cannot absorb the rainwater anymore, or during storm weather at springtides, as extremely high water levels in the Scheldt river can be reached when (north)western storm winds impel extra sea water into the estuarine funnel. This resulted in massive floods in the Netherlands (1953), which were acted upon with the ambitious Delta-plan. In Belgium, the Sigma-plan was an answer to the major floods along the Scheldt river in 1976.

Controlled flooding areas are important components of these river management plans. The construction of controlled flooding areas along the river is aimed to provide additional temporary water storage during events of high rainfall and discharge. This allows to attenuate high peak water levels, and thus reduces the chances of flooding. By permanently keeping the groundwater levels high in parts of these flooding areas instead of just flooding them occasionally, the government also hopes to restore valuable wetland habitats. However, suspended particles in the river water, especially the smallest ones, tend to bind heavy metals (Millward & Liu, 2003). In that way, sedimentation of suspended matter in these wetlands could create a significant risk for metal accumulation in these valuable habitats, especially as metals are not biodegradable. The Scheldt river has indeed already been found to be highly polluted, receiving industrial and domestic wastewaters (Paucot & Wollast, 1997). Several former and current wetlands along the river Scheldt are contaminated by metals because of either flooding and overbank sedimentation or land disposal of dredged sediments (Vandecasteele *et al.*, 2002). This contamination constitutes an important obstacle when managing, restoring or creating intertidal wetlands. Removal and cleaning of all accumulated polluted sediments quickly may prove to be an unrealistic option in the short term because of the large costs involved. On the other hand, several processes occurring in wetland sediments (such as sulphide precipitation) could also significantly reduce the bioavailability of metals which have a nutritional role to play, such as copper (Cu) and zinc (Zn)

(Gambrell, 1994). As the range between essentiality and toxicity is quite narrow for most metals and organisms, wetlands can only be created and sustainably managed if processes affecting metal mobility and availability are thoroughly understood and metal fate can be predicted. We therefore aimed to identify the most important factors that affect metal accumulation and mobility in the intertidal reed beds of the Scheldt estuary. The results of this study are summarised below and entirely published by Du Laing (2006).

11.2   Factors Affecting Metal Concentrations of the Intertidal Scheldt Sediments

Metal accumulation in wetland sediments is determined not only by input coming from discharge of industrial and urban sewage or by atmospheric deposition in the wetland ecosystems, but also by the extent by which the substrate is capable to bind and release metals. This is governed by soil and sediment characteristics such as pH, cation exchange capacity, organic matter contents, redox conditions and chloride contents. These properties determine the type and stability of metal sorption or precipitation, which are also related to the metal mobility, bioavailability and potential toxicity (Schierup & Larsen, 1981; Du Laing et al., 2002). The factors affecting metal concentrations of the upper 0–20 cm sediment layer of the intertidal reed beds along the river Scheldt have been discussed by Du Laing et al. (2007a). These sediments were found to be significantly contaminated with trace metals. The cadmium (Cd) concentrations often exceed the Flemish remediation values for nature areas (Vlarebo, 1996), whereas the sediments are moderately contaminated by chromium (Cr), Cu and Zn. Lead (Pb) concentrations are occasionally high. Nickel (Ni) concentrations however lean towards background values. Significant correlations were observed between metals, organic matter and clay contents, as has also been observed for other wetlands (Callaway et al., 1998; Shriadah, 1999; Gallagher et al., 1996; Suzuki et al., 1989). So metal concentrations can be predicted from clay or organic matter contents in this upper sediment layer. They were somewhat higher than predicted at sites within a range of a few kilometres from specific point-sources of metals (shipyards, industrial areas, major motorways). They were lower than predicted at sites which are regularly subjected to flooding by high salinity water. In that way, the ratio between observed and predicted concentrations seems to be a valuable tool to identify areas which are specifically impacted by point sources.

   The metal accumulation due to association of metals with organic matter and clay was found to be supplemented by an accumulation as sulphides at higher sampling depths within the sediments. Under deeply reducing conditions, sulphides are indeed formed as a result of the reduction of sulphates. During this sulphidisation process, metals can adsorb or co-precipitate with sulphide-containing minerals or can precipitate directly as discrete/separate solid phases (Billon et al., 2001). We found that the sediment depth at which metal sulphide precipitation significantly contributes to metal accumulation in the intertidal Scheldt marshes depends on the sampling location. It varies from a few centimetres in clayey, organic sediments to more than 1 m in sandy sediments. Temporal variations of Cu, Cd, Pb and Zn concentrations could only be linked to newly formed sulphides or sulphide oxidation at the sites with the lowest sulphide contents. At sampling sites containing high sulphide amounts, variations of total metal concentrations should be primarily attributed to metal exchange and the presence of mobile metal complexes. Litter decomposition at the end of the reed plants' growing season could hereby play a significant role (see Section 11.3.3).

11.3 Factors Affecting Metal Mobility and Availability in the Intertidal Scheldt Sediments

There is a general agreement that the metal mobility and metal uptake by organisms is related to particular metal fractions in the pore water rather than the total concentration in the soils or sediments (Sauvé *et al.*, 2000). Most recent studies examining relationships between availability of heavy metals in soils, toxicological impacts on soil fauna and flora, and the potential transfer of heavy metals through the food chain have therefore focused on extracting the soil solution (Tye *et al.*, 2003). In most of our experimental setups, we therefore also monitored metal concentrations in the pore water of intertidal Scheldt sediments under field conditions or subjected soils and sediments to varying conditions in a greenhouse, simultaneously monitoring metal concentrations in the pore water as a measure for the metal mobility and availability. Rhizon soil moisture samplers were used to extract and simultaneously filter the pore water, as described by Meers *et al.* (2006). One of the experimental setups, which were often used, is illustrated in Fig. 11.1.

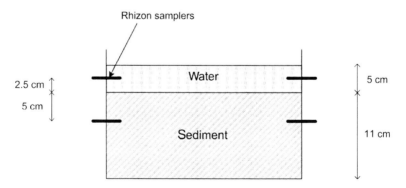

Fig. 11.1 Experimental setup used to study factors affecting metal mobility and availability in the intertidal Scheldt sediments under flooded conditions (Rhizon samplers used to sample pore water in the sediment and surface water above the sediment)

11.3.1   Effect of the Hydrological Regime

The hydrological regime to which a metal-polluted overbank sedimentation zone is subjected has a pronounced influence on the metal concentrations in the pore water of the upper soil or sediment layer. The flooding regime primarily affects the supply of oxygen. When soils or sediments are flooded, biological and microbiological activity combined with limited oxygen diffusion under these conditions causes oxygen depletion and thus establish reducing conditions. In shortage of oxygen, micro-organisms start to use other electron acceptors such as nitrate, Mn and Fe. In highly reducing conditions, sulfate reduction and methanogenesis will occur (Hadas et al., 2001). The dominating mineralisation process mainly depends on the availability of all products and micro-organisms involved. Although different electron acceptors theoretically will be oxidised in a sequence as redox potential decreases, overlapping processes have been observed (Peters & Conrad, 1996).

Typical for intertidal sediments, they will be re-aerated during emerged periods. These alternating aerobic and anaerobic conditions will have specific influences on most of the processes regulating the speciation, mobility and bioavailability of metals (Calmano et al., 1993, in Charlatchka & Cambier, 2000) such as:

- Sorption/desorption onto different solid components
- Adsorption/co-precipitation onto hydrous oxides of Fe and Mn
- Formation/decomposition of soluble and insoluble metal inorganic complex compounds
- Dissolution of carbonates, metal oxides or hydroxides
- Precipitation as insoluble sulphides under highly reducing conditions and their dissolution of sulphates under oxic conditions

An overview of these possible interactions is given in Fig. 11.2.

From a greenhouse experiment, as described by Du Laing et al. (2007b), we could conclude that flooding conditions lead to an increase of the Fe, Mn, Ni and Cr mobility and a decrease in the Cd, Cu and Zn mobility in the upper layer of a calcareous dredged-sediment-derived soil. Keeping the soil at field capacity resulted in a low pore water concentration of Fe, Mn and Ni, while the Cd, Cu, Cr and Zn pore water concentrations increased. Alternating hydrological conditions resulted in fluctuating metal concentrations in the pore water. When the flooding lasts for a maximum of 2 days and is followed by a longer emerged period or when the soil is kept continuously on field capacity, release of most metals appears primarily related to calcium (Ca) release. The metals could be mobilised as a result of Ca release, or they might be subject to the same factors which affect Ca release, such as the dissolution of carbonates. In conditions that allow sufficiently strong reducing conditions to be established, there is evidence that the mobility of Cd, Cu, and to a lesser extent Zn, is primarily affected by formation and re-oxidation of sulphides, whereas Ni would mainly be mobilised as a result of Fe/Mn oxide reduction.

Another greenhouse experiment also revealed the way the water table level affects the metal availability in the saturated and the unsaturated zones of the upper

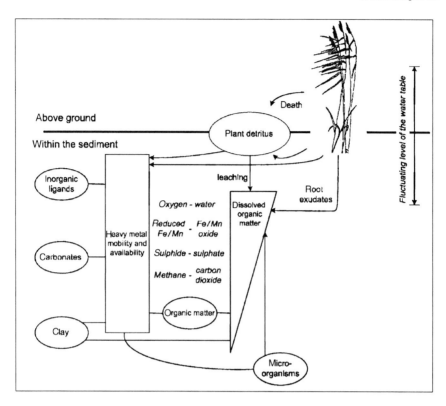

Fig. 11.2 Overview of possible interactions and factors affecting metal mobility and availability in intertidal sediments

1 m profile of intertidal sediments within the Scheldt estuary. Nickel and Cr release was found to be facilitated by reductive conditions, and was slightly promoted by increasing salinities. These reducing conditions can occur below the water table, or above the water table, but at a sufficiently high sampling depth (more than 30 cm below the sediment surface). Chromium was however also mobilised under more oxic conditions, as it can also form complexes with organic substances. Copper, Zn and especially Cd are all primarily released above the water table under high salinity conditions. Cadmium, Cu and Zn are also released below or just above the water table when organic matter is being decomposed, resulting in Ca, Mn, Ni and/or Fe release upon CO_2 accumulation and Fe/Mn oxide reduction without simultaneous sulphide production. Under greenhouse conditions, it takes about 2 months before the Flemish soil sanitation threshold levels for groundwater are reached for Ni or Cd concentrations in the pore water, upon the creation of waterlogged or emerged conditions, respectively. However, the total Ni concentrations in the sediments did not exceed any soil sanitation threshold for the solid soil fraction. Slight water level fluctuations in short term might thus result in metals becoming locally available just above or below the water table. These fluctuations are not expected to create a significant overall metal mobilisation, but they might affect the bioavailability.

11.3.2   Effect of Salinity

Increasing salinity often reduces bioavailability and toxicity towards aquatic organisms of metals that are present in the water phase (e.g. Verslycke *et al.*, 2003). Increasing salinity however will also enhance metal mobilisation from the solid sediment fraction towards the water phase. This can be attributed to chlorocomplexation (Paalman *et al.*, 1994) and cation exchange (Tam & Wong, 1999). It was especially observed for Cd as the stability and solubility of cadmium chloride complexes are relatively high and the affinity for sorption to the solid soil phase is low (Doner, 1978; Comans & Van Dijk, 1988). As tidal variations in the Scheldt estuary result in varying salinities of the river water and pore water of the sediments, we studied the evolution of Cd concentrations in the pore water of intertidal Scheldt sediments upon flooding with water of varying salinities.

Cadmium concentrations in pore water of sediments and surrounding surface waters were found to significantly exceed sanitation thresholds and quality standards during flooding of oxidised sediments. The effects were observed already at lower salinities of 0.5 g NaCl l^{-1} and increased with increasing salinity. They were however not observed in deeply reduced, sulphide-rich sediments, which always showed a low Cd availability. Figure 11.3 illustrates the Cd concentrations in pore water of two different sediments flooded with water of three different salinities, as a function of sampling time after flooding. The "Kijkverdriet" sediment was initially sulphide-rich, whereas the "Lippensbroek" sediment was initially oxidised and reducing conditions were established during the first weeks of flooding.

This implies that risks related to Cd leaching can be created when constructing flooding areas in the metal-polluted zones in the brackish part of estuaries. These risks can however be reduced by inducing sulphide precipitation because Cd is then immobilised as sulphide and does not react anymore to changes in salinity. This could be achieved by permanently flooding the polluted sediments, because

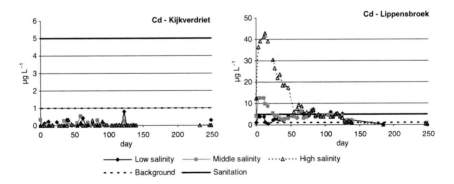

Fig. 11.3 Cd concentrations in pore water (µg l^{-1}) of two different sediments flooded with water of three different salinities, as a function of sampling time; the Kijkverdriet sediment was initially sulphide-rich whereas the Lippensbroek sediment was oxidised; background levels and sanitation thresholds for groundwater according to Vlarebo (1996)

sulphates are sufficiently available in the river water of the brackish part of the estuary. Microbial sulphate reduction is not expected to be limited by a low organic matter availability as biomass turnover is very high in intertidal reed beds. Stimulation of sulphide formation should however be done with great caution as accidental oxidation of the sulphide-rich sediments could also create a significant risk for metal mobilisation (Singh *et al.*, 1998), which is in turn facilitated by increasing salinity (Gerringa *et al.*, 2001). Thus, a "chemical time bomb" might indeed be created (Stigliani, 1992). Low sulphate availabilities might be limiting sulphide formation especially in the freshwater sediments and soils.

11.3.3   *Effect of Litter Decomposition on Metal Mobility and Availability*

When mudflats get banked up or when controlled flooding areas are being constructed, emergent vegetation can colonise the marshes. The marshes in the brackish and freshwater part of the Scheldt estuary are mainly vegetated by reed plants and willows. Uptake of metals by these plants directly reduces the metal input into the adjacent waters. Moreover, as these plants are characterised by a very high biomass production and turnover, the upper soil layer will be enriched by organic matter originating from rooting and the decay and decomposition of plant parts. As such, these plants can especially provide a sink if during decomposition metals are bound to the litter by passive sorption on the organic surfaces or by physiological mechanisms of microbial colonizers of the litter. Several studies suggest that metals in litter are available to deposit feeders and thus can enter estuarine food webs (Weis & Weis, 2004). Consumption of metal-laden detritus can cause metal accumulation and deleterious effects in higher trophic levels (Dorgelo *et al.*, 1995; Du Laing *et al.*, 2002; Weis *et al.*, 2002). The role of plant sequestration of metals into long-term sinks depends on the rate of uptake by the plant, rates of translocation and retention within individual tissue types, and the rate and mode of tissue decomposition (Catallo, 1993; Kadlec & Knight, 1996).

We therefore studied the association of metals with reed litter upon litter decomposition under field conditions on a Scheldt marsh (Du Laing *et al.*, 2006). During decomposition, most metal contents in the reed litter increased considerably. There are indications that fungal activity is an important factor determining metal accumulation in decomposing stem tissue. This could also be the case for leaf blades, but for this tissue type the effect of fungal activity on metal concentrations is found to be overruled by passive metal sorption and trapping of sediment particles and associated metals. Both factors are of intermediate importance for leaf sheaths.

The seasonal supply of organic matter can however not only act as a metal sink, but also as a source. As mentioned above, changing redox potentials as a result of ongoing aerobic and anaerobic decomposition processes can induce shifts between dissolved and solid element species in the sediments. Organic matter can be a limiting factor for these processes. Supply of easily biodegradable organic matter can thus result in a faster consumption of oxygen in aerobic sediments and nitrate reduction, reductive dissolution of oxic Mn and Fe minerals and sulphate reduction in anaerobic sediments, affecting metal mobility (Tanji *et al.*, 2003). Some organic fractions can also directly act as soluble or insoluble complexing agents, increasing or decreasing the metal mobility. The interaction with metals and solubility of these complexes in pore water depends on several factors, e.g. pH, Ca contents and cation exchange capacity of the sediments (Kalbitz & Wennrich, 1998). Moreover, decomposition of organic matter can affect metal speciation by release of CO_2 and lowering of the pH (Charlatchka & Cambier, 2000). The dominating type of inter-action between organic matter and metals is expected to depend upon the amount and type of the organic matter (Kashem & Singh, 2001), which is affected by the decomposition status.

We therefore also studied the effect of supplying grinded litter to the upper 0–10 cm sediment layer on the metal concentrations in the pore water of this sediment layer. This was done under greenhouse conditions. During the first weeks after supplying the litter to the sediments, Cd, Cu, Ni and Zn concentrations in pore water of this sediment layer were found to exceed the Flemish sanitation thresholds for ground water significantly (Vlarebo, 1996). The Ni concentrations however also significantly exceeded its sanitation threshold during a longer period of time upon addition of any type and amount of organic matter, whereas this was also the case for Zn upon addition of reed stem material. Willow leaf material was rapidly decomposed, which resulted in high Fe and Mn oxide reduction rates, subsequently resulting in high metal concentrations in the pore water. Upon addition of willow leaves, also large amounts of sulphates were released, which induced the formation of sulphides and precipitation of metal sulphides. As sufficient sulphide amounts were produced, all metal concentrations in the pore water can be reduced to low concentrations within the first weeks. Metal release also increased with increasing added amounts of reed stems. However, lower sulphates amounts were released by the reed stems. As a result, insufficient amounts of sulphides could be produced to precipitate all metals and reduce Ni, Zn, Fe and Mn concentrations to within acceptable limits in a reasonable amount of time. This in turn seemed to reduce sulphate reduction rates. As such, a vicious circle seemed to be produced which is characterised by only slowly decreasing Zn, Ni, Fe and Mn concentrations, when high amounts of low-sulphate releasing reed stem material were added to a permanently flooded soil. The latter is illustrated in Fig. 11.4. When the soils were alternately flooded, metals were exported through the percolates. Total metal export by the percolates significantly increased when organic matter was added. The export was concentrated in the first weeks after organic matter addition. As a result, concentrations of most metals in the pore water were more rapidly reduced to within acceptable limits compared to permanently flooded soils.

Taking into account all effects mentioned above, it could be concluded that the dominating plant species in intertidal wetlands or floodplains affects the metal fate upon litter decomposition at the end of the growing season. Willow leaves seem to immobilise metals more significantly upon decomposition compared to reed leaves and stems. To confirm this, the factors affecting sulphur uptake and sulphate release

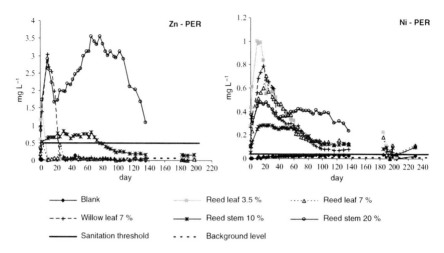

Fig. 11.4 Evolution of Zn and Ni concentrations in the pore water (mg l^{-1}) of a permanently flooded soil to which different concentrations and types of grinded litter material were added (reed leaves added to 3.5% and 7%, willow leaves to 7% and reed stems to 10% and 20%), as a function of time upon flooding (blank = soil without litter supply)

by these plant species should however still be identified. As decomposing reed plant material seems to mobilise metals, harvesting reed plants at the end of the growing season in metal-polluted zones within newly created wetland areas indirectly might lead to a reduction of the distribution of metals within the wetland.

11.3.4   *Validation Under Field Conditions*

Overall, the same factors could be shown to affect metal fate in field conditions. However, the effects were often less clear, as the resolution of measurements in the field was much lower compared to the greenhouse experiments due to sampling restrictions (e.g. samples could only be taken at low tides). As a result, some variations in metal mobility might not have been observed in the field. Moreover, several counteracting factors determine the ultimate metal fate in the field. The sediments and soils were subjected to more moderate conditions in the field compared to the greenhouse. The temperatures in the field are, e.g. not as high as under greenhouse conditions. These are expected to primarily affect the microbial reduction rates. In contrary, the sediments in the field are subject to more dynamic variations compared to the greenhouse, such as daily flooding. At most of the sampling sites, the level of the water table fluctuated only slightly during fall, winter and spring. However, the water table level significantly decreased during summer at most of the sites, especially those with more sandy sediments. As a result, the highest sulphide concentrations are found at the sites where the water table level never decreased significantly. These sulphides primarily determine Cd, Cu, Ni and Zn concentrations in the pore water. Organic complexation was also found to result in the mobilisation of Cu, Ni and Cr under field conditions. Soluble organic molecules probably originated from the incomplete decomposition of organic matter at lower temperatures during the winter months. The concentrations of Cd, Ni and Zn in the pore water were moreover affected by Fe/Mn oxide reduction, whereas Cd and Zn concentrations were also affected by the salinity. Moreover, we found increasing Cd and Zn contents in ground-dwelling spiders with increasing salinity, despite markedly lower total contents in the high-salinity sediments (Du Laing *et al.*, 2002). The latter confirms the dominating effect of the salinity on the mobility and availability of these metals in the oxic upper intertidal sediment layer.

References

Billon, G., Ouddane, B., Laureyns, J., & Boughriet, A. (2001). Chemistry of metal sulfides in anoxic sediments. *Physical Chemistry Chemical Physics, 3*, 3586–3592.

Callaway, J.C., Delaune, R.D., & Patrick, W.H. (1998). Heavy metal chronologies in selected coastal wetlands from Northern Europe. *Marine Pollution Bulletin, 36*, 82–96.

Calmano, W., Hong, J., & Förstner, U. (1993). Binding and mobilization of heavy-metals in contaminated sediments affected by pH and redox potential. *Water Science & Technology, 28*, 223–235.

Catallo, W.J. (1993). Exotoxicology and wetland ecosystems: Current understanding and future needs. *Environmental Toxicology and Chemistry, 12*, 2209–2224.

Charlatchka, R., & Cambier, P. (2000). Influence of reducing conditions on solubility of trace metals in contaminated soils. *Water, Air, and Soil Pollution, 118*, 143–167.

Comans, R.N.J., & Van Dijk, C.P.J. (1988). Role of complexation processes in cadmium mobilization during estuarine mixing. *Nature, 336*, 151–154.

Doner, H. (1978). Chloride as a factor in mobilities of Ni(II), Cu(II) and Cd(II) in soil. *Soil Science Society of America Journal, 42*, 882–885.

Dorgelo, J., Meester, H., & Vanvelzen, C. (1995). Effects of diet and heavy metals on growth rate and fertility in the deposit-feeding snail *Potamopyrgus jenkinsi* (Smith) (Gastropoda: Hydrobiidae). *Hydrobiologia, 316*, 199–210.

Du Laing, G. (2006). Dynamics of heavy metals in reedbeds along the banks of the river Scheldt. Dissertation, Ghent University, Ghent, Belgium, 284 p.

Du Laing, G., Bogaert, N., Tack, F.M.G., Verloo, M.G., & Hendrickx, F. (2002). Heavy metal contents (Cd, Cu, Zn) in spiders (*Pirata piraticus*) living in intertidal sediments of the river Scheldt estuary (Belgium) as affected by substrate characteristics. *Science of the Total Environment, 289*, 71–81.

Du Laing, G., Van Ryckegem, G., Tack, F.M.G., & Verloo, M.G. (2006). Metal accumulation in intertidal litter through decomposing leaf blades, sheaths and stems of *Phragmites australis*. *Chemosphere, 63*, 1815–1823.

Du Laing, G., Vandecasteele, B., De Grauwe, P., Moors, W., Lesage, E., Meers, E., Tack, F.M.G., & Verloo, M.G. (2007a). Factors affecting metal concentrations in the upper sediment layer of intertidal reedbeds along the river Scheldt. *Journal of Environmental Monitoring, 9*, 449–455.

Du Laing, G., Vanthuyne, D.R.J., Vandecasteele, B., Tack, F.M.G., & Verloo, M.G. (2007b). Influence of hydrological regime on pore water metal concentrations in a contaminated sediment-derived soil. *Environmental Pollution, 147*, 615–625.

Gallagher, K.A., Wheeler, A.J., & Orford, J.D. (1996). An assessment of the heavy metal pollution of two tidal marshes on the north-west coast of Ireland. *Biology and Environment: Proceedings of the Royal Irish Academy, 96B*, 177–188.

Gambrell, R.P. (1994). Trace and toxic metals in wetlands – a review. *Journal of Environmental Quality, 23*, 883–891.

Gerringa, L.J.A., Baar, H.J.W., Nolting, R.F., & Paucot, H. (2001). The influence of salinity on the solubility of Zn and Cd sulphides in the Scheldt estuary. *Journal of Sea Research, 46*, 201–211.

Hadas, O., Pinkas, R., Malinsky-Rushans, N., Markel, D., & Lazar, B. (2001). Sulfate reduction in Lake Agmon, Israel. *Science of the Total Environment, 266*, 203–209.

Kadlec, R.H., & Knight, R.L. (Eds.) (1996). *Treatment wetlands*. Boca Raton, FL: CRC Press.

Kalbitz, K., & Wennrich, R. (1998). Mobilization of heavy metals and arsenic in polluted wetland soils and its dependence on dissolved organic matter. *Science of The Total Environment, 209*, 27–39.

Kashem, M.A., & Singh, B.R. (2001). Metal availability in contaminated soils: I. Effect of flooding and organic matter changes in Eh, pH and solubility of Cd, Ni and Zn. *Nutrient Cycling in Agroecosystems*, 61, 247–255.

Meers, E., Du Laing, G., Unamuno, V.G., Lesage, E., Tack, F.M.G., & Verloo, M.G. (2006). Water extractability of heavy metals from soils: Some pitfalls. *Water, Air, and Soil Pollution, 176*, 21–35.

Millward, G.E., & Liu, Y.P. (2003). Modelling metal desorption kinetics in estuaries. *Science of the Total Environment, 314–316*, 613–623.

Paalman, M.A.A., van der Weijden, C.H., & Loch, J.P.G. (1994). Sorption of cadmium on suspended matter under estuarine conditions: Competition and complexation with major seawater ions. *Water, Air, and Soil Pollution, 73*, 49–60.

Paucot, H., & Wollast, R. (1997). Transport and transformation of trace metals in the Scheldt estuary. *Marine Chemistry, 58*, 229–244.

Peters, V., & Conrad, R. (1996). Sequential reduction processes and initiation of CH_4 production upon flooding of oxic upland soils. *Soil Biology & Biochemistry, 28*, 371–382.

Sauvé, S., Norvell, W.A., McBride, M., & Hendershot, W. (2000). Speciation and complexation of cadmium in extracted soil solutions. *Environmental Science & Technology, 34*, 291–296.

Schierup, H.H., & Larsen, V.J. (1981). Macrophyte cycling of zinc, copper, lead and cadmium in the littoral zone of a polluted and a non-polluted lake. I: Availability, uptake and translocation of heavy metals in *Phragmites australis* (Cav.) Trin. *Aquatic Botany, 11*, 197–210.

Shriadah, M.M.A. (1999). Heavy metals in mangrove sediments of the United Arab Emirates shoreline (Arabian Gulf). *Water, Air, and Soil Pollution, 116*, 523–534.

Singh, S.P., Tack, F.M., & Verloo, M.G. (1998). Heavy metal fractionation and extractability in dredged sediment derived surface soils. *Water, Air, and Soil Pollution, 102*, 313–328.

Stigliani, W.M. (1992). *Overview of the chemical time bomb problem in Europe*. Paper presented at the European State-of-the-Art Conference on Delayed Effects of Chemicals in Soils and Sediments, 2–5 September 1992.

Suzuki, T., Moriyama, K., & Kurihara, Y. (1989). Distribution of heavy metals in a reed marsh on a riverbank in Japan. *Aquatic Botany, 35*, 121–127.

Tam, N.F.Y., & Wong, Y.S. (1999). Mangrove soils in removing pollutants from municipal wastewater of different salinities. *Journal of Environmental Quality, 28*, 556–564.

Tanji, K.K., Gao, S., Scardaci, S.C., & Chow, A.T. (2003). Characterizing redox status of paddy soils with incorporated rice straw. *Geoderma, 114*, 333–353.

Tye, A.M.S., Young, D., Crout, N.M.J., Zhang, H., Preston, S., Barbosa-Jefferson, V.L., Davidson, W., McGrath, S.P., Paton, G.I., Kilham, K., & Resende, L. (2003). Predicting the activity of Cd^{2+} and Zn^{2+} in soil pore water from the radio-labile metal fraction, *Geochimica et Cosmochimica Acta, 67*, 375–385.

Vandecasteele, B., Vos, B., & Tack, F.M.G. (2002). Heavy metal contents in surface soils along the Upper Scheldt river (Belgium) affected by historical upland disposal of dredged materials. *Science of the Total Environment, 290*, 1–14.

Verslycke, T., Vangheluwe, M., Heijerick, D., De Schamphelaere, K., Van Sprang, P., & Janssen, C.R. (2003). The toxicity of metal mixtures to the estuarine mysid *Neomysis integer (Crustacea: Mysidacea)* under changing salinity. *Aquatic Toxicology, 64*, 307–315.

Vlarebo, Vlaams Reglement Bodemsanering (1996). Flemish Soil Remediation Decree ratified by the Flemish government on 22/2/1995. (Translated from Dutch)

Weis, J.S., & Weis, P. (2004). Metal uptake, transport and release by wetland plants: implications for phytoremediation and restoration. *Environment International*, *30*, 685–700.

Weis, J.S., Windham, L., Santiago-Bass, C., & Weis, P. (2002). Growth, survival, and metal content of marsh invertebrates fed diets of detritus from *Spartina alterniflora* Loisel. and *Phragmites australis* Cav. Trin. ex Steud. from metal-contaminated and clean sites. *Wetlands Ecology and Management*, *10*, 71–84.

Chapter 12
Reed Bed Sewage Treatment and Community Development/Participation

Sean O'Hogain(✉)

Abstract This chapter will review the various theories of community development, particularly the concept of community participation. It will review various
terms and will explain and evaluate the practical approaches underlying the termi-
nology. The history of the Community Participatory Approach (CPA) will be
presented. The chapter will also present the basic principles underlying this
approach. It will also present the tools and techniques of CPA and their concern
with reinforcing community structures, promoting community involvement, and
building community capabilities and developing sustainability. The chapter will
then proceed to review the technology of reed bed treatment systems (RBTS).
Hybrid systems will be examined and the suitability of hybrid RBTS to community
wastewater treatment will be presented. The principle considerations of reed bed
design, construction and maintenance will be examined with mention of the newer
developments and design modifications. The implementation of a RBTS in a small
community will be examined from the point of view of CPA. The principles of
RBTS design will be combined with CPA tools and techniques to show the useful-
ness of CPA to RBTS design, construction and maintenance.

Keywords Community development, constructed wetlands, horizontal flow,
vertical flow

12.1 Introduction

The methodology of development/aid is topical. Recent natural disasters have
thrown into sharp relief the problem of community development. The concept
emerged in the 1960s and is a process of identifying community leaders, organizing

School of Civil, Structural and Building Services Engineering, Dublin Institute
of Technology, Bolton Street, Dublin 1, Ireland

(✉) Corresponding author: e-mail: sean.ohogain@dit.ie

groups or building on existing groups and training these groups and individuals to assess their needs and resources; prioritizing a list of problems that can be addressed; planning a project or an activity; obtaining resources to implement the plan; taking action; and evaluating their impact using the lessons learned, to begin the cycle again. Community development takes into account the external environment including macroeconomic and political realities and global trends (Chand, 2006). It centers upon the planning and implementation of projects that improve a community's well-being (Castelloe *et al.*, 2002). While community action and system changes are beneficial, community development has sometimes failed to provide the support that grassroots groups need to address community issues on their own without the help of community organizers (Castelloe *et al.*, 2002).

Studies have pointed out that people in low-wealth communities are not the target for community development and have stressed the need to center community development on these people, as it is they who determine, drive and control the entire development process (Chambers, 1997; Prokopy & Castelloe, 1999). This has lead to the concept of community participation, which is a social process whereby specific groups with shared needs, often but not always living in a defined geographic area, actively pursue identification of their needs, make decisions and establish mechanisms to meet these needs.

Community members' participation in a program or activity can be thought of in terms of a continuum from minimal to very high. At the low end, community members may attend an event such as a health fair that has been planned and carried out by health service providers. At the higher end, community members may identify the need for planning and information, petition the ministry of health to request services and supplies, train local community members to distribute and manage their own supplies fund and inventory, etc. (Chand, 2006).

12.2 Participatory Development

12.2.1 History

In the 1970s and early 1980s, a desire by decision-makers to more effectively incorporate the perspectives and priorities of the local people in decision-making, policy development and project implementation led to the emergence of a number of "participatory approaches" to development (Duraiappah *et al.*, 2005). This reorientation towards greater participation in development by individuals was motivated by a desire to move from an emphasis on top-down, technocratic and economic interventions towards greater attention to bottom-up, community-level interventions (Kanji & Greenwood, 2001).

Participatory approaches to development quickly evolved throughout the 1980s and into the early 1990s with the introduction of methods such as Rapid Rural Appraisal, Participatory Action Research, Participatory Poverty Assessments, Appreciative Inquiry, and, particularly, Participatory Rural Appraisal. Development of the latter approach spawned the emergence of new tools and principles for implementing and understanding participatory development. Throughout this period, researchers and community organizers sought to improve their understanding of insider/local knowledge as a balance to the dominance of outsider/western scientific knowledge (Kanji & Greenwood, 2001).

This has continued to such an extent that participation has become a mainstream component of development. Engagement of local stakeholders, involvement of poor members of communities, responsiveness to the outcomes of consultations—these have become central tenets of development and (typically) conditions for funding (Duraiappah *et al.*, 2005).

The growing adoption of a participatory approach to development reflects a continuing belief in a bottom-up approach in which participants become agents of change and decision-making. Participation is seen as providing a means to enable meaningful involvement of the poor and voiceless in the development process, allowing them to exert greater influence and have more control over the decisions and institutions that affect their lives.

12.2.2   Definitions of Participation

Three definitions of participation are given below.

Participation implies "empowering people to mobilize their own capacities, be social actors, rather than passive subjects, manage the resources, make decisions, and control the activities that affect their lives" (Cernia, in IIED, 1994).

Participation is the process through which stakeholders' influence and share control over priority setting, policymaking, resource allocations and access to public goods and services (World Bank, 2006).

Participation is the organized efforts to increase control over resources and regulative institutions in given social situations on the part of groups and movements hitherto excluded from such control (Pearse and Stiefel, in Kanji & Greenwood, 2001).

Participation thus involves a shift in power in the process of development away from those who traditionally defined the nature of the problem and how it may be addressed (governments, outside donors) to the people immediately impacted by the issue (Duraiappah *et al.*, 2005). At its pinnacle, participation involves a transformation of the traditional development approach, resulting in the enhancement of the capabilities of the local people and communities to define and address their own needs and aspirations (Sen, 1999). Over the years, a number of participatory approaches have been developed to meet the needs of different disciplines, settings and objectives. Five main approaches have been used:

Rapid Rural Appraisal – used to obtain information in a timely, cost-effective, accurate and insightful manner as a basis for development planning and action.
Participatory Rural Appraisal – a series of exercises that emphasize local knowledge for rural planning.

Participatory Poverty Assessments – used to understand poverty from the perspective of a range of stakeholders, particularly the poor.

Participatory Action Research – used to empower participants and enhance collaboration and expedite knowledge acquisition and social change.

Appreciative Inquiry – a philosophy that the past successes of individuals, communities and organizations are the basis for future success.

The approach which is under consideration in this chapter, The Community Participatory Approach (CPA), is an amalgam of all of these approaches, taking the positive aspects of each. CPA recognizes, organizes and utilizes the experience, wisdom and power of the community members working in groups to identify, prioritize, plan, implement, monitor and evaluate community-based reconstruction and development activities.

As it utilizes local knowledge and wisdom, CPA can be an effective tool for disaster prevention, emergency preparedness, postdisaster reconstruction and all community-based development programs. CPA emphasizes community capacity building, empowerment and community mobilization that contribute significantly to improve the socioeconomic status of the most disadvantaged communities. It is an open-ended and longer-term process and therefore needs longer-term partnership with the communities. It encourages building on community strengths and social readiness to accomplish tasks or improve community living standards (Duraiappah *et al.*, 2005).

Emphasis is also laid on understanding the community's culture, social structure, demographics, political structure, major issues/concerns on the minds of residents and natural and human resources available within the community to address them. The five basic principles of CPA are summarized in Table 12.1.

It follows a participatory approach where many if not all the decisions are made by local communities themselves (not external agencies or their agents) on the basis of shared responsibility. It is a "bottom-up" rather than a "top-down" approach. CPA complements community resources where necessary. External development agencies may "top-up" community resources with materials, skills or funds where this is necessary. A large share of the resources are provided by the local community or groups. It is a gender- and age-sensitive process. Women are normally disadvantaged compared to men in our society. Similarly children's voices are rarely heard during the planning process. Development agencies must recognize this imbalance and ensure that it supports initiatives which, whenever possible, will reduce such inequality. CPA also ensures that activities are sustainable (i.e., they will last after external support has ended); in particular, only those activities that do not damage the environment and have sound cost recovery and maintenance plans are supported (Chand, 2006). Enhanced community capacity to meet the future

Table 12.1 Basic principles of CPA (After Chand, 2006)

1. Community participation in decision-making
2. Complementary to community resources
3. Gender & age sensitivity
4. Sustainability
5. Holistic

needs and a sense of ownership among its members are essential for the long-term sustainability of any development activity. CPA recognizes interlinks among various needs of the community. While emphasis may be placed on one or other sectors during the implementation phase, a thorough analysis and research is carried out for the wide range of community needs. People and their lives can not be dealt with by one or a few sector-specific interventions. It requires holistic thinking to deal with the real issues of a real community (Chand, 2006).

12.2.3   Community Participatory Approach Tools and Techniques

CPA tools and techniques are designed to suit the needs of multisectoral integrated community-based activities. The tools can easily and effectively be used in communities in order to facilitate the process. These tools are:

- Organize community members, strengthen their capacity and mobilize them.
- Gather information on the community's resources.
- Organize the community's collective knowledge and wisdom in a logical framework and utilize it for their development.
- Analyze community problems, and identify and prioritize local development needs.
- Develop community organizational structures and operating procedures.
- Develop community vision; long-term development strategy and Community Action Plans (CAP) which respond to these needs.
- Design and manage the community-based projects.
- Monitor and evaluate the successes and failures the better to plan for future.
- Establish links with external donor agencies for support if needed.

12.2.4   Community Participatory Approach Outcomes

In a community where CPA is utilized, an ever-increasing number of people will participate in community activities and in the decision-making process. This is one of the main outcomes of using this approach. Others are shown in Table 12.2.

Successful use of CPA in community development ensures increased community control over the decision-making process, and especially over how funds are used. It also provides an opportunity for community members to acquire leadership skills and to practice these in a real-world situation. CPA is considered successful when community leaders bring new people into decision-making and thus expand the leadership base within the community. As individuals develop new skills and expertise, the level of volunteer service is raised, ultimately benefiting the community at large. Formal or informal community-based groups/organizations are the main stakeholders of the reconstruction and community development

Table 12.2 CPA outcomes (After Chand, 2006)

1. Community participation expanded at all levels of the development process
2. Increased community control over decision-making process
3. Expanded leadership base
4. Strengthened individual base
5. Effective community organizations and institutions
6. Increased access to the resources and better resource utilization by the community
7. A widely shared vision and a community strategic plan
8. Consistent, tangible progress toward goals

efforts. All types of civic societies and traditional institutions – such as schools, youth clubs, women's societies, NGOs, local government/municipality councils – are the mainstay of the CPA process. CPA contributes to strengthening these institutions and helps them run well and efficiently and ultimately make the community stronger. CPA believes in the theory that the poverty of underdevelopment is a result of the lack of access to resources and therefore helps communities to bridge the gulf between rural communities and the outside resources.

However, the primary focus of CPA is to identify and use local resources first. Communities that balance local self-reliance with the use of outside resources can face the future with confidence. A well-thought-out village strategy and a vision of the best community future is an important part of planning, but in CPA the emphasis is on those who participate in developing that strategy and how widely that vision is shared among community residents. CPA presents a community response to the future in the form of a well-documented strategic plan which is a way to understand and manage change. CPA not only helps people to develop vision and plans, but it also builds a community capacity that turns plans into results. The momentum and bias for action come through as a community gets things done.

12.3   Reed Bed Treatment Systems

12.3.1   Introduction

Reed bed treatment systems (RBTS) are engineered wastewater systems designed and constructed to use the natural processes that occur in natural wetlands, but within a more controlled environment (Vymazal, 2005). Most RBTS for wastewater treatment are planted with emergent macrophytes with design variations in the flow of the effluent through the bed and also the choice of media. The most common systems are referred to as vertical flow (VF) systems and horizontal subsurface flow (HF) systems, a classification based on the movement of the effluent through the bed. Reed beds have been used for decades mostly for the treatment of domestic or municipal sewage. However, recently they have also been used to treat other effluents, besides domestic wastewater. These applications include treating industrial and agricultural wastewaters, landfill leachate, stormwater runoff and acid mine drainage. As many of these wastewaters are difficult to treat in a single-stage system, hybrid systems, which consist of combinations of VF and HF reed beds staged in series, have been introduced.

12.3.2 Hybrid Reed Bed Treatment Systems

The most common hybrid systems are VF beds followed by HF beds arranged in a staged manner (Vymazal, 2005). Horizontal reed beds have the effluent passing through the media horizontally. The media, usually gravel, is planted with reeds, and the rhizomes grow horizontally and vertically and help to keep the bed media open for water flow. The media is sealed by an impervious material. The rhizomes also serve as sites for large populations of bacteria which colonize the rhizosphere. Oxygen is transmitted to the rhizosphere via the leaves and stems of the reeds. However, there is not enough oxygen transfer in the HF reed beds to achieve nitrification. The media is of a uniform grading, and most horizontal beds are planted with *Phragmites australis*, though bulrush and iris have also been used.

Vertical reed beds have the effluent passing through the media vertically, after being distributed over the full surface of the bed. As with HF beds, the rhizomes of the reeds serve to open up the media and increase the hydraulic conductivity. The rhizosphere serves the same function as in the HF flow beds and the media is sealed by an impervious material. However, oxygen transfer is enhanced in VF beds due to the fact that they are dosed intermittently (Blix and Schierup, 1989). As a result the liquid drains from the bed, and the bed refills with air which is then trapped by the next dose of effluent. In VF beds the media is graded gravel with a sand on top, though this has changed in the more recent designs. VF beds are also built to be operated in parallel, as they need to be rested after a period of dosing.

HF systems cannot achieve a fully nitrified effluent because of their limited oxygen-transfer capacity. VF systems, on the other hand do provide good conditions for nitrification but no denitrification occurs in these systems. Therefore, the advantages of the HF and VF systems can be combined to complement each other. It is possible to produce an effluent low in biochemical oxygen demand (BOD), which is fully nitrified and partly denitrified and hence has a much lower total-N concentrations. Phosphorus removal is not noted to any degree except where special media is used (Cooper, 1999, 2001).

12.3.3 Suitability of Reed Bed Treatment Systems to Community Participatory Approach

RBTS are a suitable technology for community wastewater treatment for a number of reasons (Table 12.3). It is an efficient and economical form of wastewater treatment and produces a high-quality effluent. It is a green process and is therefore sustainable. The system itself can be constructed from locally available materials.

Table 12.3 Suitability of RBTS to community wastewater treatment

Low construction cost
Low operating costs
Low maintenance requirement
Ideal for village systems
Up to 2,000 PE
Treatment of small flows/remote locations
Low Ecological Footprint that provides wildlife habitat
Can be constructed with local materials

Construction costs are low as are maintenance requirements and operating costs. The ecological footprint of any RBTS is low and it also serves as a wildlife habitat. It is a robust technology and can deal with varying flows, from low to high and is especially suited to remote locations with a minimum of infrastructure.

12.4   Design and Construction of Hybrid Systems

Various studies, manuals and guidelines have dealt with the design and construction of hybrid RBTS (Cooper, 1990; Cooper *et al.*, 1996; Vymazal *et al.*, 1998; O'Hogain, 2001; Brix and Arias, 2005). The main considerations of hybrid systems are: characterization of wastewater, choice of design, pretreatment, construction, layout and system operation, liner, filter media, distribution, recirculation/willows/agricultural use of effluent, planting, management/maintenance requirements.

Characterization of the wastewater involves identifying the type of wastewater generated, identifying the inputs into the system and possible alternatives to including them in the RBTS. Characterization of the wastewater may involve sampling and analysis and treatability studies.

Choice of design involves determining the population equivalent (PE) accurately, survey of the land and available space, topography, power availability and requirements, end use of effluent, sludge generation, etc. Also required here is a detailed study of the existing infrastructure in the community and whether it is sufficient to support a central wastewater treatment plant or whether latrines are required with the reed beds sited nearby.

Pretreatment requirements are for a sedimentation or a septic tank prior to discharge into the vertical flow wetland to minimize the risk of clogging. The sludge must be removed from the tank once a year to maintain a sufficient sludge-removal capacity. However, systems have been installed without pretreatment, most notably in France (Boutin & Lienard, 2004). Here a siphon is used to dose the beds and no sludge is generated. This design has a useful cost and maintenance payoff and is worthy of consideration in any community situation.

Construction, layout and system operation involves the planning, laying out and preparation of the site for construction. Materials and labor and other logistical problems would come under this section also. Arranging teams of workers and also local materials can make this a complicated task.

All RBTS are lined to ensure that the pollutants in the wastewater do not percolate through the ground and into the water. The impervious material used can range from puddle clay to a polyethylene sheet. The liner plays an important role in that without it pollution will continue to take place and if not present the final effluent will be of reduced volume. Local knowledge will identify any local sources of impervious materials.

The filter medium is probably the most important part of the project, especially in the VF beds. The conventional VF beds consist of layers of graded gravel topped by a sand layer. The porosity of this sand layer is critical to the successful operation of the system. If the sand is too porous, the wastewater will penetrate through it too quickly and the full surface of the bed may not be used; also treatment may be reduced as contact time with the media is reduced. If the sand is not porous enough, ponding will occur, resulting in little or no treatment taking place. Striking the right balance entails choosing the right sand for the bed.

This is particularly critical if the new generation of Compact Vertical Beds (CVF) are being used. CVF is a concept introduced in the last few years (Weedon, 2003). The basic difference between CVF and VF beds is that the media in CVF is sand alone, to a depth of 0.8–1 m. An important advantage of the CVF beds is that no parallel beds are needed and therefore maintenance is reduced. However, proper sizing and the correct sand selection is critical. It is recommended that the sand have a d_{10} between 0.25 and 1.2 mm, a d_{60} between 1 and 4 mm, and that the uniformity coefficient ($U = d60/d10$) should be less than 3.5 (Brix $et\ al.$, 2005). The contents of clay and silt (particles less than 0.125 mm) must be less than 0.5%. In practice, only washed sand materials can be used. The effective filter depth is 1.0 m and the surface of the filter should be level. It is important not to compact the sand during construction. There is also a sand test with may prove very advantageous in field conditions, where such accuracy of sand analysis is not always possible (Griggs & Grant, 2000).

The media in HF beds is of a uniform diameter, laid on the media and supported if required. Local knowledge will help source these materials.

In all VF beds it is necessary that sewage is distributed evenly over the surface of the bed by a network of pressurized distribution pipes. It is important that the whole distribution system is placed under pressure for a period that is long enough to secure an even distribution of water over the entire bed surface. Recent studies have recommended pipes with a diameter of 32–45 mm and with 5–7 mm holes placed in the bottom of the pipes for every 0.4–0.7 m (Brix & Arias, 2005). In practice, the volume pumped should be at least three times the volume of the distribution pipe system. The type of distribution system is likely to be a variant on locally available materials, cost and functionality.

The final use of the effluent and the quality required are the main factors in deciding whether recirculation of the effluent, discharge to a willow bed or irrigation for agricultural production is chosen as part of the design. In an area where fuel is an issue or where materials such as willows can be beneficial to the local community, the final effluent can be discharged to a willow bed, and the willows harvested and used locally. Where there are irrigation problems, the finished effluent may, after careful analysis, be approved as irrigation water for agricultural work.

Beds are usually planted with the common reed, *P. australis*, in a density of approximately four plants per square meter. The best planting time is April–May in Europe, but planting can be done all year round except in periods with risks of severe frost. The best results are obtained if potted seedlings are used, but pieces of rhizomes might also be used. Similar precautions may be taken when planting locally available reeds. Particular care should be paid to the plants in the first 2 years of growth. Locally available species will be identified through the CPA.

Management and maintenance requirements will depend on decisions taken earlier. Pretreatment will require that de-sludging takes place annually. The use of parallel VF beds will require a weekly program of rotation. Recirculation, willows or irrigation use of the final effluent will also have their own maintenance implications. The distribution pipes should be cleaned and flushed once per year to remove sludge and biofilm that might have blocked some of the holes. During the first growing season, it is important to remove any weeds that might compete with the planted reeds. The reeds should not be harvested during autumn because the plant material will help insulate the filter against frost during winter.

12.5   Combining Community Participatory Approach and Reed Bed Treatment Systems

The challenge in using CPA at a community level is engaging the poor in the project. A similar problem exists in designing and constructing a RBTS that can be designed, built and maintained by the people of the community after the development workers are gone. Therefore, different techniques are required to achieve these goals. CPA utilizes tools and techniques easily understood. These tools include maps, flow diagrams, seasonal calendars and visual techniques. Some of these techniques are listed in Table 12.4.

The first five techniques are suitable for the gathering of information required for wastewater characterization. They involve walking and mapping the community,

Table 12.4 CPA tools and techniques

Transect walk
Community mapping
Village history
Seasonal/daily calendar
Trend line
Focus group discussions
Village institution diagram
Gender and age role analysis
Problem analysis
Project planning for suitable solutions
Problem ranking
Electing community committees
Community strategic plan
Community action planning

which also serves as a survey of community land, land use and topography. Accurate determination of the local PE would also result from this. The trend line will help to identify changes in community life and how problem-solving was applied in the past and how it may be applied to the present situation.

Focus group discussions allow members to share ideas and identify and understand the issues. These serve an important role in explaining the need for a sanitation project in the community, and also serve as forums where hygiene and sanitation issues can be raised, explained and demonstrated. A village institution diagram allows community members to identify local, national and international organizations who might be of help and also to identify past relationships and successes and failures. Gender and age role analysis identifies the different roles in each sector and can help target groups with special projects and assistance. These functions are important in building up the teamwork and the understanding in the community to mobilize the people and resources necessary to carry out the first five techniques. These techniques are not mutually exclusive, and where RBTS is concerned, will go on simultaneously and possibly seamlessly. These tools will be the building blocks for gathering the necessary information to choose a design and even to decide whether a RBTS is the best solution to the sanitation problem identified in the previous eight techniques.

This leads to problem analysis, which analyses the root causes of the problems and starts the dialogue on finding sustainable solutions. Here, and in the focus group discussions mentioned above, is where the theoretical educational side of RBTS and sanitation generally have to be dealt with, through discussions and possibly theater. This would entail portraying sanitation, disease transmission, hygiene and other health topics in an understandable and clear manner. Education alone is never enough and therefore the practical application of this data reinforces the educational point. Popular education lacks the action component that is needed to translate critical consciousness into community change. Learning is best acquired not through training or education sessions, but through the process of building indigenous grassroots organizations and carrying out locally controlled community improvement projects (Chand, 2006)

Project planning, problem ranking and electing committees involve design, construction and layout of the RBTS. Here the problems of the locality come to the fore. Materials, design restrictions due to local factors, ability to carry out the work and local responsibility can be dealt with. These arrangements go hand in hand with the strategic community plan and the community action plan, which is a comprehensive implementation plan. These can follow the points in Table 12.4 that remain to be dealt with. By incorporating the points in Section 12.4 into Table 12.4, the design, construction and maintenance of a RBTS can be carried out in a CPA.

12.6 Conclusion

"Due to the complexity of community dynamics as a human process there are no blueprints, nor ready made recipes for participatory processes that can be applied to promote participatory development" (Botes & van Rensburg, 2000).

Each community situation is unique and different. As facilitators, development workers should foster the principle of minimum intervention and respect the indigenous knowledge of the disadvantaged groups in the community. Experience with CPA has demonstrated that the manner in which community members are included in a process sets the context for the results generated (Duraiappah *et al.*, 2005). The aim of this chapter was to take a methodology, CPA, and outline the tools and techniques involved in this approach. By analyzing the installation of a RBTS it was shown how the CPA can be applied to this process, the better to promote community participation in their own development and to achieve sustainable and self-sufficient sanitation.

References

Botes, L., & van Rensburg, D. (2000). Community participation in development: Nine plagues and twelve commandments. *Community Development Journal, An International Forum, 35*, 41–58.

Boutin, C., & Lienard, A. (2004). Reed bed filters for wastewater treatment in France. *Water, 21*, April 2004.

Brix, H., & Schierup, H.H. (1989). The use of macrophytes in water pollution control. *Ambio, 18*, 100–107.

Brix, H., & Arias, C. (2005). The use of vertical flow constructed wetlands for on-site treatment of domestic wastewater: New Danish guidelines. *Ecological Engineering, 25*, 491–500.

Castelloe, P., Watson, T., & White, C. (2002). *Participatory change: An integrative approach to community practice.* Asheville, NC: Center for Participatory Change.

Chambers, R. (1997). *Whose reality counts?: Putting the first last.* London: Intermediate Technology Publications.

Chand, P.B. (2006). Community Participatory Approaches, Tools & Techniques. RedR/IHE, Learning Support & Capacity Building Programme, 22 Palmyrah Avenue, 5th Floor, Colombo - 03, Sri Lanka.

Cooper, P.F. (Ed.) (1990). European Design and Operations Guidelines for Reed Bed Treatment Systems. Prepared for the European Community/European Water Pollution Control Association Emergent Hydrophyte Treatment System Expert Contact Group. Swindon, UK: WRc Report UI 17.

Cooper, P.F. (1999). A review of the design and performance of vertical flow and hybrid reed bed treatment systems. *Water Science & Technology, 40*(3), 1–9.

Cooper, P.F. (2001). Nitrification and denitrification in hybrid constructed wetlands systems. In J. Vymazal (Ed.), *Transformations on nutrients in natural and constructed wetlands* (pp. 257–270). Leiden, The Netherlands: Backhuys.

Cooper, P.F., Job, G.D., Green, M.B., & Shutes, R.B.E. (1996). *Reed beds and constructed wetlands for wastewater treatment.* Medmenham, Marlow, UK: WRc Publications.

Duraiappah, A.K., Roddy, P., & Parry, J. (2005). *Have participatory approaches increased capabilities?* Manitoba, Canada: International Institute for Sustainable Development.

Griggs, J., & Grant, N. (2000). *Reeds beds: Application, specification, design and maintenance. Good Building Guide*, Vol. 42, Part 1 & 2. Watford, Garston, UK: BRE.

International Institute for Environment and Development (IIED). (1994). *Whose Eden?: An overview of community approaches to wildlife management.* Nottigham, UK: Russell Press.

Kanji, N., & Greenwood, L. (2001). *Participatory approaches to research and development in IIED: Learning from experience.* London: IIED.

O'Hogain, S. (2001). The Application of Hybrid Reed Bed Sewage Treatment Technology. PhD. Thesis, Trinity College Dublin, Ireland.

Prokopy, J., & Castelloe, P. (1999). Participatory development: Approaches from the global south and the United States. *Journal of the Community Development Society, 30*, 213–231.

Sen, A. (1999). *Development as freedom*. New York: Knopf.

Vymazal, J. (2005). Horizontal sub-surface flow and hybrid constructed wetlands systems for wastewater treatment. *Ecological Engineering, 25*, 478–490.

Vymazal, J., Brix, H., Cooper, P.F., Green, M.B., & Haberl, R. (Eds.). (1998). *Constructed wetlands for wastewater treatment in Europe*. Leiden, The Netherlands: Backhuys.

Weedon, C.M. (2003). Compact vertical flow constructed wetland systems—first two years' performance. *Water Science & Technology, 48*(5), 15–23.

World Bank. (2006). Participation web site. http://www.worldbank.org/participation/participation/participation.htm

Chapter 13
The Constructed Wetland Association's Database of Constructed Wetland Systems in the UK

Paul Cooper(✉)

Abstract There are now believed to be more than 1,200 constructed wetland (CWs) systems in the UK. The first database of the CWs in the UK was put together by Water Research Centre (WRc) and Severn Trent Water Ltd. in 1996 to accompany a book on the design and performance of these systems. In that database, constructed by Job *et al.* (1996), only 154 beds were listed, most of which were tertiary sewage treatment sites owned and operated by Severn Trent Water. The Constructed Wetland Association (CWA) was formed in 1999 as a UK water industry body in response to problems caused by unscrupulous constructors. A group of experienced, reputable designers and constructors formed the CWA to bring to together best UK practice in order to counteract this problem. The group contains major water companies, designers, constructors, academics, plant growers and operators. They decided that one of the best ways of countering the problem was to assemble a database of design and performance from well-designed systems in order to demonstrate clearly what good systems could achieve for common treatment requirements. After negotiation the CWA group took over responsibility for the database from WRc. The CWA has produced nine updates of the database, which now contains information from more than 1,000 beds. It contains examples of the different variants of CWs in use in the UK. Most of these sites treat sewage/domestic wastewater but the database also includes examples of systems for the treatment of mine water, sludge, landfill leachate, industrial effluents, surface run-off and road run-off. Particular treatment applications are illustrated by case studies which are summary articles describing design, construction and performance.

Keywords Case studies, constructed wetlands, database, design, models, performance data, reed beds, sewage treatment, tertiary treatment, wastewater treatment

Independent Consultant, PFC Consulting, The Ladder House, Cheap Street, Chedworth, Cheltenham, GL54 4AB, United Kingdom

(✉) Corresponding author: e-mail: paul.cooper@ukonline.co.uk

13.1   Introduction

During the 1990s many constructed wetland systems were installed in the UK and they became accepted as conventional technology. Unfortunately this attracted a number of less-scrupulous constructors who had little or no experience. They thought that they could make a quick profit fitting reed bed Treatment systems but they installed systems which often did not perform the treatment required. Some of these constructors went into bankruptcy meaning that their clients had (a) no treatment and (b) no way of recovering their money. This situation was causing a problem to the experienced, reputable designers and constructors of reed bed systems who were beginning to suffer because of these "black sheep". The Constructed Wetland Association (CWA) was formed in October 1999 in response to this problem. A group of constructors, designers, water company operators, researchers and plant suppliers got together to form an industry association with the following aims:

- To promote the perception, appreciation and application of constructed wetland (CWs) systems for wastewater treatment.
- To act as a focal point, and to establish a coherent, comprehensive and up-to-date picture of constructed wetland and reed bed treatment technology for water pollution control in the UK.

The CWA organises meetings, training courses and maintains a web site but its main contribution is a database of the CW systems in the UK. The website contains a 'buyers' guide' containing advice for those who wish to purchase design, construction, plants or CW-related equipment.

The membership of the CWA comprises:

- Multi-site users (water companies, water utilities and commercial companies)
- Designers/consultants
- Researchers
- Suppliers of wetland plants and other relevant materials
- Constructors

The UK's environmental regulators (the Environment Agency (EA) for England and Wales and the Scottish Environmental Protection Agency (SEPA)) are prevented from being members of a trade association by national policy but they are supportive of the aims of the CWA and have supplied information on the CW sites in their areas.

Applicants, to become members of CWA, have to show evidence of technical and financial competence by providing information by way of design information and testimonials from satisfied clients and/or academic publication references. One of the ways that was identified to help maintain the reputation of CW systems (and that of the CWA group members) was the collection of design and *performance data* to give confidence to potential users. As a consequence of membership of the CWA, members are asked to share at least some of their performance data. It is appreciated that some data is commercially sensitive and members do not have to share this with other CWA members but most performance data can be shared.

One of the obvious routes to providing confidence to clients and potential users was to create a CWA database.

13.2 The Constructed Wetland Association United Kingdom Database

The first UK reed bed database was the idea of Gareth Job, and he built it whilst working at Water Research Centre (WRc) with the present author and Ben Green from Severn Trent Water on the book *Reed Beds and Constructed Wetlands for Wastewater Treatment* (Cooper *et al.*, 1996), which was published by WRc on behalf of Severn Trent Water and themselves. The database (Job *et al.*, 1996) was produced as a supplement to the book and contained information on 154 sites, most of which were Severn Trent Water sites, some of which had performance data. It was produced using Microsoft Access 95 and was initially available on four floppy disks or a CD. In 1999, the CWA wanted to update the UK database and since Severn Trent Water were one of the founder members of the CWA and had provided much data of the data for the original database, they were able to persuade WRc to allow them to take responsibility for updating the UK database. Since 2000, the author and staff members of ARM Ltd. have updated the CWA UK database on nine occasions on a CD using data provided by CWA members and the UK environmental regulators. The database is still constructed with Windows Access (using the version 2003) but now contains information on 1,012 sites/beds (CWA, 2006; Cooper, 2006).

A number of researchers from outside the UK showed interest in the CWA database and asked to become members in order to have access to the data. A decision was taken in 2005 to admit members from outside the UK at a reduced rate since their main benefit would be the use of the CWA database. However, a stipulation of membership (both for UK and non-UK members) is that they must not take named data from the database and put it into other commercial or publicly available databases. Summarised information can be used elsewhere. This restriction is because of the danger of potential infringement of UK and EU data protection laws. The association now has members in the Australia, the Czech Republic, Italy and the USA.

13.3 Contents of the Database

Figure 13.1 is a screenshot showing what is contained in the database. The separate sections of the database are accessed by a series of buttons on the main menu. The **Member/Contact** database shows the contact details of all the CWA members and their specialist interests and expertise. It also shows some non-members, such as regulators, who have contributed information to the database.

One key feature of the database is a series of **Case Studies**. These are Microsoft Word documents ranging in size from 4 to 16 pages typically containing process

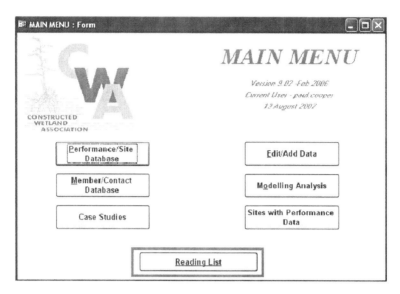

Fig. 13.1 Opening screen showing the sections of the database

descriptions of the sites, performance data tables, site diagrams, photographs and references. These are aimed at demonstrating to potential users how a particular wastewater treatment problem has been solved by a CW design. These cover all the types of design used to treat the most common UK *Treatment Applications*. They can be accessed directly from the *Case Study* button on the main menu but also through individual buttons inserted into the Performance/Site Database which is described later. The case studies included are shown in Table 13.1. A three-page *Reading List* covers the sources of information on wastewater treatment with a particular bias towards the UK and European systems and practice. The papers presented at the 10 IWA conferences, which were published in *Water Science and Technology*, are one of the main sources of the literature cited. The heart of the database is accessed by the *Performance/Site Database* button. This brings up the screen shown in Fig. 13.2. There are 1,012 beds listed from more than 900 sites in the UK. Of these, 201 beds have some performance data but the amount varies widely from data over several years to just the occasional record. This *Performance Data Selection* screen allows the user to select sites to view based on *Treatment Application*, *Bed type* and *Geographical area*.

It is also possible to use the *Performance Data Selection* screen to search to see which systems have been built by a particular designer or constructor as shown in Fig. 13.3. It is also possible to drill down and see their bed type, treatment application, location, etc.

In the UK the regional water companies are based upon river basins. It is possible to identify all the CWs in a water company area (Fig. 13.4 using the Severn Trent Water, the company which has the largest number of CWs, as the example) by searching under the *Geographical Area* button. Not all the CWs shown will be

Table 13.1 List of the *Case Studies* included in the CWA database

Secondary sewage treatment in horizontal flow (HF) beds
Tertiary treatment of sewage in HF beds
Hybrid VF beds followed by HF beds system
Tertiary nitrification in vertical flow (VF) beds
Sludge-drying VF reed beds
Separate storm sewage overflow treatment in HF beds
Combined storm sewage overflow + tertiary treatment in HF beds
Compact VF beds for secondary treatment

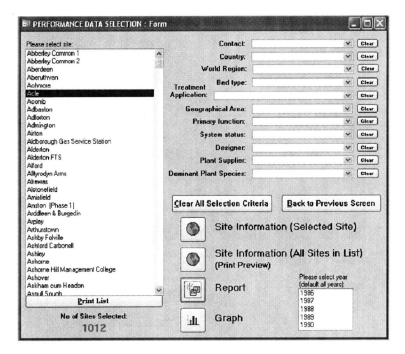

Fig. 13.2 The main screen of the *Performance Data Selection* form

owned and operated by the Severn Trent Water company but the vast majority will
be. The UK government's environmental regulator, the Environment Agency, is
organised on a similar regional setup. Going on a further step it is possible to identify
those sites in the Severn Trent region that are used for a particular **Treatment
Application**, e.g. tertiary sewage treatment (Fig. 13.5). It is also possible to do further
filtering and identify beds built by a particular **Designer**/constructor, **Bed type**, etc.

Figure 13.6 shows a selection done, looking for all the **Vertical downflow systems**
in the database. It is possible to tighten the filtering process to select the systems
built by a constructor of a particular bed type for a specific treatment application.
The **Site Details** are accessed by clicking on the **Site Information (selected site)**
button. The **Performance Data** are accessed directly by clicking on the site name

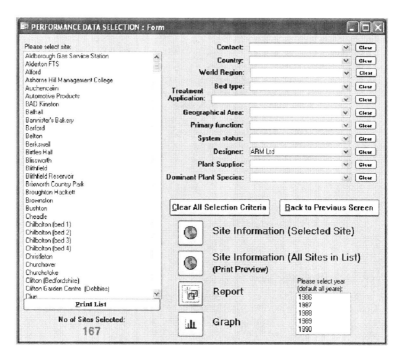

Fig. 13.3 Search shows the total number of beds a particular designer/constructor has built

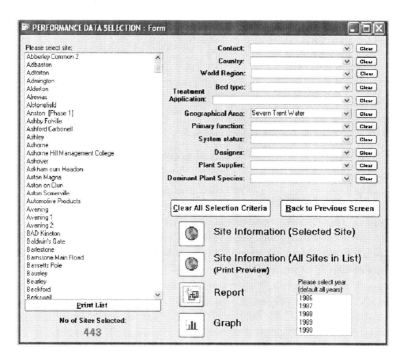

Fig. 13.4 Shows the total number of beds in the Severn Trent Water region

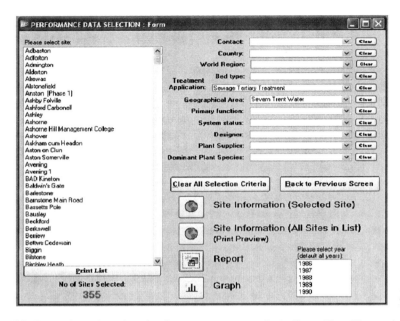

Fig. 13.5 Shows the total number of tertiary treatment systems in the Severn Trent Water region

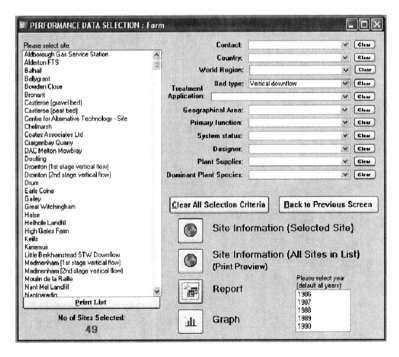

Fig. 13.6 Screen showing selection of sites with vertical downflow systems

in the dropdown list after selection on the *Performance Data Selection* form and then clicking the *Report* button. This brings up a form of the type part of which is shown in Fig. 13.7.

If we select one of these vertical downflow sites, Medmenham (first-stage vertical flow (VF)), we can we see a little more of the details contained within the database. Fig. 13.7 shows the standard format of the *Site Details* screen, giving details such as address, bed type and treatment application. Not shown in Fig. 13.7 are additional site details such as area, length, width, depth together with media type and size and full details of the consent (the effluent standard which has to be achieved). Some sites also have photographs and site drawings.

This *Site Details* screen also gives access to the *Case Study*, if there is one for the site, via the button contained therein. Figure 13.8 shows a screen from within the case study for VF tertiary nitrification at WRc Medmenham.

When a particular site has been selected, it is then possible to click on the *Report* button which brings up a spreadsheet showing the performance data (Fig. 13.9).

Clicking on the *Graph* button allows the performance data to be shown against time (see Fig. 13.10 where a graph of data from another VF site at Oaklands Park is shown). The main menu screen also has a button which allows a user to access those *Sites with Performance Data*. There are 201 sites with some performance data but the amount for each site is somewhat variable. Some sites have several

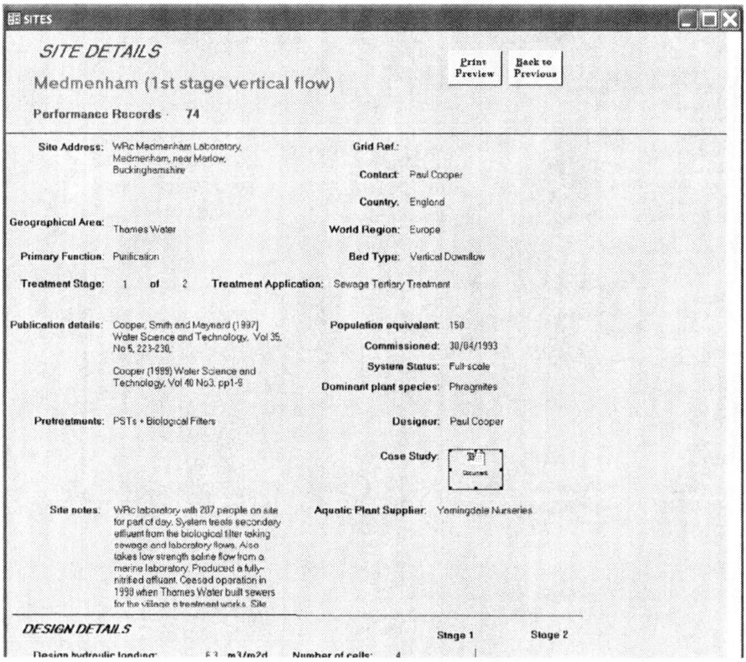

Fig. 13.7 Example of a screen showing the basic details of an individual site

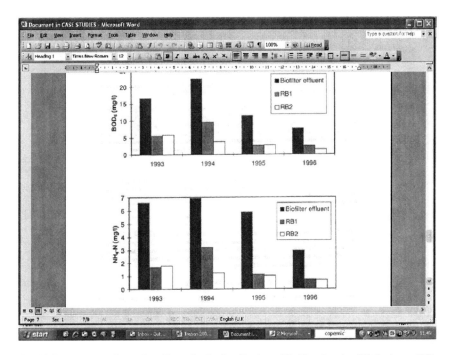

Fig. 13.8 A screen from the *Case Study* for *Tertiary Nitrification in VF beds* at WRc, Medmenham

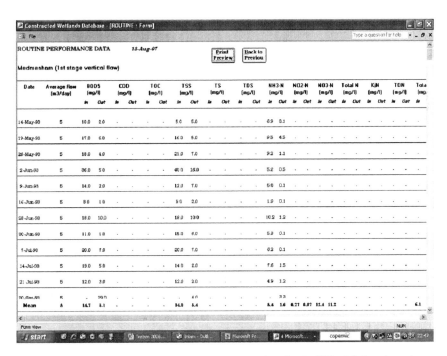

Fig. 13.9 An example of part of the Performance Data sheet for the WRc, Medmenham site

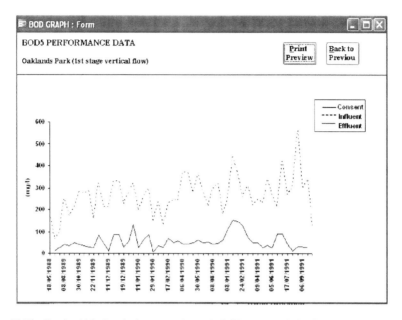

Fig. 13.10 Graph of biochemical oxygen demand (BOD_5) removal for Oaklands Park first VF stage produced directly in the database. The key within the figure shows three lines which are common to all the graphs in the database. In this particular example there is no consent (standard for the effluent) applicable because this is the first stage of a five-stage system and the consent would apply to the final stage prior to discharge

years' data recorded, while others have only a few analyses from a small number of days. The other information missing from most sites is flow rate.

In the late 1980s and early 1990s, when CWs were still somewhat new and experimental, both influent and effluent composition and flow rates were measured in order to develop the design and operating information. The situation now is that the CWs are regarded as conventional, safe solutions to many treatment applications and only effluent analysis is carried out (needed for regulatory compliance) because of the expense of analysing influent concentrations and measuring flow rates. The *Add/Edit data* screen allows users to add their own data if they have some new material to contribute. The *Modelling Analysis* function allows a user to model data from a particular site using four standard models.

13.4   Summary of the Current Situation in the United Kingdom

The author has been recording the growth of reed beds in the UK over the past 20 years and has written five papers over the period detailing this development. The UK database came into action in 1996, when it only contained 154 beds. It is

Table 13.2 Breakdown of the *Bed Type* in the UK database in June 2007

1,012 beds recorded
107 secondary sewage treatment beds
698 tertiary treatment HF beds
70 vertical flow (VF) beds
49 VF beds
21 compact VF
19 hybrid systems
46 storm sewage overflow treatment
6 Separate storm sewage overflow treatment beds
40 combined storm sewage overflow and tertiary treatment beds

Table 13.3 Breakdown of *Treatment Applications* in the database in June 2007

Types of wastewater	Number of systems
Sewage treatment	874
Minewater treatment	50
Landfill leachate treatment	24
Industrial wastewater	19
Surface run-off treatment	16
Agricultural run-off treatment	13
Road run-off treatment	16

believed that there are now between 1,100 and 1,200 beds in total in June 2007 and the CWA database has some form of record for 1,012 of these. The breakdown of the bed types is shown in Table 13.2.

The use of reed bed treatment systems originated in the wastewater treatment industry with the water companies (which are based on river basins) and the applications are still primarily for sewage treatment (Table 13.3), and most of the systems are used to provide tertiary treatment for small village treatment plants in order to avoid consent (standards) failure.

13.5   Conclusions

1. The CWA was formed as a coordinating organisation within the UK water industry in response to poor design by some constructors.
2. The CWA database has proved a useful tool to the UK water industry as a method of demonstrating what is possible with good design.
3. The database has collected together basic information from more than 1,000 sites of which about 201 have some performance data.
4. Case studies have been provided for nine of the most common design/treatment applications as a method of building confidence in constructed wetland systems.

References

Cooper, P.F. (2006). *The Constructed Wetland Association UK constructed wetland database.* Paper presented at the 10th IWA conference Wetland Systems for Water Pollution Control, Lisbon, Portugal.

Cooper, P.F., Job, G.D., Green, M.B., & Shutes, R.B.E. (1996). *Reed beds and constructed wetlands for wastewater treatment.* Swindon, UK: Water Research Centre.

CWA. (2006). *Constructed wetland interactive database* Version 10.01, June, 2007, available from the Constructed Wetland Association, UK as part of membership. www.constructedwetland.co.uk.

Job, G.D., Green, M.B., & Cooper, P.F. (1996). *Reed beds and constructed wetlands for wastewater treatment.* WRc, Swindon, UK. CD Rom containing data from 154 sites.

Chapter 14
Nitrogen Removal by a Combined Subsurface Vertical Down-Flow and Up-Flow Constructed Wetland System

Suwasa Kantawanichkul[1]($\boxtimes$), Kiattisak Pingkul[1], and Hiroyuki Araki[2]

Abstract A subsurface vertical up-flow followed by down-flow constructed wetland system was used to treat anaerobic digester effluent from pig-farm wastewater. The system consists of two beds connected in series and have equal size of $1.2 \times 1.2 \times 0.6$ m (W × L × D). The first bed was planted with *Typha angustifolia* and the second bed was planted with *Cyperus alternifolius* L. The wastewater with NH_3-N and chemical oxygen demand (COD) concentration of approximately 300 mg l^{-1} was fed to the first bed which performed as a vertical up-flow bed and then passed to the second vertical down-flow bed and recirculated to the first bed with the recycle ratio of 1:1 to enhance nitrogen removal. The system was operated with the hydraulic loading rate (HLR) of 10, 20, 30 and 40 cm day^{-1} including recycled water to evaluate the capability of the system under high HLR. The results showed that the removal efficiencies of N and COD decreased when the system operated with higher HLR. The removal efficiency for NH_3-N and COD was maximum at 10 cm day^{-1} (79 and 74% respectively). Nitrification was significant in vertical down-flow bed, denitrification in the up-flow bed at every HLR. The two-stage constructed wetland removed 96–98% of incoming total suspended solids (TSS) and the removal was not affected by high HLR. The removal efficiency of total phosphorus (TP) was low (24–27%) at all HLR.

Keywords Combined constructed wetland, hydraulic loading rate, pig-farm wastewater, vertical down-flow bed, vertical up-flow bed

[1] Department of Environmental Engineering, Chiang Mai University, Chiang Mai 50202, Thailand.

[2] Institute of Lowland Technology, Saga University, Saga, Japan

($\boxtimes$) Corresponding author: e-mail:suwasa@eng.cmu.ac.th

14.1   Introduction

Wastewater pollution problem is rather serious in Thailand. The major source is the nonpoint source pollution and especially wastewater from agriculture and animal farms has polluted many rivers and streams. There are several large-scale pig farms (more than 5,000 pigs/farm) in Thailand due to the increasing demand for food. These farms produce massive loads of wastewater each day, although most of them treat their wastewater by anaerobic system and biogas is valuable product from the treatment process. However, the effluent quality from anaerobic system channel digester followed by Upflow Anaerobic Sludge Blanket (UASB) is still beyond the national effluent standard, especially for ammonia and chemical oxygen demand (COD) (Ministry of Science, Technology and Environment, 2001[1]). The constructed wetland is an appropriate posttreatment system for pig farm due to low cost and simple technology compared to conventional systems. Cooper (1999) reviewed different options for horizontal and vertical flow systems and for the combination of both beds together. Horizontal flow systems (HF) have high capacity for biochemical oxygen demand (BOD) and total suspended solids (TSS) removal but for nitrification, HF has limited oxygen-transfer capacity. Vertical flow systems (VF) are considerably smaller but have higher oxygen-transfer capacity than HF. Vymazal (2001) also stated that combined subsurface flow constructed wetland system has been accepted for high nitrogen-removal efficiency. Therefore, a combination of up-flow and down-flow systems is one of a promising system for nitrification, denitrification and COD and suspended solids removal (Perfler *et al.*, 1999). However, more research for the application in tropical climate is still required. The objectives of the study were:

1. To evaluate the optimum hydraulic loading rate (HLR) and nitrogen-removal efficiency of pig-farm wastewater by using a combined constructed wetland system: a vertical up-flow bed followed by a vertical down-flow bed with continuous feeding.
2. To estimate the nitrification reaction in the vertical down-flow bed and denitrification reaction in the vertical up-flow bed.

14.2   Materials and Methods

Two experimental beds of $1.2 \times 1.2 \times 1.0$ m (W × L × D) were lined with PVC sheet and filled with gravel (3–6 cm) at the bottom (15 cm) and sand (0.2–0.4 cm) on top (45 cm). The first bed (up-flow) was planted with *Typha angustifolia* and the second bed (down-flow) was planted with *Cyperus alternifolius*, with initial density of 16 rhizomes per bed. Pig-farm wastewater effluent from channel digester followed by UASB was continuously fed to the bottom of the first bed and

---

[1] Thai standards for medium-to-large pig farm (from 60–600 LU, 1 LU (livestock unit) = 500 kg pig weight): TKN < 200 mg l^{-1}, pH = 5.5–9, BOD_5 < 100 mg l^{-1}, COD < 400 mg l^{-1}, SS < 200 mg l^{-1}

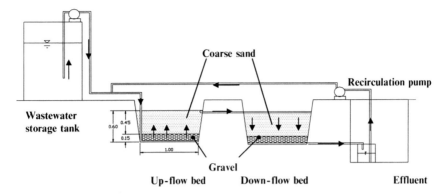

Fig. 14.1 Experimental setup

flowed upward to the surface. The wastewater was collected and then distributed to the second bed via perforated pipes laid on the surface of the bed. The water trickled vertically downward and was collected at the bottom of the second bed as shown in Fig. 14.1. The effluent was then recirculated to the first bed with the ratio of 1:1 which is the ratio of wastewater inflow (L/hr) and recirculation water flow (L/hr) and the flow rate of both is equal. To evaluate the optimum HLR of the system, the HLR was controlled at 5, 10, 15 and 20 cm day^{-1} or 10, 20, 30 and 40 cm day^{-1}, including recirculation water.

The influent and the effluent from both beds were sampled and analyzed for COD, TKN, NH_3-N, NO_x-N, SS, TP, alkalinity, pH and temperature according to the American Public Health Association's (1995) *Standard Method for Water and Wastewater*. Nitrogen uptake by plants was also measured.

14.3   Results and Discussion

14.3.1   Effect of HLR on Nitrogen Removal

The wastewater temperature and pH ranged from 27.1 to 29.9°C and 6.8 to 7.2, respectively. The system was operated under TKN and NH_3-N loading rates of 297–1,092 and 252–952 kg ha^{-1} d^{-1}, depending on the flow rate. The concentration of total Kjeldahl nitrogen (TKN) and NH_3-N in the raw wastewater was 285 mg l^{-1} and 245 mg l^{-1} in average. The oxidized nitrogen concentration was very low, 0.04 mg l^{-1} on average. The average COD concentration was 383 mg l^{-1}.

The concentrations of TKN, NH_3-N and COD were diluted after mixing with the recirculating water and all four HLRs had the same pattern for nitrogen removal. Concentrations of oxidized nitrogen in the raw wastewater increased after mixing with the recycled effluent, decreased in the first bed due to saturation and anoxic condition within the bed and then increased again as a result of high nitrification in the second bed. More than half of oxidized nitrogen had been denitrified. The removal of TKN and NH$_3$-N was around 60–62% at HLR of 10 and 20 cm day^{-1} and around 50–52% when the hydraulic loading increased to 30 and 40 cm day^{-1} as shown in Tables 14.1–14.4.

Table 14.1 Nitrogen removal at HLR of 10 cm day^{-1}

Parameter	TKN	NH$_3$-N	Oxidized N	TN
Raw wastewater (mg l^{-1})	302	252	0.04	302
	($\pm$ 24)[a]	($\pm$ 14)[a]	($\pm$ 0.00)[a]	($\pm$ 24.0)[a]
Mixed water (mg l^{-1})	182	152	23.4	205.4
	($\pm$ 13)[a]	($\pm$ 8)[a]	($\pm$ 2.3)[a]	($\pm$ 11.9)[a]
Effluent (up-flow) (mg l^{-1})	120	94	10.5	130.5
	($\pm$ 5)[a]	($\pm$ 5)[a]	($\pm$ 0.7)[a]	($\pm$ 5.6)[a]
Removal efficiency (%)	60.0	62.7	55.2	56.8
Effluent (down-flow) (mg l^{-1})	63	52	46.7	109.7
	($\pm$ 5)[a]	($\pm$ 5)[a]	($\pm$ 4.6)[a]	($\pm$ 5.3)[a]
Removal efficiency (%)	47.9	44.6	–	15.9
Total removal efficiency (%)	79.1	79.3	–	63.7

[a] SD of n = 6 (at steady state)

Table 14.2 Nitrogen removal at HLR of 20 cm day^{-1}

Parameters	TKN	NH$_3$-N	Oxidized N	TN
Raw wastewater (mg l^{-1})	273	238	0.04	273
	($\pm$ 16)[a]	($\pm$ 17)[a]	($\pm$ 0.01)[a]	($\pm$ 13.3)[a]
Mixed water (mg l^{-1})	170	148	20.6	190.6
	($\pm$ 12)[a]	($\pm$ 12)[a]	($\pm$ 2.63)[a]	($\pm$ 12.4)[a]
Effluent (up-flow) (mg l^{-1})	109	92	5.5	114.5
	($\pm$ 5)[a]	($\pm$ 5)[a]	($\pm$ 0.8)[a]	($\pm$ 4.9)[a]
Removal efficiency (%)	60.2	61.4	73.5	58.1
Effluent (down-flow) (mg l^{-1})	66	59	41.2	107.2
	($\pm$ 10)[a]	($\pm$ 7)[a]	($\pm$ 5.3)[a]	($\pm$ 9.8)[a]
Removal efficiency (%)	38.9	36.1	-	6.4
Total removal efficiency (%)	75.7	75.3	-	60.7

[a] SD of n = 5 (at steady state)

Table 14.3 Nitrogen removal at HLR of 30 cm day^{-1}

Parameter	TKN	NH$_3$-N	Oxidized-N	TN
Raw wastewater (mg l^{-1})	302	252	0.04	302
	($\pm$ 24)[a]	($\pm$ 14)[a]	($\pm$ 0.00)[a]	($\pm$ 24.0)[a]
Mixed water (mg l^{-1})	194	161	21.6	215.6
	($\pm$ 14)[a]	($\pm$ 8)[a]	($\pm$ 1.5)[a]	($\pm$ 14.3)[a]
Effluent (up-flow) (mg l^{-1})	151	123	12.3	163.3
	($\pm$ 15)[a]	($\pm$ 7)[a]	($\pm$ 1.2)[a]	($\pm$ 14.1)[a]
Removal efficiency (%)	50.2	51.3	43.2	45.9
Effluent (down-flow) (mg l^{-1})	86	69	43.2	129.2
	($\pm$ 7)[a]	($\pm$ 5)[a]	($\pm$ 2.9)[a]	($\pm$ 6.8)[a]
Removal efficiency (%)	42.7	43.4	-	20.9
Total removal efficiency (%)	71.4	72.4	-	57.2

[a] SD of n = 6 (at steady state)

Table 14.4 Nitrogen removal at HLR of 40 cm day[-1]

Parameter	TKN	NH3-N	Oxidized- N	TN
Raw wastewater (mg l[-1])	273	238	0.04	273
	(± 16)[a]	(± 17)[a]	(± 0.01)[a]	(± 16.4)[a]
Mixed water (mg l[-1])	180	159	21.8	201.8
	(± 13)[a]	(± 12)[a]	(± 4.7)[a]	(± 12.9)[a]
Effluent (up-flow) (mg l[-1])	129	116	7.4	136.4
	(± 9)[a]	(± 8)[a]	(± 0.7)[a]	(± 9.3)[a]
Removal efficiency (%)	52.7	51.4	65.9	50.0
Effluent (down-flow) (mg l[-1])	87	79	43.5	130.5
	(± 11)[a]	(± 7)[a]	(± 9.3)[a]	(± 10.6)[a]
Removal efficiency (%)	32.8	31.3	-	4.3
Total removal efficiency (%)	68.2	66.6	-	52.2

[a] SD of $n = 5$ (at steady state)

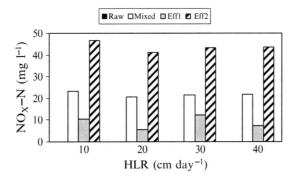

Fig. 14.2 Influent and effluent concentrations of oxidized nitrogen at four different HLRs. Raw = raw wastewater (close to zero), mixed = raw mixed with recycled effluent, Eff1 = effluent from up-flow bed, Eff2 = effluent from down-flow bed

After percolating through the second bed, 31–47% of TKN and NH$_3$-N were oxidized resulting in 79, 75, 72 and 68% total removal efficiencies of the system at HLR of 10,20,30 and 40 cm day[-1], respectively. The aerobic condition in the down-flow bed enhanced nitrification and oxidized nitrogen increased again to 41–46 mg l[-1].

It was shown that the maximum removal efficiency was achieved at low HLR (10 cm d[-1]) and declined with higher HLRs. Thus, oxidized nitrogen concentration increased after mixing with the recycled effluent and was denitrified in the saturated up-flow bed. Oxidized nitrogen concentration increased again in vertical down-flow bed as nitrification occurred. Figure 14.2 shows the concentration of oxidized nitrogen as a result of nitrification and denitrification reactions in the up-flow and down-flow beds. The effect of HLR was not clearly notified for oxidized nitrogen as for TKN and NH$_3$-N but the nitrification in the down-flow and the denitrification in the up-flow bed are substantial.

14.3.2   Effect of HLR on COD Removal

COD concentration of the raw wastewater varied in the range of 370–383 mg l^{-1} (96–370 kg COD ha^{-1} day^{-1}) average. Most of the COD was removed in the down-flow beds and the removal efficiency decreased with increasing HLR. The reduction and pattern of COD reduction with time from both beds at HLR of 10 cm day^{-1} is shown in Fig. 14.3 and Tables 14.5–14.8.

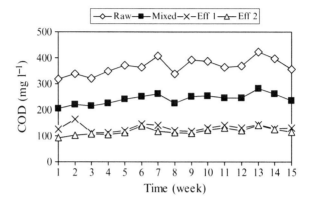

Fig. 14.3   COD concentration in the inlet and outlet at HLR of 10 cm day^{-1}

Table 14.5   Removal efficiencies at HLR of 10 cm day^{-1}

Parameter	COD	SS	TP
Raw wastewater (mg l^{-1})	383	387	87.2
	(± 24.2)	(± 38.5)	(± 4.3)
Mixed water (mg l^{-1})	241	97	75.6
	(± 13.3)	(± 19.5)	(± 3.7)
Effluent (up-flow) (mg l^{-1})	121	8.0	69.6
	(± 5.8)	(± 1.9)	(± 2.4)
Removal efficiency (%)	68.4	95.7	20.1
Effluent (down-flow) (mg l^{-1})	100	4.2	64
	(± 3.9)	(± 0.8)	(± 3.3)
Removal efficiency (%)	17.7	48.1	8
Total removal efficiency (%)	74	97.8	27

Table 14.6   Removal efficiencies at HLR of 20 cm day^{-1}

Parameter	COD	SS	TP
Raw wastewater (mg l^{-1})	370	253	92
	(± 33.9)	(± 43.8)	(± 11.07)
Mixed water (mg l^{-1})	242	129	79
	(± 21)	(± 22.5)	(± 11.6)
Effluent (up-flow) (mg l^{-1})	135	6.8	71
	(± 17.4)	(± 2.4)	(± 10.9)
Removal efficiency (%)	63.7	97.3	23
Effluent (down-flow) (mg l^{-1})	109	5.2	67
	(± 10.7)	(± 1.4)	(± 18.4)
Removal efficiency (%)	19.3	24	27
Total removal efficiency (%)	70.4	97.6	27

Table 14.7 Removal efficiencies at HLR of 30 cm day^{-1}

Parameter	COD	SS	TP
Raw wastewater (mg l^{-1})	383	187	87.2
	($\pm$ 24.2)	($\pm$ 38.5)	($\pm$ 4.3)
Mixed water (mg l^{-1})	25.4	95.4	76.5
	($\pm$ 15.6)	($\pm$ 19.5)	($\pm$ 3.5)
Effluent (up-flow) (mg l^{-1})	135	10.7	72.1
	($\pm$ 6.6)	($\pm$ 1.9)	($\pm$ 8.03)
Removal efficiency (%)	64.8	94.2	17.3
Effluent (down-flow) (mg l^{-1})	126	6.1	65.8
	($\pm$ 8.9)	($\pm$ 1.1)	($\pm$ 3.7)
Removal efficiency (%)	6.7	42.9	8.9
Total removal efficiency (%)	67.2	96.7	24.6

Table 14.8 Removal efficiencies at HLR of 40 cm day^{-1}

Parameter	COD	SS	TP
Raw wastewater (mg l^{-1})	370	253.4	91.3
	($\pm$ 33.9)	($\pm$ 43.7)	($\pm$ 11)
Mixed water (mg l^{-1})	252	129.8	80.3
	($\pm$ 22)	($\pm$ 22.6)	($\pm$ 11)
Effluent (up-flow) (mg l^{-1})	150	8.2	71.9
	($\pm$ 17.7)	($\pm$ 3.3)	($\pm$ 10.6)
Removal efficiency (%)	59.5	96.7	21.6
Effluent (down-flow) (mg l^{-1})	134	6.3	68.9
	($\pm$ 11.8)	($\pm$ 2.4)	($\pm$ 11.1)
Removal efficiency (%)	10.7	2.3	4.1
Total removal efficiency (%)	6	97.5	24.9

14.3.3   Effect of HLR on TP and TSS Removal

HLRs did not influence TP and TSS removal efficiencies (Fig. 14.4 and Tables 14.5–14.8). Low water velocities coupled with the presence of vegetation and gravel substrate promoted the filtration of solid materials (Kadlec & Knight, 1996). The up-flow of wastewater enhanced the sedimentation and there was no clogging during the experimental period of 4 months for each HLR.

Phosphorus is an important nutrient required for plant growth and is frequently a limiting factor for vegetative productivity. Phosphorus is transformed in the wetland by a complicated biogeochemical cycle and wetlands are not efficient in phosphorus reduction (Kadlec & Knight, 1996). Phosphorus can also be retained in the substrate via adsorption and precipitation, depending on the properties of the substrate. However, the effluents at four HLRs are below the national standard of pig-farm effluent in Thailand. The concentration of nitrate (which is not referred in the standard) was still high and posttreatment for denitrification is essentially required.

Fig. 14.4 Effect of HLR on removal efficiencies of TKN, COD, TP and TSS

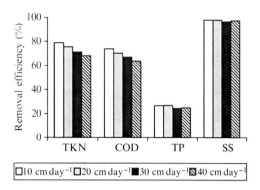

14.3.4 Plant Growth Rate and Nitrogen Accumulation

From previous experiences (Kantawanichkul *et al.*, 2001) it has been seen that *Typha* can grow well in water-saturated bed and *Cyperus* can survive well in more dry bed. In this study, *Typha* was selected for the up-flow bed and *Cyperus* was planted in the down-flow bed. Before starting each HLR, plants were cut to the same height (70 cm) and at the end of the experiment of each HLR, plants were randomly sampled to evaluate dry mass and nitrogen accumulation.

Typha had higher growth rate than *Cyperus* at every HLR. After 14 weeks, dry mass production rate were 6.5, 4.5, 7.5 and 6.3 g m^{-2} day^{-1} for *Typha* and 5.7, 2.9, 6.7 and 2.9 g m^{-2} day^{-1} for *Cyperus* at 10, 20, 30 and 40 cm day^{-1}, respectively (Fig. 14.5). Nitrogen accumulation were 0.29, 0.11, 0.31 and 0.14 g m^{-2} day^{-1} for *Typha* and 0.12, 0.049, 0.16 and 0.059 g m^{-2} day^{-1} for *Cyperus* at 10, 20, 30 and 40 cm day^{-1}, respectively (Fig. 14.6).

Typha had higher mass increment and nitrogen accumulation than *Cyperus* at every HLR. Both plants had minimum dry mass at hydraulic loading of 20 cm day^{-1}; this was caused by rats attack and most of leaves were damaged. Nitrogen accumulation in both plants were very low, only 0.51–3.84% of total nitrogen loading in the up-flow bed by *Typha* and 0.43–3.68% of total nitrogen loading in the down-flow bed by *Cyperus*, respectively.

14.4 Conclusions

The subsurface up-flow followed by down-flow constructed wetland system is suitable for treating wastewater with high ammonium concentration. The denitrification in the up-flow bed and the nitrification in the down-flow bed were noticeable. However, the higher HLR caused the reduction in retention time and impeded the removal efficiency of the system. From this study, the maximum nitrogen- and COD-removal efficiencies were achieved at the minimum HLR (10 cm day^{-1}).

Fig. 14.5 Dry mass at different HLRs

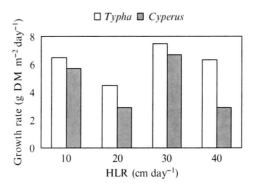

Fig. 14.6 Nitrogen accumulation in plants at different HLRs

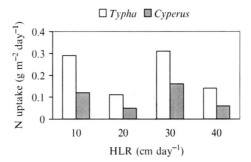

The removal of SS and TP were not dependent on HLR. However, all four HLRs could produce the effluent below national effluent standards for pig farms. However, for economic reasons, higher HLRs (20–40 cm day^{-1}) could be considered.

Nitrogen uptake by plants was very low compared to total nitrogen loading to the system. Further treatment was also needed for nitrate reduction in the final effluent. The system could be operated by continuous feeding without clogging throughout the experiment.

Acknowledgements The authors would like to thank the Institute of Lowland Technology, Saga University, Saga, Japan, for the financial support of this project.

References

Cooper, P. (1999). A review of the design and performance of vertical-flow and hybrid reed bed treatment systems. *Water Science & Technology*, *40*(3), 1–9.

Kadlec, R.H., & Knight, R.L. (1996). *Treatment wetlands*. Boca Raton, FL: Lewis.

Kantawanichkul, S., Neamkam, P., & Shutes, R.B.E. (2001). Nitrogen removal in a combined system: Vertical vegetated bed over horizontal flow sand bed. *Water Science & Technology*, *44*(11/12), 138–142.

Ministry of Science, Technology and Environment. (2001). Notification of the Ministry of Science Technology and Environment (BE2544). Published in the Royal Government Gazette, Vol. 118, Special Part 18d, dated February 23, BE2544.

Perfler, R., Laber, J., Langergraber, G., & Haberl, R. (1999). Constructed wetlands for rehabilitation and reuse of surface waters in tropical and subtropical areas-first results from small-scale plots using vertical flow beds. *Water Science & Technology*, 4(3), 155–162.

American Public Health Association. (1995). *Standard methods for the examination of water and wastewater* (19th edn.). Washington, DC: USA American Public Health Association/American Water Works Association/Water Environment Federation.

Vymazal, J. (2001). Types of constructed wetlands for wastewater treatment: Their potential for nutrient removal. In J. Vymazal (Ed.), *Transformation of nutrient in natural and constructed wetlands* (pp. 1–94). Leiden, The Netherlands: Backhuys.

Chapter 15
Statistical Analysis of Treatment Performance in Aerated and Nonaerated Subsurface Flow Constructed Wetlands

Scott Wallace[1]($\boxtimes$), Jaime Nivala[1], and Troy Meyers[2]

Abstract This study compares the treatment performance of aerated and nonaerated horizontal flow constructed wetlands. Results are presented in a series of performance charts that relate effluent concentrations to influent mass loadings. A statistical approach has been developed that allows the application of confidence intervals to system performance. Performance data is presented for predicted median (50th-percentile) and 90th-percentile effluent concentrations. Aeration was found to improve removal of biochemical oxygen demand (BOD), total suspended solids (TSS), and to a lesser extent, total Kjeldahl nitrogen (TKN). Benefits of aeration include lower median effluent concentrations at a given influent mass loading, less variability in effluent quality, and lower background concentrations (C*). In contrast, aeration was observed to have no impact on total phosphorus (TP) or fecal coliform bacteria removal.

Keywords Aeration, coliform bacteria, constructed wetland, suspended solids

15.1 Introduction

Treatment performance in constructed wetlands is a combined result of the internal mechanisms that store, transform, and remove organic matter and associated pollutants. Understanding why treatment performance varies between individual wetland systems requires a detailed knowledge of the mechanisms within each wetland system. Unfortunately, internal mechanisms have only been quantified for a very small percentage of treatment wetlands. These data are scarce because of the time

[1] North American Wetland Engineering LLC, 4444 Centerville Road, Suite 140, White Bear Lake, Minnesota 55127, USA

[2] Mathematics Department, Luther College, Decorah, Iowa, USA

($\boxtimes$) Corresponding autor: e-mail: swallace@nawe-pa.com

J. Vymazal (ed.) *Wastewater Treatment, Plant Dynamics and Management in Constructed and Natural Wetlands*,
© Springer Science + Business Media B.V. 2008

commitment and high cost associated with intensive sampling and analysis. Due to the limited amount of information available, a mechanistic approach (e.g. one that is based on knowledge of internal mechanisms) cannot be used to predict system performance with a high degree of accuracy. By necessity, current design tools use "lumped" parameters (such as first-order rate coefficients) to represent the aggregate impact of the internal mechanisms in wetland treatment systems.

Variability between systems is generally much greater than that within an individual wetland system (Wallace & Knight, 2006). Because the combined effect of internal treatment processes varies from one system to the next, using the performance of a single wetland system to predict the performance of a proposed wetland introduces an unquantified level of risk into system design. As a result, design engineers need robust tools that enable them to account for intersystem variability.

Statistics offers an alternate approach to treatment wetland design. Rather than attempting to separate and quantify each of the internal mechanisms (for which there is currently insufficient data anyway), statistics can be used to assess the aggregate result of the internal mechanisms within a system. Assessment of intra- and intersystem variability makes it possible to develop confidence intervals that quantify the factor of safety associated with a given wetland design. By utilizing appropriate time intervals (such as system-months), seasonal and stochastic variabilities can also be accounted for in the engineering design.

15.2   Methods

The treatment performance of 22 conventional (nonaerated) horizontal subsurface-flow (HSSF) constructed wetlands was statistically analyzed by the authors as part of a recent study conducted for the Water Environment Research Foundation (WERF) (Wallace & Knight, 2006). Performance charts were created based on an assumed mass-in, concentration-out relationship for both free water surface (FWS) and HSSF wetland types. Each data point on the resulting performance charts represents a "system-month". One system-month represents the average influent mass loading and average effluent concentration for one system during a particular month. Performance charts for FWS and HSSF systems were created for biochemical oxygen demand (BOD), total suspended solids (TSS), total Kjeldahl nitrogen (TKN), and total phosphorus (TP) for both system types.

For this study, the performance of 17 aerated HSSF constructed wetlands (designed by North American Wetland Engineering LLC) (NAWE) was compared to the nonaerated HSSF wetlands in the WERF study. The aerated wetlands analyzed in this study were all located in the cold temperate climate of Minnesota, used similar gravel bed material, and were insulated with a 15 cm mulch layer to prevent freezing during wintertime operation. Hydraulic loading rates were between 0.1 and 4.9 cm d^{-1}. A typical schematic of a NAWE-aerated wetland systems is shown in Fig. 15.1.

Aerated subsurface-flow wetlands, also known as Forced Bed Aeration (FBA) wetlands (Wallace, 2002), have been widely used in North America to improve

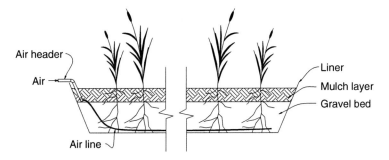

Fig. 15.1 Aerated horizontal subsurface flow wetland schematic

Table 15.1—Equations used in performance charts

Parameter		Aerated HSSF Wetlands	Non-aerated HSSF Wetlands
BOD	50th	$f(x)=\begin{cases}-0.34625x^2+4.02x & x<4.0\\ 10.54+1.25(x-4.0) & x\geq4.0\end{cases}$	$f(x)=\begin{cases}-0.998125x^2+9.235x+3.0 & x<4.0\\ 23.97+1.25(x-4.0) & x\geq4.0\end{cases}$
	90th	$f(x)=\begin{cases}-0.34625x^2+4.02x+22.33 & x<4.0\\ 32.87+1.25(x-4.0) & x\geq4.0\end{cases}$	$f(x)=\begin{cases}-0.998125x^2+9.235x+44.99 & x<4.0\\ 65.96+1.25(x-4.0) & x\geq4.0\end{cases}$
TSS	50th	$f(x)=\begin{cases}-0.758125x^2+7.315x+1.63 & x<4.0\\ 18.76+1.25(x-4.0) & x\geq4.0\end{cases}$	$f(x)=\begin{cases}-0.758125x^2+7.315x+6.0 & x<4.0\\ 23.13+1.25(x-4.0) & x\geq4.0\end{cases}$
	90th	$f(x)=\begin{cases}-0.758125x^2+7.315x+26.24 & x<4.0\\ 43.37+1.25(x-4.0) & x\geq4.0\end{cases}$	$f(x)=\begin{cases}-0.758125x^2+7.315x+38.91 & x<4.0\\ 56.04+1.25(x-4.0) & x\geq4.0\end{cases}$
TKN	50th	$f(x)=\begin{cases}-31.02x^2+63.29x & x<1.0\\ 32.27+1.25(x-1.0) & x\geq1.0\end{cases}$	$f(x)=\begin{cases}-36.88x^2+75.01x+1.5 & x<1.0\\ 39.63+1.25(x-1.0) & x\geq1.0\end{cases}$
	90th	$f(x)=\begin{cases}-31.02x^2+63.29x+31.9 & x<1.0\\ 64.17+1.25(x-1.0) & x\geq1.0\end{cases}$	$f(x)=\begin{cases}-36.88x^2+75.01x+31.17 & x<1.0\\ 69.3+1.25(x-1.0) & x\geq1.0\end{cases}$
TP	50th	$f(x)=\begin{cases}-58.4375x^2+4.8x & x<4.0\\ 9.85+1.25(x-0.4) & x\geq4.0\end{cases}$	$f(x)=\begin{cases}-55.3125x^2+4.55x & x<4.0\\ 9.35+1.25(x-0.4) & x\geq4.0\end{cases}$
	90th	$f(x)=\begin{cases}-58.4375x^2+4.8x+8.73 & x<4.0\\ 18.58+1.25(x-0.4) & x\geq4.0\end{cases}$	$f(x)=\begin{cases}-55.3125x^2+4.55x+9.05 & x<4.0\\ 18.4+1.25(x-0.4) & x\geq4.0\end{cases}$

treatment performance by increasing oxygen transfer and maintaining aerobic conditions in the rhizosphere. This design approach has been utilized to sustain high levels of nitrification in subsurface-flow wetlands and to treat high-strength or industrial wastewaters, including dairy waste, petroleum hydrocarbons, and landfill leachate (Wallace, 2001a, b; Wallace & Kadlec, 2005).

Performance charts (Table 15.1) were created for the aerated dataset using the same approach that was used for the WERF report. The data plots give bound lines that are polynomial curve fits based on an assumed first-order background concentration k-C^* model. The polynomial curve fits bound 50% and 90% of the data points for each dataset. The 50% curve represents median performance, whereas the 90% curve provides more conservative design criteria.

15.3   Results

15.3.1   *Biochemical Oxygen Demand*

BOD influent mass loading and effluent concentration for aerated (184 system-months) and nonaerated (217 system-months) HSSF wetland data is summarized in Fig. 15.2.

As seen in Fig. 15.2, aeration of HSSF wetlands produces lower median effluent concentrations at a given influent loading rate. A commonly used sizing criteria for HSSF wetland systems is 5 m² per population equivalent (m² PE⁻¹), resulting in an applied organic load of approximately 8 g m⁻² day⁻¹ (GFA, 1998). Figure 15.2 indicates that a nonaerated HSSF wetland would return a median effluent concentration of approximately 28 mg l⁻¹ at this loading rate, whereas an aerated HSSF wetland would return a median effluent concentration of approximately 15 mg l⁻¹.

Higher BOD removal rates in aerated HSSF wetlands can be attributed to aerobic decomposition of organic matter, which has faster kinetics than the anaerobic processes that predominantly occur in nonaerated HSSF. The higher removal rates observed in aerated HSSF wetlands allow these systems to be sized smaller than their nonaerated counterparts. Figure 15.2 indicates that a nonaerated wetland designed to achieve a median effluent concentration of 20 mg l⁻¹ would have to

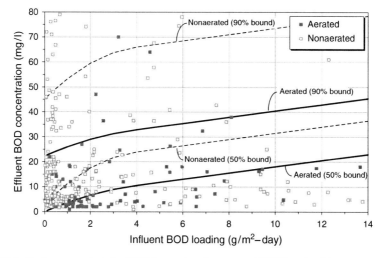

Fig. 15.2 Performance chart for BOD (aerated and nonaerated HSSF wetlands)

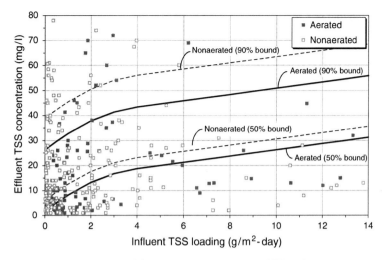

Fig. 15.3 Performance chart for TSS (aerated and nonaerated HSSF wetlands)

loaded at 2.5 g m^{-2} day^{-1} (16 m^2 PE^{-1}), whereas an aerated HSSF wetland could be loaded at up to 11 g m^{-2} day^{-1} (3.6 m^2 PE^{-1}). Comparison of the 50% and 90% bound lines for aerated and nonaerated HSSF wetlands indicates that aerated HSSF wetlands have more consistent effluent quality (less variability in treatment performance) than their nonaerated counterparts.

In order for a system to perform similar to the data and trend-lines for aerated systems in Fig. 15.2, the aeration system must be sized to meet the carbonaceous BOD. If the aeration system is undersized (or if it fails), the wetland system will function as a nonaerated system. Intermittent or inadequate aeration of a HSSF wetland will result in system performance intermediate to the curves presented in Fig. 15.2.

15.3.2 Total Suspended Solids

TSS influent mass loading and effluent concentrations for aerated (134 system-months) and nonaerated (277 system-months) HSSF wetland data is summarized in Fig. 15.3.

Figure 15.3 indicates that aerated HSSF wetlands produce, on average, lower TSS effluent concentrations than nonaerated wetlands. At a commonly used sizing criteria of 5 m^2 PE^{-1}, the resulting influent TSS loading is approximately 7 g m^{-2} day^{-1}. Data from Fig. 15.3 indicates that at this loading rate, a nonaerated HSSF wetland would return a median effluent concentration of approximately 25 mg l^{-1} whereas an aerated wetland would return a median effluent concentration of 20 mg l^{-1}.

One hypothesis for the better TSS performance of aerated HSSF wetlands is that the aerobic conditions created by the aeration system allows for the development of a grazing protozoan community to develop. Protozoa, particularly ciliated protozoa, have been documented to reduce turbidity in water by heavily grazing suspended unicellular bacteria (Curds, 1975). This grazing community of protozoans cannot develop in the anaerobic (reducing) conditions predominant in nonaerated HSSF wetlands.

This observation is further supported by the author's anecdotal evidence of TSS performance in wetland systems that have been operated both with and without aeration. Periods of nonaeration in aerated wetland systems can occur due to broken blowers, cracked aeration headers, or clogged aeration tubing orifices. TSS performance in these wetland systems substantially improved upon repair and reinstatement of the aeration system.

15.3.3   Total Kjeldahl Nitrogen

The apparent relationship between TKN influent mass loading and effluent concentrations for aerated (67 system-months) and nonaerated (120 system-months) HSSF wetland data is summarized in Fig. 15.4.

Performance data in Fig. 15.4 indicates that aerated HSSF wetlands return lower effluent TKN concentrations than nonaerated systems, although for median effluent TKN concentrations below 40 mg l^{-1}, both types of wetlands have to be loaded at less than 1 g m^{-2} day^{-1} (approximately 12 m^2 PE^{-1}). Since the typical range of nitrogen

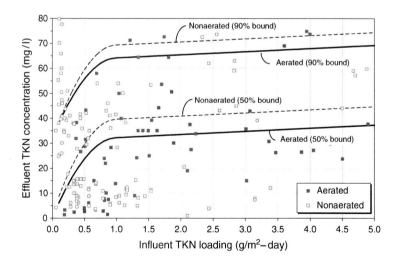

Fig. 15.4 Performance chart for TKN (aerated and nonaerated HSSF wetlands)

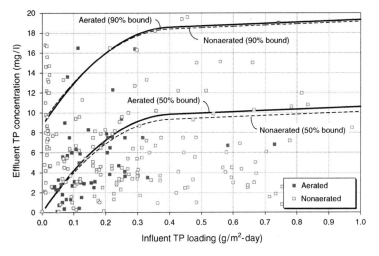

Fig. 15.5 Performance chart for TP (aerated and nonaerated HSSF wetlands)

concentration in domestic wastewater is 40–70 mg l^{-1} (Metcalf & Eddy, 1998), Fig. 15.4 indicates that little net TKN reduction occurs at loadings greater than 1 g m^{-2} day^{-1}.

This field data is not consistent with data generated by pilot systems designed to measure the effect of wetland aeration on nitrification. In laboratory studies, wetland aeration has been found to increase nitrification rates tenfold (Kinsley *et al.*, 2002). One possible explanation is that several of the aerated HSSF wetlands used in this study were retrofit systems, and the applied aeration many not have been sufficient to satisfy both the applied carbonaceous and nitrogenous oxygen demands.

15.3.4 Total Phosphorus

TP influent mass loading and effluent concentrations for aerated (79 system-months) and nonaerated (183 system-months) HSSF wetland data is summarized in Fig. 15.5. Generally speaking, VSB systems are too small (relative to their wastewater loading) to remove significant amounts of phosphorus through the accretion of organic sediments. Typical phosphorus concentrations in domestic wastewater are 4–12 mg l^{-1} (Metcalf & Eddy, 1998). Performance data presented in Fig. 15.5 indicates there is no apparent difference in phosphorus reduction between aerated and nonaerated HSSF wetlands. This is an expected result, since aeration will not affect the adsorption capacity of the HSSF bed media or plant biomass cycling. The geochemistry of the aerated HSSF wetlands included in this study is not conducive to phosphorus removal by precipitation, so aeration played no role in that removal mechanism, either.

15.3.5   Fecal Coliform Bacteria

Fecal coliform bacteria are commonly used as a pathogen indicator in the United States. Influent and effluent concentration data for fecal coliform bacteria reported for aerated (70 system-months) and non aerated (570 system-months) HSSF wetlands are presented in Fig. 15.6.

As seen in Fig. 15.6, there is no apparent difference in fecal coliform reductions between aerated and nonaerated HSSF wetlands. Both systems deliver a reduction in fecal coliform bacteria that can be approximated as a 2-log reduction. However, available data indicates that HSSF effluents are unlikely to meet a 200 CFU 100 ml^{-1} effluent standard, which is commonly applied to surface water discharges in the United States.

15.3.6   Background Concentrations

Variability within a wetland system will produce a range of effluent concentrations at a given mass loading. This concept of statistical variability can be applied to background concentrations, and is denoted as C^* (Kadlec & Knight, 1996). For the purpose of this study, C^* was defined as the most probable effluent concentration returned by the wetland at an influent mass loading of zero. Determinations of C^* were made at the median (50th-percentile) and 90th-percentile, as summarized in Table 15.2.

While there are different approaches to determining C^*, the method used here yields results that are consistent with current theory and previous literature on the

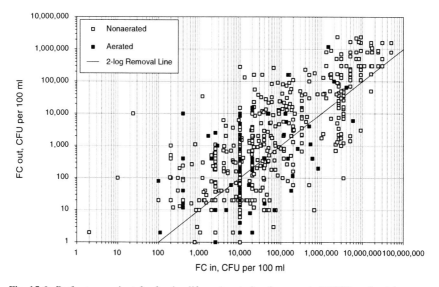

Fig. 15.6 Performance chart for fecal coliform (aerated and nonaerated HSSF wetlands)

Table 15.2 Background concentrations for aerated and nonaerated HSSF wetlands

	Aerated HSSF Wetlands		Nonaerated HSSF Wetlands	
	50th Percentile Background Concentration, C^*	90th Percentile Background Concentration, C^*	50th Percentile Background Concentration, C^*	90th Percentile Background Concentration, C^*
BOD_5	0	22	3	45
TSS	1.6	26	6	39
TKN	0	32	1.5	31
TP	0	9	0	9

subject (U.S.EPA, 1993; Kadlec & Knight, 1996; Kadlec *et al.*, 2000). Aeration of HSSF wetlands decreases variability in effluent concentrations, as can be seen by the difference between the 50th- and 90th-percentile C^* values for aerated and nonaerated system in Table 15.2. Additionally, aerated wetland systems exhibit lower median C^* values for BOD, TSS, and TKN, and lower C^* values for BOD and TSS at the 90% confidence interval. In contrast, aeration has no impact on TP for fecal coliform removal in HSSF wetlands.

15.4   Conclusions

The data presented in this study indicates that aeration of HSSF wetlands will improve removal of BOD, TSS, and to a lesser extent, TKN. Benefits of aeration include lower median effluent concentrations at a given influent mass loading, less variability in effluent quality, and lower background concentrations (C^*). In contrast, aeration was found to have no impact on total phosphorus (TP) or fecal coliform removal.

References

Curds C.R. (1975). Protozoa. In C.R. Curds. & H.A. Hawkes (Eds.), *Ecological aspects of used-water treatment, Vol. 1: Biological activities and treatment processes* (pp. 203–268). New York: Academic.

Gesellschaft zur Förderung der Abwassertechnik d. V (GFA). (1998). *Arbeitsblatt ATV – A 262 Grundsätze für Bemessung, Bau und Betrieb von Planzenbeetwn für kommunales Abwasser bei Ausbaugrößen bis 1000 Einwohnerwerte*. Germany: St. Augustin.

Kadlec, R.H., Knight, R.L., Vymazal, J., Brix, H., Cooper, P.F., & Haberl, R. (2000). *Constructed wetlands for pollution control: Processes, performance, design and operation*. IWA Specialist Group on Use of Macrophytes in Water Pollution Control. London: IWA Publishing.

Kadlec, R.H., & Knight, R.L. (1996). *Treatment wetlands*. Boca Raton, FL: CRC Press.

Kinsley, C.B., Crolla, A.M., & Higgins, J. (2002). Ammonia reduction in aerated subsurface flow constructed wetlands. In *Proceedings of the 8th International Conference on Wetland Systems for Water Pollution Control* (pp. 961–971). Dar se Salaam, Tanzania: Comprint International Limited, University of Dar Es Salaam.

Tchobanoglous, G., & Burton, F.L. (eds.), (1998). *Wastewater engineering, treatment, disposal, and reuse*, 4th edition. New York: Metcalf & Eddy Inc./McGraw-Hill.

U.S.EPA. (1993). *Guidance for design and construction of a subsurface flow constructed wetland*, US EPA Region 6 Water Management Division Municipal Facilities Branch Technical Section, Washington DC.

Wallace, S.D. (2001a). Onsite remediation of petroleum contact wastes using subsurface flow wetlands. In *Proceedings of the 2nd International Conference on Wetlands and Remediation*. Columbus, OH: Battelle Institute..

Wallace, S.D. (2001b). *Treatment of cheese-processing waste using subsurface flow wetlands*. In *Proceedings of the 2nd International Conference on Wetlands and Remediation*. Columbus, OH: Battelle Institute.

Wallace, S.D. (2002). *Patent: Method for removing pollutants from water*. Minnesota, US Patent 6,406,627. October 19, 2002.

Wallace, S.D., & Kadlec, R.H. (2005). BTEX degradation in a cold-climate wetland system. In *Proceedings of the 9th International Conference on Wetland Systems for Water Pollution Control* (pp. 165–171). Lyon: ASTEE.

Wallace S.D., & Knight R.L. (2006). *Feasibility, design criteria, and O&M requirements for small-scale constructed wetland wastewater treatment systems*. Alexandria, Virginia Water Environment Research Foundation.

Chapter 16
Constructed Wetland Břehov: Three Years of Monitoring

Lenka Kröpfelová(✉)

Abstract Constructed wetland Břehov for 100 PE was put in operation in October 2003. It is a typical horizontal subsurface flow system designed to treat municipal wastewater in the Czech Republic. Vegetated beds with a total surface area of 504 m² are filled with washed gravel (4–8 mm) and planted with bands of *Phalaris arundinacea* and *Phragmites australis*. During the period 2004–2006, the treatment performance and vegetation growth were regularly monitored. The average treatment effects for BOD_5, COD and suspended solids were high and amounted to 88%, 77% and 88%, respectively. The average outflow BOD_5 COD and TSS concentrations during the monitored period were 16.5 mg l⁻¹, 78.3 mg l⁻¹ and 19.2 mg l⁻¹, respectively. Removal of TN and TP amounted to 45% and 40%, with respective average outflow concentrations of 28 mg l⁻¹ and 4.3 mg l⁻¹. These results are quite commonly achieved in horizontal subsurface constructed wetlands in the Czech Republic. The results also indicated that during the warm period (May–October) treatment efficiency for nitrogen and phosphorus was higher than that during cold period of the year (November–April). Removal of organics and suspended solids was not affected by the season. *Phalaris* reached its aboveground biomass during the second growing season (2,265 g m⁻²), while *Phragmites* increased its biomass over the 3-year period (2,532 g m⁻²). The nutrient aboveground standing stocks were comparable with those observed in natural stands and did not exceed 5.3% and 6.2% of the inflow load of phosphorus and nitrogen.

Keywords Biomass, constructed wetland, horizontal flow, subsurface flow systems, wastewater

ENKI, o.p.s., Dukelská 145, 379 01 Třeboň, Czech Republic

(✉) Corresponding author: e-mail: kropfelova@enki.cz

J. Vymazal (ed.), *Wastewater Treatment, Plant Dynamics and Management in Constructed and Natural Wetlands,*
181

16.1   Introduction

The first constructed wetland (CW) in the Czech Republic was built in 1989 (Vymazal, 1998). By the end of 2005, at least 180 CWs were put in operation. CWs in the Czech Republic are designed with horizontal subsurface flow and most of them are designed to treat municipal or domestic sewage (Vymazal, 2006). The filtration beds are filled with washed gravel or crushed stones with grain fractions 4–8 or 8–16 mm. The beds are mostly planted with Common reed (*Phragmites australis*) or with the combination of Common reed and Reed canarygrass (*Phalaris arundinacea*). The plants used in constructed wetlands designed for wastewater treatment should (1) be tolerant to high organic and nutrient loadings, (2) have rich belowground organs even under certain level of anoxia and/or anaerobiosis in the rhizosphere in order to provide substrate for attached bacteria and oxygenation of areas adjacent to roots and rhizomes (Čížková-Končalová *et al.*, 1996; Květ *et al.*, 1999). The most important effects of the emergent macrophytes in relation to the wastewater treatment processes in constructed wetlands with horizontal subsurface flow (HF) CWs are physical effects such as erosion control, provision of surface area for attached microorganisms and insulation of the bed surface during winter (Brix, 1997; Vymazal *et al.*, 1998).

The purpose of this study was (1) to evaluate treatment performance of the constructed wetland Břehov during the period 2004–2006, (2) to evaluate growth of *Phalaris* and *Phragmites* during three growing seasons and (3) to estimate nutrient concentrations and standing stocks in the aboveground plant parts.

16.2   Material and Methods

The study was performed in CW Břehov, South Bohemia during the period 2004–2006. CW Břehov is a typical HF system designed to treat municipal wastewater (Fig. 16.1). The system was put in operation in October 2003 and the major design

Fig. 16.1   Constructed wetland Břehov, Czech Republic (Photo author)

parameters are shown in Table 16.1. *Phalaris arundinacea* and *Phragmites australis* are planted in bands parallel to the flow in both cells (Fig. 16.2).

During the period 2004–2006, the composite 24 h samples were taken on a monthly basis at three locations: raw sewage, outflow from the septic tank (= inflow to vegetated beds) and final outflow to Břehovský Creek. In the samples, BOD_5, COD, TSS, TP, TN, NH_4-N, NO_2-N, NO_3-N, TKN and SO_4^{2-} were measured in the laboratory, and concentrations of dissolved oxygen were measured in the field during the sampling.

The beds were planted in late 2003, so the first full growing season was 2004. Samples of biomass were taken from the area of $0.25 \, m^2$ in four replicates (two samples in inflow zone and two in outflow zone) during the peak standing crop, i.e., in the beginning of July for *Phalaris* and in the beginning of September for

Table 16.1 Major design parameters of constructed wetland Břehov

Start of operation	10/2003
Pretreatment	Screens, grit chamber, septic tank
Area (m²)	504
Number of beds	2
Substrate	Washed gravel, 4–8 mm
Plants	*Phragmites + Phalaris*
PE	100
Flow (m³ day⁻¹)	23ᵃ

ᵃ Average over the monitored period

Fig. 16.2 Bands of *Phragmites australis* (in the middle) and *Phalaris arundinacea* (along the sides) in CW Břehov (Photo author)

Phragmites. The stems were clipped at the ground level and separated into stems, leaves including leaf sheaths and inflorescence (flowers). Belowground biomass was completely dug out from the sampled area and divided into roots and rhizomes and thoroughly washed with pressure water. The biomass was then dried at 70°C until constant weight, and analyzed for nutrients.

16.3   Results and Discussion

16.3.1   Water Chemistry

The treatment efficiency in individual years is shown in Table 16.2. The average treatment performances for organics, suspended solids and nutrients are presented in Figs. 16.3 and 16.4. The average treatment effects for organics and suspended solids were high and amounted to 88%, 77% and 88% for BOD_5, COD and suspended solids, respectively. Removal of TN and TP was lower – 45% and 40%, respectively. However, these results are typical for horizontal flow CWs treating municipal sewage (Vymazal, 2007).

Data presented in Table 16.3 and Fig. 16.5 indicate that treatment effect is steady throughout the year for organics and suspended solids while removal of nitrogen and phosphorus is substantially higher during the summer period. The reports on the influence of the season on removal of pollutants are not unanimous. For example, Kadlec *et al.* (2003) described two adjacent HF systems in Minnesota with the average annual air and wetland water temperatures of 5°C and 9°C with air temperatures reaching below −40°C. The results revealed the highest removal of BOD_5

Table 16.2 Average annual inflow/outflow concentrations in constructed wetland Břehov

	2004			2005			2006		
	IN	A-PR	OUT	IN	A-PR	OUT	IN	A-PR	OUT
BOD_5	76	40	7.6	133	109	27	171	98	15
COD	227	123	54	ND	152	86	435	298	95
TSS	128	46	14.8	125	75	17.9	197	75	25
TP	5	5.4	2.7	10.1	6.5	6	6.6	6.3	4.3
TN	36.1	34.6	18.1	54.8	55.3	29.6	62.6	57.0	36.3
NH_4^-N	24.2	25.9	15.7	41.3	39.9	24.6	50.7	45.3	29.6
NO_x^-N	3.4	3.2	1.2	0.33	0.36	0.24	1.3	1.2	0.55
TKN	32.7	31.4	16.9	54.5	54.9	29.3	61.3	55.8	35.7
N_{org}	8.5	5.5	1.2	13.2	15	4.7	10.6	10.5	6.1
SO_4^{2-}	45.8	45.8	22.3	66.4	69.8	33.6	30.9	38.3	20.1
DO	3.1	2.5	4.1	3.9	1.4	4.1	4.9	2.6	3.2

IN = inflow (raw wastewater), A-PR = after pretreatment (inflow to vegetated beds), OUT = final outflow. All values in mg l^{-1}.

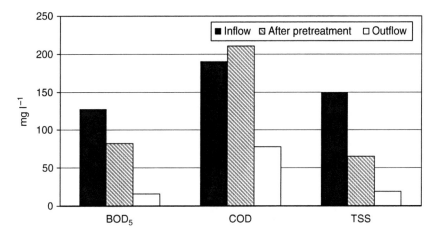

Fig. 16.3 Removal of BOD$_5$, COD and TSS in CW Břehov. Average values for the period of 2004–2006

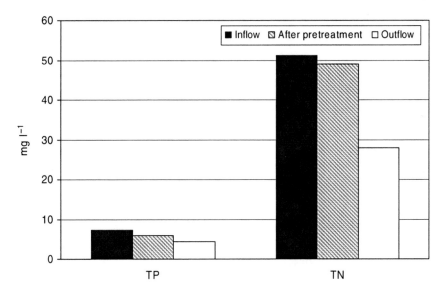

Fig. 16.4 Removal of TP and TN in CW Břehov. Average values for the period of 2004–2006

during autumn and summer with substantial decrease in treatment efficiency during winter and spring. Similar results were obtained for nitrogen and phosphorus removal, where highest mass removal occurred in summer. On the other hand, Jenssen and Mæhlum (2003) reported that there was no significant difference in efficiency between cold (<4°C) and warm (>11°C) periods for all parameters tested (TP, TN, COD, BOD$_7$, TSS) in nine constructed wetlands in Norway. Züst and

Table 16.3 Removal of pollutants in summer (May–October) and winter (January–April, November–December) during the period 2004–2006. All values in mg l^{-1}

	Period	Inflow	Outflow
BOD$_5$	Summer	148	10.9
	Winter	235	31.5
COD	Summer	408	80
	Winter	551	141
TSS	Summer	155	17
	Winter	190	24
TN	Summer	68.6	29.5
	Winter	62.7	48.1
TP	Summer	8.4	4.6
	Winter	8.6	6.2

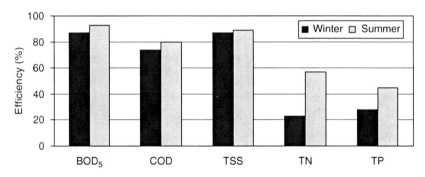

Fig. 16.5 Removal of pollutants in "summer" (May–October) and "winter" (November–April) periods in CW Břehov during 2004–2006

Schönborn (2003) could not find any influence of wastewater temperature on the removal efficiency of COD, NH$_3$-N and TP in constructed wetlands located at the altitude of 730 m with average annual temperature of 8.4°C in Switzerland. The treatment efficiency was still very good at temperatures as low as 0.5°C.

16.3.2 Nutrients in the Plant Biomass

Values of aboveground dry biomass of both *Phalaris* and *Phragmites* during 2004–2006 are presented in Tables 16.4 and 16.5. The aboveground biomass development followed the pattern reported by Vymazal and Kröpfelová (2005), i.e., that *Phalaris* reached the maximum standing crop during the second growing season (2,265 g m^{-2}), while *Phragmites* steadily increased its aboveground biomass over

Table 16.4 Dry biomass (g m^{-2}) and stem number of *Phalaris arundinacea* in CW Břehov during the period 2004–2006

Phalaris arundinacea	2004			2005			2006		
	Inflow	Outflow	Average	Inflow	Outflow	Average	Inflow	Outflow	Average
Number of stems per 1 m^2	342	449	396	844	868	856	656	690	673
Stems	559	1,251	905	1,084	1,692	1,388	1,084	818	951
Leaves	302	590	446	788	839	814	730	514	622
Flowers	56	78	67	19	105	62	104	22	63
Aboveground mass	917	1,919	1,418	1,893	2,636	2,265	1,918	1,354	1,636
Roots	126	248	187	204	600	402	X	X	X
Rhizomes	186	403	295	459	1,200	830	X	X	X
Belowground mass	310	651	481	663	1,800	1,232	X	X	X
Total biomass	1,227	2,570	1,899	2,556	4,436	3,497	X	X	X

X – the samples of belowground mass were not taken in 2006

Table 16.5 Dry biomass (g m^{-2}) and stem number of *Phragmites australis* in CW Břehov during the period 2004–2006

Phragmites australis	2004			2005			2006		
	Inflow	Outflow	Average	Inflow	Outflow	Average	Inflow	Outflow	Average
Number of stems per 1 m^2	422	322	372	460	384	422	360	232	296
Stems	272	417	345	820	1,076	948	1,418	1,142	1,280
Leaves	408	462	435	1,256	940	1,098	1,396	1,018	1,207
Flowers	0	0	0	0	0	0	54	36	45
Aboveground mass	680	879	780	2,077	2,015	2,046	2,868	2,196	2,532
Roots	106	186	146	60	90	75	X	X	X
Rhizomes	491	507	499	640	1,020	830	X	X	X
Belowground mass	597	693	645	700	1,110	905	X	X	X
Total biomass	1,277	1,572	1,425	2,777	3,125	2,951	X	X	X

X – samples of belowground mass were not taken in 2006

3-year period (2,532 g m^{-2} in the third growing season). Vymazal and Kröpfelová (2005) reported that *Phragmites* aboveground biomass varied widely between 1,652 and 5,070 g m^{-2}, with an average value of 3,266 (±1,050) g m^{-2} in 13 HF constructed wetlands in the Czech Republic. The aboveground biomass of *Phalaris* in seven systems varied between 345 g m^{-2} and 1,902 g m^{-2} with an average value of 1,286 (±477) g m^{-2}. These results indicate that *Phragmites* biomass is still well within the range found in the Czech CWs, while *Phalaris* biomass in Břehov was very high.

The results indicated that the biomass of both plants were approximately twice higher during the second growing season. For *Phragmites*, the biomass increased gradually during the third growing season while *Phalaris* biomass decreased during this period. However, the decrease was not significant.

Belowground biomass of both plants substantially increased between the first and second growing seasons. However, contrary to most natural stands where belowground/aboveground biomass ratio is commonly > 1 (usually between 2 and 5), in our study the aboveground to belowground ratio was < 1 for both plants. This has also been observed in other systems as a consequence of continuous sufficient supply of nutrients and/or of a stress caused by heavy loads of pollutants (e.g., Gries & Garbe, 1989; Haberl & Perfler, 1990; Behrends *et al.*, 1994; Bernard & Lauve, 1995; Vymazal & Kröpfelová, 2005).

16.3.3 Nutrient Standing Stocks

Nutrient standing stocks are presented in Table 16.6. They are within the range commonly found in natural wetlands. Brix and Schierup (1989) suggested that standing stock in aboveground biomass of emergent macrophytes, and thus available for harvesting, is roughly in the range 3–15 g P m^{-2} year^{-1} and 20–250 g N m^{-2}. Vymazal (1995) reported aboveground phosphorus standing stock in the range of 0.1–11 g P m^{-2} year^{-1} and 22–88 g N m^{-2} for 29 various emergent species. Johnston (1991) gives the range for nitrogen standing stock in emergent species between 0.6 and 72 g N m^{-2} with an arithmetic mean of 20.7 g N m^{-2}. Mitsch and Gosselink (2000) reported that the aboveground stock of nitrogen in freshwater marsh plants ranges from as low as 3 to 29 g N m^{-2}. Vymazal *et al.* (1999) reported nitrogen standing stock in aboveground biomass of *Phragmites australis* and *Phalaris arundinacea* growing in natural stands in the range of 0.04–63.4 g N m^{-2} and 2.0–15.5 g N m^{-2}, respectively. For constructed wetlands, Vymazal (2005) reported the aboveground nitrogen standing stock for *Phragmites australis* and *Phalaris arundinacea* growing in 29 constructed treatment wetlands in Europe, North America, Australia and Asia in the range of 3.8–62.1 g N m^{-2}. The aboveground phosphorus

Table 16.6 Nutrient standing stocks in *Phragmites australis* and *Phalaris arundinacea* growing in constructed wetland Břehov during the period 2004–2006

| | Phragmites | | Phalaris | | Phragmites | | Phalaris | |
| | Phosphorus (g P m^{-2}) | | | | Nitrogen (g N m^{-2}) | | | |
	ABG	**BG**	**ABG**	**BG**	**ABG**	**BG**	**ABG**	**BG**
2004	1.71	1.64	1.89	0.73	18.4	12.3	18.9	7.6
2005	4.40	2.43	3.7	2.82	47.7	17.3	32.3	20.5
2006	5.16	X	3.98	X	45.6	X	25	X

ABG – aboveground, BG – belowground
X – samples of belowground mass were not taken in 2006

standing stock in these plants growing in 23 constructed wetlands varied between 0.4 and 6.1 g P m^{-2} (Vymazal, 2004).

The results of this study are in agreement with the literature information on constructed wetlands that it would be misleading to expect higher standing stocks due to higher inflow concentrations. The results also indicated that the amount of nitrogen and phosphorus sequestered in the aboveground biomass during the peak standing crop was quite low as compared to the inflow loading. The standing stocks formed only 1.2%, 5.3% and 3.5% of the inflow phosphorus load and 1.9%, 6.2% and 2.7% of the inflow nitrogen load during 2004, 2005 and 2006, respectively.

16.4   Conclusions

The average treatment effects for BOD$_5$, COD and suspended solids were high and amounted to 88%, 77% and 88%, respectively. The average outflow BOD$_5$, COD and TSS concentrations during 2004–2006 were 16.5 mg l^{-1}, 78.3 mg l^{-1} and 19.2 mg l^{-1}, respectively. Removal of TN and TP amounted to 45% and 40%, with respective average outflow concentrations of 28 mg l^{-1} and 4.3 mg l^{-1}. These results are quite commonly achieved in horizontal subsurface constructed wetlands in the Czech Republic. The results also indicated that warm period (May–October) treatment efficiency for nitrogen and phosphorus was higher than that in cold period of the year (November–April). Removal of organics and suspended solids was not affected by the season.

Phalaris reached its aboveground biomass during the second growing season (2,265 g m^{-2}) while *Phragmites* increased is biomass over the 3-year period (2,532 g m^{-2}). The nutrient aboveground standing stocks were comparable with those observed in natural stands and did not exceed 5.3% and 6.2% of the inflow load of phosphorus and nitrogen.

Acknowledgement   The study was supported by grants No. 206/06/0058 "Monitoring of Heavy Metals and Selected Risk Elements during Wastewater Treatment in Constructed Wetlands" from the Czech Science Foundation and No. 2B06023 "Development of Mass and Energy Flows Evaluation in Selected Ecosystems" from the Ministry of Education, Youth and Sport of the Czech Republic.

References

Behrends, L.L., Bailey, E., Bulls, M.J., Coonrod, H.S., & Sikora, F.J. (1994). Seasonal trends in growth and biomass accumulation of selected nutrients and metals in six species of emergent aquatic macrophytes. In *Proceedings of the 4th International Conference on Wetland Systems for Water Pollution Control* (pp. 274–289). Guangzhou, P.R., China: ICWS'94 Secretariat.

Bernard, J.M., & Lauve, T.E. (1995). A comparison of growth and nutrient uptake in *Phalaris arundinacea* L. growing in a wetland and a constructed bed receiving landfill leachate. *Wetlands, 15,* 176–182.

Brix, H. (1997). Do macrophytes play a role in constructed treatment wetlands?. *Water Science and Technology, 35*(5), 11–17.

Brix, H., & Schierup, H.-H. (1989). The use of aquatic macrophytes in water pollution control. *Ambio, 18,* 100–107.

Čížková-Končalová, H., Květ, J., & Lukavská, J. (1996). Response of *Phragmites australis, Glyceria maxima* and *Typha latifolia* to addition of piggery sewage in a flooded sand culture. *Wetlands Ecology and Management, 4,* 43–50.

Gries, C., & Garbe, D. (1989). Biomass, nitrogen, phosphorus and heavy metal content of Phragmites australis during the third growing season in a root zone waste water treatment. *Archiv für Hydrobiologie, 117,* 97–105.

Haberl, R., & Perfler, R. (1990). Seven years of research work and experience with wastewater treatment by a reed bed system. In P.F. Cooper & B.C. Findlater (Eds.), *Constructed wetlands in water pollution control* (pp. 205–214). Oxford: Pergamon.

Jenssen, P.D., & Mæhlum, T. (2003). Treatment performance of multistage wastewater constructed wetlands in Norway. In Ü. Mander, C. Vohla, & A. Poom (Eds.), *Proceedings of International Conference on Constructed and Riverine Wetlands for Optimal Control of Wastewater at Catchment Scale* (pp. 11–16). Tartu, Estonia: University of Tartu, Institute of Geography.

Johnston, C.A. (1991). Sediments and nutrient retention by freshwater wetlands: Effects on surface water quality. *CRC Critical Reviews in Environmental Control, 21,* 491–565.

Kadlec, R.H., Axler, R., McCarthy, B., & Henneck, J. (2003). Subsurface treatment wetlands in the cold climate in Minnesota. In Ü. Mander & P. Jenssen (Eds.), *Constructed wetlands for wastewater treatment in cold climates* (pp. 19–52). Southampton, UK: WIT Press.

Květ, J., Dušek, J., & Husák, Š. (1999). Vascular plants suitable for wastewater treatment in temperate zones. In J. Vymazal (Ed.), *Nutrient cycling and retention in natural and constructed wetlands* (pp. 101–110). Leiden, The Netherlands: Backhuys.

Mitsch, W.J., & Gosselink, J.G. (2000). *Wetlands,* 3rd ed. New York: Wiley.

Vymazal, J. (1995). *Algae and element cycling in wetlands.* Chelsea, MI: Lewis.

Vymazal, J. (1998). Czech Republic. In J. Vymazal, H. Brix, P.F. Cooper, M.B. Green, & R. Haberl (Eds.), *Constructed wetlands for wastewater treatment in Europe* (pp. 95–121). Leiden, Netherlands: Backhuys.

Vymazal, J. (2004). Élimination du phosphore par faucardage de la fraction aérienne de la végétation dans les marais artificiels pour le traitement des eaux usées. *Ingénieries eau agriculture territoires, Special issue,* 13–21.

Vymazal, J. (2005). Removal of nitrogen via harvesting of emergent vegetation in constructed wetlands for wastewater treatment. In J. Vymazal (Ed.), *Natural and constructed wetlands: Nutrients, metals and management* (pp. 209–221). Leiden, The Netherlands: Backhuys.

Vymazal, J. (2006). The use constructed wetlands for wastewater treatment in the Czech Republic. In A.R. Burk (Ed.), *Focus on ecology research* (pp. 175–196). New York: Nova Science.

Vymazal, J. (2007). Removal of nutrients in various types of constructed wetlands. *Science of the Total Environment, 380,* 48–65.

Vymazal, J., & Kröpfelová, L. (2005). Growth of *Phragmites australis* and *Phalaris arundinacea* in constructed wetlands for wastewater treatment in the Czech Republic. *Ecological Engineering, 25,* 606–621.

Vymazal, J., Brix, H., Cooper, P.F., Haberl, R., Perfler, R., & Laber, J. (1998). Removal mechanisms and types of constructed wetlands. In J. Vymazal, H. Brix, P.F. Cooper, M.B. Green & R. Haberl (Eds.), *Constructed wetlands for wastewater treatment in Europe* (pp. 17–66). Leiden, The Netherlands: Backhuys.

Vymazal, J., Dušek, J., & Květ, J. (1999). Nutrient uptake and storage by plants in constructed wetlands with horizontal sub-surface flow: a comparative study. In J. Vymazal (Ed.), *Nutrient cycling and retention in natural and constructed wetlands* (pp. 85–100), Leiden, The Netherlands: Backhuys.

Züst, B., & Schönborn, A. (2003). Constructed wetlands for wastewater treatment in cold climates: Planted soil filter Schattweid – 13 years' experience. In Ü. Mander & P. Jenssen (Eds.), *Constructed wetlands for wastewater treatment in cold climates* (pp. 53–68). Southampton, UK: WIT Press.

Chapter 17
Factors Affecting the Longevity of Subsurface Horizontal Flow Systems Operating as Tertiary Treatment for Sewage Effluent

David Cooper[1], Paul Griffin[2], and Paul Cooper[1](✉)

Abstract Compared with other forms of wastewater treatment, horizontal flow reed beds require very little by way of operational and maintenance input. Unfortunately, in practice this frequently results in them receiving little or no attention at all. Tertiary treatment reed beds are not a "fit and forget" solution, but they are often treated this way because they are very forgiving and abuse-tolerant. Severn Trent Water Ltd. installed their first tertiary reed beds in the late 1980s and the company now has almost 350 units, mainly at village works, to ensure that the effluent from existing secondary treatment units is polished to a produce a high-quality effluent well within the standard. ARM Ltd. has been designing and constructing reed beds for 20 years and has units on more than 250 sites, a large number of these are on Severn Trent Water sites. After a number of years a few of these tertiary reed beds have deteriorated to the extent that they are close to failing to comply with the regulator's requirements. Severn Trent Water Ltd. recognised this situation and has committed a budget for a programme of reed bed maintenance. Initially a budget was allocated to refurbish the reed beds in the poorest condition. Resources are now being directed to regular maintenance of all reed bed sites. Between November 2002 and October 2005, ARM personnel carried out more than 300 primary surveys of reed bed sites in order to determine the condition of the reed beds and to establish what if any remediation was necessary. Since September 2004, when the first part of this chapter was presented at the IWA Avignon conference, more than 200 secondary surveys have been carried out when staff visited sites to carry out remedial work. The work is being further extended by use of ground penetrating radar (GPR) to try to assess the degree of clogging in different beds and in different parts of the beds. The preliminary results from this study are included. A series of tertiary surveys will be made in the future when the sites are revisited as part of a rolling maintenance contract.

[1] ARM Ltd, Rydal House, Colton Road, Rugeley, Staffordshire, WS15 3HF, United Kingdom

[2] Severn Trent Water Ltd., Technology and Development, Avon House, Coventry, CV3 6PR, United Kingdom

(✉) Corresponding author: e-mail: paul.cooper@ukonline.co.uk

J. Vymazal (ed.) *Wastewater Treatment, Plant Dynamics and Management in Constructed and Natural Wetlands,*
© Springer Science + Business Media B.V. 2008

Keywords Constructed wetland systems, ground penetrating radar; horizontal flow beds' long-term deterioration, maintenance

17.1   Introduction

Horizontal subsurface flow reed beds are now widely used for the treatment of sewage in the United Kingdom, especially for the final 'polishing' of secondary effluents to provide tertiary effluents. Severn Trent Water Ltd. has been at the forefront in developing this form of treatment. Their first reed beds were installed in 1986 and the company now has more than 350 units, mainly on small rural works. Most of their systems are horizontal flow systems for tertiary treatment to improve the quality of the secondary effluent produced by rotating biological contactors but also from biological filters, small activated sludge plants or submerged aerated filters (SAFs). They are used to protect the consent (standard) against any malfunction in the secondary treatment process primarily by removing total suspended solids (TSS) and providing some oxidation capability which will provide some additional biochemical oxygen demand (BOD_5) removal and nitrification. Severn Trent also has a few vertical flow beds for tertiary treatment and a small number of horizontal flow beds for secondary treatment. The company treats the wastewater from and supply potable water to about 8 million people in central England centred on Birmingham. ARM Ltd. has been designing and constructing reed beds for than 20 years and has units on almost 250 sites, a large number of which are Severn Trent sites. An earlier version of this study was presented at the IWA conference in Avignon in 2004 (Cooper *et al.*, 2005).

17.2   Reed Bed Maintenance Survey

During the period November 2002 to June 2004 inclusive, Severn Trent and ARM personnel visited 126 sites, which had been identified by their operations managers as being in most need of remediation. The company produced an interim report [after the first survey of 126 beds] for Severn Trent Water (Cooper, 2004) upon which the results and conclusions in the previous publication (Cooper *et al.*, 2005) were based. The survey has continued over the past 2 years and the refurbishment of some sites has been completed. Inspection and routine maintenance has been carried out on the other sites. At the present time a total of 683 visits have been made to 329 sites and 255 reed beds have been evaluated. Some reed beds still have to be evaluated.

The first author has developed a reed bed survey technique. This consists of visiting each site and completing a standard form, which the company had developed from their experience of designing and monitoring reed beds over the past 20 years. The data from this is logged in a Microsoft Access database. The form contained fields relating to the following items:

- Name and location plus contact details
- Treatment application, i.e. 1ry, 2ry or 3ry treatment
- Type of bed – HF, VF, etc.
- Pre-treatment prior to the reed beds
- Water levels – inlet and outlet
- Sludge levels – inlet and outlet
- Weed infestation – types of plant and the extent
- Condition of reeds
- Distribution system
- Collection system
- Recommendations

All this information is logged on the first (survey) visit. After the survey visit a judgement is made on the level of effort needed in the next year at the different sites. There are three levels of effort involved in the different types of visit to the sites for monitoring and maintenance:

1. *Service*, which will involve two men for up to 1 day
2. *Maintenance*, greater than 1 day's effort from two or more men
3. *Refurbishment*, involving taking some or all of the gravel out for cleaning or replacement and the amount of time will depend upon the size of the beds.

ARM Ltd. aims to visit every newly commissioned or newly refurbished bed to remove any weeds before they have chance to take over the bed and thus inhibit reed growth. Figure 17.1 shows the distribution of construction dates for 280 beds in Severn Trent Water including the 255 beds so far examined in the survey. The other 25 beds are not yet included in this evaluation because some details are missing. Almost half of the beds are more than 10 years old and some are nearly 20 years old. Two of the beds have been rebuilt.

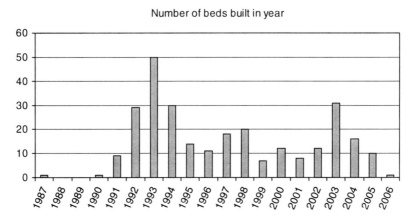

Fig. 17.1 The distribution of beds in the survey of Severn Trent sites

Table 17.1 summarises the number of beds in the different types of treatment applications that have been evaluated in the survey to date. All beds were horizontal flow subsurface systems, except for two vertical flow beds. The comparison of the earlier part of the study (when only 126 sites had been looked at) and the more recent situation (after the total reached 255) is shown in Table 17.1.

The main problems that were revealed in the earlier survey of 126 sites continue to be the most significant in the extended survey for 255 sites. These are shown in Table 17.3. Table 17.4 shows the different types of weeds that have been noted in the survey.

One of the biggest problems identified in the survey of the first 126 sites was that of water on the surface of most of the beds. Table 17.5 shows that this is still the case with these systems, but there was an even higher percentage of these in the second part of the survey. Water on the surface leads to bypassing, leading to reduced retention time in the system, and hence inadequate treatment. The opinion formed during the earlier survey that a bed with a depth of sludge >150 mm at the inlet and >40 mm at the outlet would lead to water on the whole surface is still firmly held. Water on the surface may result from clogging due to sludge build-up; however, there were some sites where the outlet water level control had been set at too high a position leading to surface flooding. This may be a result of poor communication between operators and management. Those sites with water on surface at the outlet *almost invariably* had water over the whole bed since the beds started out flat at commissioning. The problem highlighted in the earlier study regarding

Table 17.1 Type of treatment provided by the reed beds in the survey

Function of reed bed	Number of beds evaluated[a]
Secondary treatment	13 (6)
Tertiary treatment only	209 (104)
Combined tertiary + storm treatment	60 (15)
Storm sewage treatment only	3 (1)
Total	255 (126)

[a]Numbers in brackets refer to the first survey when 126 sites had been inspected.
The form of secondary treatment for the effluents flowing to the tertiary treatment beds is shown in Table 17.2.

Table 17.2 Types of secondary treatment ahead of all the *tertiary* beds

Type of pre-treatment	Number of sites
Rotating biological contactor	159
Biological filter	77
Activated sludge (oxidation ditch)	2
Submerged aerated filter (SAF)	4
Total	242

Table 17.3 Frequency of problems encountered

Problem	Frequency at 255 sites
Weed infestation (>25% cover)	130 (47)[a]
Growth of saplings	34 (24)
Inlet distributor problem	34 (30)
Outlet collector problem	21 (16)
Water on most of the bed	76 (21)

[a] Numbers in brackets refer to the first survey when 126 sites had been inspected in 2004.

Table 17.4 Types and frequency of weeds and trees found in the survey

Plants	Frequency in 255 sites with at least 25% cover
Willowherb/*Epilobium*	41
Woody night shade /*Solanum dulcamosa*	22
Nettle/*Urtica dioica*	91
Dock/*Rumex*	1
Bindweed/*Polygonum convolvulus*	2
Willow/Salix and Alder /*Alnus* saplings	34
Brambles/*Rubus*	1
Mixture of weeds	135
Grass	2

Table 17.5 No. of sites with surface flooding/relationship to sludge depth

	Number of reports from 255 sites
Sludge depth > 150 mm at inlet	111 (27)[a]
Sludge depth > 40 mm at outlet	48 (9)
Bed flooded at Inlet on 1st visit	132 (24)
Bed flooded at Outlet on 1st visit	76 (21)

[a] Numbers in brackets refer to the first survey when 126 sites had been inspected in 2004.

clogging again showed up strongly in the more recent site surveys. In some cases it may be possible to solve the problem by removing a small amount of media from near the surface in the inlet area, but in some cases it is necessary to remove media to a substantial depth. A study is being carried out at ARM Ltd. looking at potential methods for cleaning the clogged bed media (i) outside the bed and (ii) in situ.

The other main problem highlighted in the previous study was problems with the inlet distribution arrangements. This is illustrated by Fig. 17.2 which was used in the earlier paper (Cooper *et al.*, 2005; Horton, 2003). Sludge build-up on the surface at Knightcote is much higher down the right-hand side of the bed. It is believed that this is because the upflow riser pipes were fed from the right-hand end and so the flow was going preferentially to the riser pipes at that end, leading to preferential flow to that side and greater sludge deposition there. Sludge solids and sewage debris may also have deposited in the pipe furthest from the inlet, thus

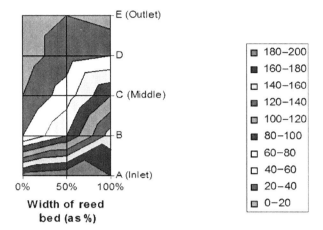

Fig. 17.2 Distribution of sludge on the tertiary bed at Knightcote

further reducing the flow to the left-hand side of the bed. It appears that maintenance has not been enough to prevent the clogging of the distribution further away
from the inlet.

The authors believe that remediation should be triggered when surface accumulation of sludge and debris gets to within 1 or 2 m of the end of the bed. The
Severn Trent Water beds are typically 12–15 m long and are always wider than
they are long. There are a number of 'triggers', which indicate the need for
maintenance/remediation:

1. Inlet distribution system adversely affected by solids accumulation
2. Standing water on 40% of the bed surface
3. Sludge accumulation to within 1–2 m of the outlet of the bed
4. Reeds only covering 40% of the bed surface
5. Surface flow across the bed or operation of the overflow during 'normal' rainfall
   events

The background to points 1, 2, 3 and 5 are fairly obvious from the previous com-
mentary. The significance of point 4 relating to poor reed growth is that the primary
function of the reeds is to alleviate/delay clogging of the bed. Gravel beds without
Phragmites australis will function for a while before clogging after several weeks/
months dependant upon the type of waste and loading rate. The reeds delay clogging by several mechanisms including movement of the gravel as the roots and
rhizomes grow and movement caused by Windrock. With respect to point 2 it was
generally found that when there was water on the surface of beds at the outlet
end, it extended back to the inlet since the bed started out as flat.

17.3   Ground Penetrating Radar Study

A preliminary study (Fogg, 2005) has recently been carried out by ADAS Ltd. (formerly the Agricultural Development and Advisory Service, part of the UK government's Department of Agriculture) for Severn Trent Water of the use of ground penetrating radar (GPR), in an attempt to try to determine where the build-up of organic matter (sludge and biomass) was taking place. This was aimed at finding a non-invasive method of determining where the sludge build-up/clogging of the bed was taking place.

Figure 17.3 shows a GPR survey along the *length* of a bed at Lighthorne Heath. (i) The dark areas are an indication of organic matter build-up at the inlet end of the bed and (ii) the lighter areas at the base of the bed indicate areas of low organic build-up, i.e. void spaces. Figure 17.4 shows a scan across *width* of the outlet of a

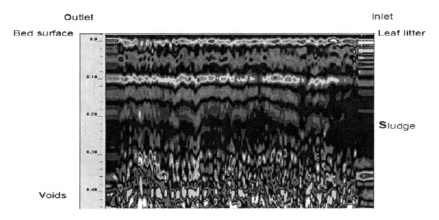

Fig. 17.3 GPR survey along the length of a bed at Lighthorne Heath. The dark areas indicate organic matter, i.e. sludge/biomass

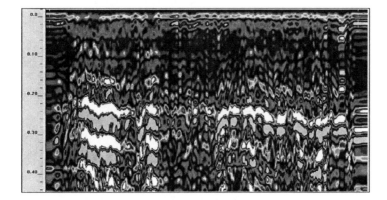

Fig. 17.4 GPR scan across width of the outlet at Ashorne. Pale areas indicate void spaces

bed at Ashorne. This shows higher a percentage of void spaces at the outlet end of a bed. The study showed that it was possible to differentiate between organic matter and the void spaces, and it was also possible to clearly see the civil structures of the bed. The second phase of the study will commence in mid 2006 with the aim of calibrating the system. It will seek to identify more closely what the different colours relate to, i.e. sludge, roots/rhizomes, bed medium and void spaces.

17.4   Conclusions

Many of the sites had the outlet water-level control set at the wrong position and this led to water on the surface. There is a need for more regular inspection of the outlet water-level control arrangement. This emphasises the need for better local communication.

1. There is also a need for greater attention to be paid to maintaining the inlet distributor to ensure that it is not clogged with sewage debris.
2. Surface flooding was mainly associated with sludge depth at the inlet. This is a problem when the depth of sludge is >150 mm.
3. Where flooding occurs it may be necessary to inspect the gravel at depth to check for clogging.
4. Remove and clean gravel where clogged at depth. On-site cleaning is being investigated.
5. Weed infestation should be dealt with in spring/early summer.
6. Ground penetrating radar has been used for preliminary tests. It may be useful in future as a method assessing the depth of sludge penetration in a non-invasive manner. This technique is still in the development phase.
7. Despite problems with lack of maintenance, tertiary reed beds have played a significant part in maintaining compliance with standards at a large number of small treatment works.

Acknowledgement   The authors wish to thank Severn Trent Water Ltd. for the permission to present information derived from the survey and report prepared for them by ARM Ltd.

References

Cooper, D.J. (2004). Report on the maintenance of Reed Beds in Severn Trent Water Ltd. ARM Ltd report to Severn Trent Water Ltd. Rugeley Staffs., UK, 3 February 2004.

Cooper, D.J., Griffin, P., & Cooper, P.F. (2005). Factors affecting the longevity of sub-surface horizontal flow systems operating as tertiary treatment for sewage effluent. *Water Science and Technology, 51*(9), 127–135.

Fogg, P. (2005). A novel approach to assess organic material accumulation in constructed reed beds: Phase 1. Report to Severn Trent Water Ltd by ADAS UK Ltd, Mansfield, Nottinghamshire, December 2005.

Horton, D. (2003). Review of tertiary and storm reed bed operations in Severn Trent Water. MSc thesis, Department of the Built Environment, Anglia Polytechnic University, UK, May 2003.

Chapter 18
Investigations on Nitrogen Removal in a Two-Stage Subsurface Vertical Flow Constructed Wetland

Günter Langergraber[1](✉), Christoph Prandtstetten[2], Alexander Pressl[1], Kirsten Sleytr[1], Klaus Leroch[2], Roland Rohrhofer[2], and Raimund Haberl[1]

Abstract A two-stage constructed wetland (CW) system consisting of two vertical flow beds with intermittent loading operated in series has been investigated. The first stage uses a grain size of 1–4 mm for the main layer and has a drainage layer that is operated under saturated conditions; the second stage uses a grain size of 0.06–4 mm and a drainage layer with free drainage. The two-stage system was operated with an organic load of 80 g COD m^{-2} day^{-1} for the first stage (1 m^2 per person equivalent), i.e. 40 g COD m^{-2} day^{-1} for the whole system (2 m^2 per person equivalent). The system has been investigated using indoor pilot-scale CWs. The results of the indoor experiments are currently verified outdoor. Organic matter and ammonia nitrogen (NH$_4$-N) removal has been very high and has been shown to be the same as for a single-stage CW system (grain size for main layer 0.06–4 mm, free drainage, organic load 20 g COD m^{-2} day^{-1}, i.e. 4 m^2 per person). A total nitrogen elimination rate of 58% could be reached for the two-stage indoor system compared to almost no nitrogen elimination for a single-stage system. The first results of the outdoor experiments are similar compared to the indoor experiments regarding organic matter, nitrogen, and microbial parameters. The nitrogen elimination reached in the first 3 months of operation was 63%. By using the two-stage CW system it is possible to obtain a higher effluent quality compared to a standard single-stage CW system with only half of the specific surface area requirement. Nitrogen removal efficiencies of about 60% and nitrogen elimination rates of more than 1,400 g N m^{-2} year^{-1} can be reached without recirculation.

Keywords Nitrogen removal, subsurface two-stage system, vertical flow constructed wetlands

[1] Institute of Sanitary Engineering and Water Pollution Control, University of Natural Resources and Applied Life Sciences, Vienna (BOKU), Muthgasse 18, A-1190 Vienna, AUSTRIA

[2] ÖKOREAL GmbH, Carl Reichert-Gasse 28, A-1170 Vienna, AUSTRIA

(✉) Corresponding author: e-mail: guenter.langergraber@boku.ac.at

J. Vymazal (ed.) *Wastewater Treatment, Plant Dynamics and Management in Constructed and Natural Wetlands*,
© Springer Science + Business Media B.V. 2008

18.1   Introduction

Constructed wetlands (CWs) designed to improve water quality, use the same processes that occur in natural wetlands but have the flexibility of being constructed. CWs are used worldwide to treat different qualities of water (e.g. Kadlec *et al.*, 2000; Langergraber & Haberl, 2001; Haberl *et al.*, 2003).

Using subsurface vertical flow constructed wetlands (SSVFCWs) with intermittent loading, it is possible to fulfil the stringent Austrian effluent standards regarding nitrification (Langergraber & Haberl, 2001). For small plants (less than 500 persons) the maximum ammonia nitrogen (NH_4-N) effluent concentration allowed is 10 mg l^{-1}. This has to be met at water temperatures higher than 12°C. For organic matter, effluent concentrations (90 mg COD l^{-1} and 25 mg BOD_5 l^{-1}) and treatment efficiencies (85% and 95% for COD and BOD_5, respectively) have to be met the whole year around. For small plants there is no standard for total nitrogen and phosphorus (1.AEVkA, 1996).

In this chapter results of the Austrian research project "*Optimization of subsurface vertical flow constructed wetlands (Bepflanzte Bodenfilter)*" are shown. The main goal of the project was the optimization of the specific area requirement for intermittently loaded vertical flow beds in Austria (Langergraber *et al.*, 2007). An additional goal was the optimization of nitrogen removal. The contribution shows results of indoor experiments and preliminary results of outdoor experiments regarding nitrogen removal.

18.2   Methods

18.2.1   Indoor System

In the technical laboratory of the Institute of Sanitary Engineering and Water Pollution Control, indoor experiments have been carried out at pilot-scale constructed wetlands (PSCWs). Each PSCW had a surface area of 1 m^2. The 50 cm main layer consisted of sandy substrate with a grain size of 0.06–4 mm (d_{10} = 0.2 mm; d_{60} = 0.8 mm) for eight PSCWs and 1–4 mm (d_{10} = 1.0 mm; d_{60} = 3.0 mm) for two PSCWs, respectively. The grain size distribution curves of the two sandy filter materials used are shown in Fig. 18.1. The PSCWs have been planted with *Miscanthus gigantea* and have been loaded intermittently with mechanically pretreated municipal wastewater. The organic load applied was 20 and 80 g COD m^{-2} day^{-1} for the grain size of 0.06–4 mm and 1–4 mm, respectively, corresponding to a specific surface area requirement of 4 and 1 m^2 per person equivalent, respectively.

For enhanced nitrogen removal, a two-stage system was tested (Fig. 18.2). The first stage consisting of a PSCW with a grain size of 1–4 mm for the main layer and a drainage layer operated under saturated conditions (water level 20 cm). The second

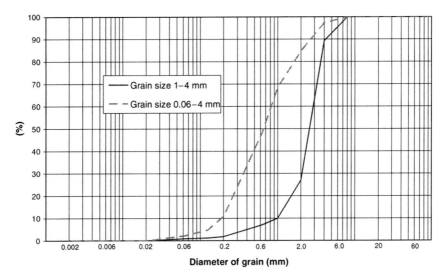

Fig. 18.1 Grain size distribution of the two sandy main layers used

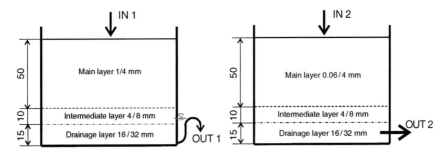

Fig. 18.2 Cross-section of the first (left) and second (right) stage of the two-stage system

stage was a PSCW with a grain size of 0.06–4 mm and free drainage. The system was operated with an organic load of 80 g COD m^{-2} day^{-1} for the first stage, i.e. 1 m^2 per person. The results of the two-stage system are compared with a "standard" system. The standard system is a one-stage PSCW with a grain size of 0.06–4 mm for the main layer, free drainage, and an organic load of 20 g COD m^{-2} day^{-1}, i.e. 4 m^2 per person. Table 18.1 compares the basic design figures of the standard and the two-stage system.

Influent and effluent samples have been taken monthly for analysis of organic matter (BOD$_5$, COD, TOC), total suspended solids (TSS), and nitrogen compounds (NH$_4$-N, NO$_2$-N, NO$_3$-N, TN), according to standard methods in the laboratory of the Institute. Furthermore, microbial parameters such as heterotrophic plate counts (HPC), *Escherichia coli*, total coliforms and enterococci have been determined.

Table 18.1 Comparison of the basic design figures of the standard and the two-stage system

System	Standard system (single-stage)	Two-stage system	
		First stage	Second stage
Main layer (50 cm)	0.06–4 mm	1–4 mm	0.06–4 mm
Organic load	20 g COD m^{-2} day^{-1}	80 g COD m^{-2} day^{-1} (first stage) 40 g COD m^{-2} day^{-1} (total)	–
Specific surface area demand	4 m^2 per person	1 m^2 per person (first stage) 2 m^2 per person (total)	–
Drainage layer	Free drainage	Saturated	Free drainage

18.2.2 Outdoor System

The experimental site for the outdoor experiments is located at the municipal wastewater treatment plant in Ernsthofen (Lower Austria), located at about 150 km west of Vienna. A three-chamber septic tank is used for mechanical pretreatment. The experimental plant consists of three SSVFCWs with intermittent loading operated in parallel with a surface area of 20 m^2 each. One bed has been converted to a two-stage system as described earlier. The filter beds are planted with common reed (*Phragmites australis*).

Samples have been taken weekly and analysed for BOD_5, COD, NH_4-N, NO_2-N, NO_3-N in the laboratory of the wastewater treatment plant using Dr. Lange cuvette tests (HACH-LANGE, Germany). Organic N (Norg) was calculated using the Norg to COD ratio from reference samples analysed in the laboratory of the Institute. The Norg to COD ratio has been shown very stable over the investigation period and its median values were 0.025, 0.034, and 0.078 for the influent, the effluent of stage one and the effluent of stage two, respectively. Furthermore, the microbial parameters HPC, *E. coli*, total coliforms and enterococci have been analysed from the reference samples in the laboratory of the Institute.

18.2.3 Data Evaluation

'STATGRAPHICS Plus for Windows 4.0' was used for drawing 'Box-and-Whisker Plots'. The box encloses the middle 50% of the data, and the median value is drawn as a horizontal line inside the box; the mean value is given as a cross. Vertical lines (whiskers) extend from each end of the box. The lower/upper whisker is drawn from the lower/upper quartile to the smallest/largest value within 1.5 interquartile ranges. The dots represent outliers. The length of the notch around the median value represents a 95% confidence interval for the median. If the notches of 2 median values from different data groups do not overlap, there is a statistically significant difference amongst the median values at the 95% confidence level.

18.3   Results and Discussion

18.3.1   Indoor System

Table 18.2 shows the influent and effluent concentrations for COD, the organic load applied and the resulting specific surface area. Although the concentrations varied the median values are close to the target loads. The organic load for the second stage of the two-stage system is far below $20\,g$ COD m^{-2} day^{-1} and therefore no overload with organic matter and no clogging problems can be expected (Winter & Goetz, 2003; Fehr *et al.*, 2003). Table 18.3 gives the TSS and TOC influent and effluent concentrations and TOC elimination rates. There is no statistically significant difference in the results for the effluent concentrations of the standard and the two-stage system.

Table 18.2 COD influent and effluent concentrations, organic load and specific surface area

Parameter Unit	COD mg l^{-1}				Organic load g COD m^{-2} day^{-1}			Specific surface area m^2 PE^{-1}	
Location	0	1	2	3	0.06–4 mm	1–4 mm	Second stage	0.06–4 mm	1–4 mm
No. of analysis	15	11/11[a]	4	6/6[a]	15	15	4	15	15
Median value	387	<20	30	<20	19.3	77.1	3.7	3.85	0.96
Mean value	442	<20	28	<20	18.9	75.4	4.6	3.77	0.94
95% C.I.	77	–	4	–	3.1	12.3	1.9	0.62	0.15
Standard dev.	151	–	4	–	6.1	24.3	1.9	1.22	0.30
Maximum	814	<20	31	<20	32.8	131.3	7.4	6.56	1.64
Minimum	253	<20	22	<20	10.5	42.0	3.5	2.10	0.53

0: influent; 1: effluent of the standard system; 2: effluent of the first stage of the two-stage system; 3: effluent of the second stage of the two-stage system, [a]Number of analysis below the limit of detection (20 mg COD l^{-1})

Table 18.3 TSS and TOC influent and effluent concentrations (mg l^{-1}) and TOC elimination rates

Parameter	TSS concentration				TOC concentration				TOC elimination rate		
Location	0	1	2	3	0	1	2	3	1	2	3
No. of analysis	7	11/4[a]	5	6/1[a]	7	11	9	6	11	9	6
Median value	123	2	5	2	151	4.3	9.4	4.8	96.8%	94.5%	96.7%
Mean value	133	2	8	2	163	4.7	9.2	5.5	96.9%	94.3%	96.4%
95% C.I.	33	<1	6	<1	22	0.4	0.7	1.1	0.3%	0.7%	1.1%
Standard dev.	44	<1	6	<1	29	0.8	1.1	1.4	0.5%	1.1%	1.4%
Maximum	223	3	19	3	204	5.7	11.0	8.1	97.6%	95.9%	97.7%
Minimum	95	<1	4	<1	134	3.8	8.0	4.6	96.1%	92.5%	94.0%

0: influent; 1: effluent of the standard system; 2: effluent of the first stage of the two-stage system; 3: effluent of the second stage of the two-stage system; [a]Number of analysis below the limit of detection (1 mg TSS l^{-1})

Table 18.4 shows the NH_4-N influent and effluent concentrations and elimination rates. Full nitrification can be reached with both systems. There is statistically significant difference for the effluent concentrations of the standard and the two-stage system. The NO_2-N, NO_3-N and Norg influent and effluent concentrations are given in Table 18.5 and the total nitrogen influent and effluent concentrations and elimination rates in Table 18.6. Compared to the standard single-stage system, the

Table 18.4 NH_4-N influent and effluent concentrations (mg l^{-1}) and elimination rates

Parameter	NH_4-N concentration				NH_4-N elimination rate		
Location	0	1	2	3	1	2	3
No. of analysis	7	11/9[a]	9	6/5[a]	11	9	6
Median value	49.1	0.030	8.0	0.03	99.94%	82.9%	99.94%
Mean value	48.1	0.032	11.0	0.10	99.94%	77.2%	99.78%
95% C.I.	2.7	0.002	5.3	0.13	0.01%	10.6%	0.31%
Standard dev.	3.7	0.004	8.1	0.16	0.02%	16.3%	0.38%
Maximum	53.8	0.040	25.3	0.42	99.98%	95.4%	99.94%
Minimum	41.9	<0.030	2.5	<0.03	99.92%	48.6%	99.00%

0: influent; 1: effluent of the standard system; 2: effluent of the first stage of the two-stage system; 3: effluent of the second stage of the two-stage system, [a]Number of analysis below the limit of detection (0.03 mg NH_4-N l^{-1})

Table 18.5 NO_2-N, NO_3-N and Norg influent and effluent concentrations (mg l^{-1})

Parameter	NO_2-N concentration				NO_3-N concentration				Norg concentration			
Location	0	1	2	3	0	1	2	3	0	1	2	3
No. of analysis	7/3[a]	11/5[a]	5/1[a]	6/4[a]	7/7[a]	11	5	6	7	11/8[a]	5/1[a]	6/4[a]
Median value	0.007	0.003	0.026	0.003	<0.1	52.3	4	26.1	9.5	1.0	1.2	1.0
Mean value	0.006	0.005	0.035	0.019	<0.1	51.7	18	30.5	10.3	1.1	1.4	1.0
95% C.I.	0.002	0.002	0.033	0.032	–	5.4	20	11.9	3.3	0.1	0.5	0.1
Standard dev.	0.003	0.004	0.038	0.040	–	9.2	23	14.9	4.4	0.2	0.6	0.1
Maximum	0.010	0.016	0.100	0.100	<0.1	67.2	49	55.5	18.2	1.7	2.5	1.2
Minimum	0.003	<0.003	<0.003	<0.003	<0.1	38.9	0.3	15.9	3.8	<1.0	<1.0	<1.0

0: influent; 1: effluent of the standard system; 2: effluent of the first stage of the two-stage system; 3: effluent of the second stage of the two-stage system, [a]Number of analysis below the limit of detection (0.003 mg NO_2-N l^{-1}, 0.1 mg NO_3-N l^{-1}, and 1 mg Norg l^{-1})

Table 18.6 TN influent and effluent concentrations (mg l^{-1}) and elimination rates

Parameter	TN concentration				TN elimination rate		
Location	0	1	2	3	1	2	3
No. of analysis	7	11	5	6	11	5	6
Median value	56.3	52.8	29.0	26.9	2%	57%	58%
Mean value	58.5	52.5	32.5	31.9	8%	46%	46%
95% C.I.	4.0	5.3	13.6	12.5	8%	22%	21%
Standard dev.	5.4	9.0	15.5	15.6	13%	25%	27%
Maximum	67.5	67.7	52.9	56.0	30%	67%	69%
Minimum	53.2	39.4	17.5	16.4	–9%	15%	10%

0: influent; 1: effluent of the standard system; 2: effluent of the first stage of the two-stage system; 3: effluent of the second stage of the two-stage system

two-stage system as described above showed a significantly higher nitrogen removal of 58% (only 2% for the standard single-stage system).

Nitrogen elimination in the two-stage CW system could be reached due the following facts:

1. Due to about 80% nitrification in the first stage the availability of NO_3-N in the saturated drainage layer is ensured.
2. By using a grain size of 1–4 mm for the main layer of the first stage, there is no complete mineralization of the organic matter, but there is enough organic matter left for denitrification in the saturated drainage layer.
3. The saturated drainage layer increases the hydraulic retention time and therefore the contact time between the denitrifying micro-organisms and the medium in the first stage.

The second stage of the two-stage CW system is mainly to guarantee full nitrification for the final effluent.

Figure 18.3 compares the TOC effluent and TN influent and effluent concentrations. No statistically significant difference can be found between the TOC effluent concentrations of the standard and two-stage system, respectively. However, the TN effluent concentrations of the two-stage CW system are significantly lower compared to the standard CW system.

Table 18.7 compares the influent, effluent and removed loads in the investigated systems. Although the same effluent concentrations could be reached, the removed loads are about twofold higher for the two-stage system compared to the standard system for TOC and NH_4-N. The removed nitrogen load in the two-stage system was very high and about 4 g N m^{-2} day^{-1}, i.e. 1,460 g N m^{-2} year^{-1}.

Table 18.8 and Fig. 18.4 show the influent and effluent concentrations and removal rates for HPC and total coliforms. The effluent concentrations and removal rates for the standard system and the two-stage system are the same although the loading of the two-stage system is twice, i.e. the retention time is only half. The same results could be found for *E. coli* and enterococci.

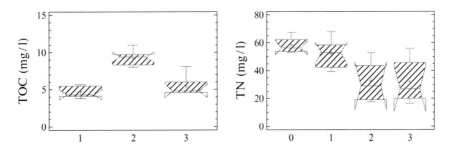

Fig. 18.3 TOC effluent and TN influent and effluent concentrations (0: influent; 1: effluent of the standard system; 2: effluent of the first stage of the two-stage system; 3: effluent of the second stage of the two-stage system)

Table 18.7 Comparison of loads in g m^{-2} day^{-1} in the two investigated systems

System	Standard system (4 m^2 per person)			Two-stage system (2 m^2 per person)		
Parameter	TOC	NH$_4$-N	TN	TOC	NH$_4$-N	TN
Influent load	8.00	2.35	2.86	16.32	5.028	6.38
Effluent load	0.20	0.0014	2.22	0.38	0.0022	2.36
Removed load	7.80	2.35	0.64	15.94	5.026	4.02

Table 18.8 Influent and effluent concentrations and removal rates for HPC and total coliforms

	HPC per ml							Total coliforms per 100 ml						
	Log concentration			Log removal rate			Log concentration				Log removal rate		
Parameter	0	1	2	3	1	2	3	0	1	2	3	1	2	3
Location	0	1	2	3	1	2	3	0	1	2	3	1	2	3
No. of analysis	8	14	7	5	14	7	5	8	14	7	5	14	7	5
Median value	6.32	3.39	5.25	3.17	2.79	0.94	3.00	7.16	2.69	5.83	2.63	4.46	1.44	5.25
Mean value	6.21	3.43	5.19	3.32	2.75	1.07	2.96	7.25	2.67	5.91	3.09	4.56	1.42	4.26
95% C.I.	0.18	0.19	0.40	0.41	0.20	0.37	0.46	0.25	0.45	0.42	0.95	0.52	0.27	1.21
Standard dev.	0.26	0.36	0.54	0.47	0.39	0.50	0.52	0.36	0.86	0.57	1.08	0.99	0.36	1.38
Maximum	6.50	4.16	6.17	4.09	3.34	1.73	3.54	7.88	4.14	6.80	4.48	6.40	1.88	5.31
Minimum	5.76	2.92	4.60	2.96	1.78	0.33	2.29	6.76	1.41	5.26	1.95	3.12	0.88	2.71

0: influent; 1: effluent of the standard system; 2: effluent of the first stage of the two-stage system; 3: effluent of the second stage of the two-stage system

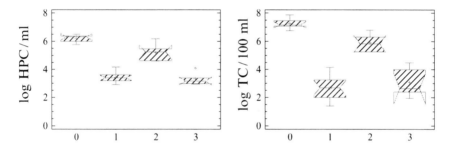

Fig. 18.4 Log HPC and total coliforms influent and effluent concentrations (0: influent; 1: effluent of the standard system; 2: effluent of the first stage of the two-stage system; 3: effluent of the second stage of the two-stage system)

18.3.2 Outdoor System

The construction work at the outdoor system took place in April 2005. In the start-up period from May until August 2005 the beds had been operated with lower organic load. From September 2005 the system was operated with the target organic load of 40 g COD m^{-2} day^{-1} for the whole system (i.e. 2 m^2 per person). Data presented here are from the period September to November 2005.

The first results of the outdoor system showed similar results as for the pilot-scale system. Table 18.9 shows the influent and effluent concentrations for BOD$_5$

and COD. The organic load and the specific surface area for the first stage have been close to the target values. The organic load for the second stage was much lower than 20 g COD m^{-2} day^{-1}. Therefore, according to Winter and Goetz (2003) and Fehr et al. (2003) no clogging problems shall be expected for the second stage. The effluent concentrations for BOD$_5$ and COD have been low.

Table 18.10 shows the measured influent and effluent concentrations for NH$_4$-N and NO$_3$-N as well as the calculated total nitrogen concentrations. The NH$_4$-N effluent concentration was below 1 mg l^{-1}, while the TN effluent concentration was around 30 mg l^{-1}. The elimination rates for BOD$_5$, COD and NH$_4$-N have been very high (Table 18.11). A stable elimination rate of 63% for total nitrogen could be reached

Table 18.9 BOD$_5$ and COD influent and effluent concentrations, organic load and specific surface area

Parameter				COD mg l^{-1}			Organic load g COD m^{-2} day^{-1}		Specific surface area m^2 PE^{-1}
		BOD$_5$ mg l^{-1}							
Location	In	Out 1	Out 2	In	Out 1	Out 2	Stage 1	Stage 2	Two-stage system
No. of analysis	11	11	11	11	11	11	11	11	11
Median value	359	48	5	578	98	20	82.4	13.5	2.06
Mean value	360	50	4	586	96	21	81.3	13.8	2.03
95% C.I.	30	11	1.6	39	23	2	5.5	3.2	0.14
Standard dev.	18	7	1.0	23	13	1	3.2	1.9	0.08
Maximum	400	70	9	627	163	23	88.2	22.9	2.20
Minimum	300	30	3	487	81	17	68.5	11.4	1.71

Table 18.10 Influent and effluent concentrations (mg l^{-1}) for nitrogen compounds

Parameter	NH$_4$-N			NO$_3$-N			TN		
Location	In	Out 1	Out 2	In	Out 1	Out 2	In	Out 1	Out 2
No. of analysis	11	11	11	11	11	11	11	11	11
Median value	65.8	21.6	0.64	0.43	5.50	27.0	80.2	31.1	29.4
Mean value	64.8	20.7	0.63	0.44	5.20	26.9	79.6	31.7	29.6
95% C.I.	6.3	2.7	0.28	0.05	2.59	2.8	7.0	2.9	2.8
Standard dev.	3.7	1.6	0.16	0.03	1.53	1.7	4.1	1.7	1.7
Maximum	73.5	27.4	0.96	0.51	9.90	30.5	89.0	35.7	32.9
Minimum	54.8	17.7	0.16	0.35	2.56	23.1	68.1	26.8	25.2

Table 18.11 Elimination rates for chemical parameters

Parameter	BOD$_5$		COD		NH$_4$-N		TN	
Location	Out 1	Out 2	Out 1	Out 2	Out 1	Out 2	Out 1	Out 2
No. of analysis	11	11	11	11	11	11	11	11
Median value (%)	86.7	98.6	83.0	96.5	66.9	99.0	61.1	63.3
Mean value (%)	86.1	98.7	83.8	96.5	65.1	99.0	61.8	63.3
95% C.I. (%)	2.6	0.4	3.7	0.2	5.3	0.47	2.9	2.0
Standard dev. (%)	1.5	0.2	2.2	0.1	3.1	0.28	1.7	1.2
Maximum (%)	91.1	99.2	86.3	96.8	75.9	99.8	65.3	66.4
Minimum (%)	82.5	97.8	72.5	96.0	57.6	98.2	55.0	59.9

Table 18.12 Log influent and effluent concentrations and removal rates for HPC and total coliforms

	HPC per ml					Total coliforms per 100 ml				
Parameter	Log concentrations			Log removal		Log concentrations			Log removal	
Location	In	Out 1	Out 2	Out 1	Out 2	In	Out 1	Out 2	Out 1	Out 2
No. of analysis	5	4	4	4	4	5	4	4	4	4
Median value	6.40	5.55	3.41	0.77	2.88	6.91	6.39	3.75	0.57	3.13
Mean value	6.34	5.54	3.41	0.76	2.89	6.83	6.30	3.72	0.56	3.15
95% C.I.	0.13	0.33	0.16	0.25	0.05	0.15	0.25	0.20	0.08	0.06
Standard dev.	0.15	0.34	0.16	0.26	0.05	0.18	0.26	0.21	0.09	0.07
Maximum	6.48	5.92	3.58	0.99	2.96	7.00	6.49	3.91	0.65	3.24
Minimum	6.10	5.14	3.23	0.49	2.83	6.59	5.94	3.45	0.46	3.09

with the two-stage outdoor system. Table 18.12 show the log influent and effluent concentrations and removal rates for HPC and total coliforms. A log removal rate of about 3 could be reached for all microbial parameters.

18.4   Conclusions

A two-stage CW system (first stage: grain size for the main layer 1–4 mm, saturated drainage layer; and second stage: grain size for the main layer 0.06–4 mm, free drainage) was operated with an organic load of 80 g COD m^{-2}day^{-1} for the first stage (1 m^2 per person), i.e. 40 g COD m^{-2} day^{-1} for the whole system (2 m^2 per person). Organic matter and NH$_4$-N removal are the same as for a single-stage standard CW system (grain size for main layer 0.06–4 mm, free drainage, organic load 20 g COD m^{-2} day^{-1}). A nitrogen elimination of 58% could be reached for the two-stage system compared to almost no nitrogen elimination for a single-stage system.

The results of the indoor experiments are currently verified outdoor mainly to secure safe operation during winter. The first result of the outdoor experiments are similar compared to the indoor experiments regarding organic matter, nitrogen, and microbiological parameters. The nitrogen elimination reached in the first 3 months of operation (September to November 2005) was 63%. However, the first promising results still have to be verified during operation of the system under winter conditions.

By using the two-stage CW system it is possible to obtain about 60% nitrogen elimination and a high nitrogen removal rate of more than 1,400 g N m^{-2}year^{-1} without recirculation. In conclusion, it can be said that a higher effluent quality can be reached with the two-stage system compared to the single-stage standard CW system. The higher effluent quality can be reached although the two-stage CW system is operated with twofold organic load resulting in half of the specific surface area requirement.

Acknowledgements The experiments are carried out within the project '*Optimization of subsurface vertical flow constructed wetlands (Bepflanzte Bodenfilter)*', funded by the Austrian Ministry for Agriculture, Forestry, Environment and Water Management. The authors are grateful for support and especially thank the mayor of the municipality Ernsthofen, Mr. Karl Huber, for making the construction of the experimental plant at the WWTP Ernsthofen possible, as well as the staff of the WWTP Ernsthofen, Mr. Franz Eglseer and Mr. Karl Hiebl, for operating the experimental plant.

References

1. AEVkA (1996): *1. Abwasseremissionsverordnung für kommunales Abwasser* (*Austrian regulation for emissions from domestic wastewater*). BGBl.210/1996, Vienna, Austria [*in German*].

Fehr, G., Geller, G., Goetz, D., Hagendorf, U., Kunst, S., Rustige, H., & Welker, B. (2003). *Bewachsene Bodenfilter als Verfahren der Biotechnologie (Endbericht des DBU-Verbundprojektes AZ 14178-01)*. Texte Nr.05/03, Umweltbundesamt, Berlin, Deutschland, ISSN 0722-186X [*in German*].

Haberl, R., Grego, S., Langergraber, G., Kadlec, R.H., Cicalini, A.R., Martins Dias, S., Novais, J.M., Aubert, S., Gerth, A., Hartmut, T., & Hebner, A. (2003). Constructed wetlands for the treatment of organic pollutants. *–Journal of Soils & Sediments, 3*, 109–124.

Kadlec, R.H., Knight, R.L., Vymazal, J., Brix, H., Cooper, P., & Haberl, R. (2000). Constructed wetlands for pollution control – processes, performance, design and operation. *IWA Scientific and Technical Report No.8*. London, UK: IWA Publishing.

Langergraber, G., & Haberl, R. (2001). Constructed wetlands for water treatment. *Minerva Biotecnologica, 13*, 123–134.

Langergraber, G., Prandtstetten, C., Pressl, A., Rohrhofer, R., & Haberl, R. (2007). Optimization of subsurface vertical flow constructed wetlands for wastewater treatment. *Water Science & Technology, 55*(7), 71–78.

Winter, K.-J., & Goetz, D. (2003): The impact of sewage composition on the soil clogging phenomena of vertical flow constructed wetlands. *Water Science & Technology, 48*(5), 9–14.

Chapter 19
Removal of Heavy Metals from Industrial Effluents by the Submerged Aquatic Plant *Myriophyllum spicatum* L.

Els Lesage[1], Charity Mundia[1], Diederik P.L. Rousseau[2], Anelies M.K. van de Moortel[1], Gijs du Laing[1], Filip M.G.Tack[1] (✉), Niels De Pauw[3], and Marc G. Verloo[1]

Abstract The potential use of *Myriophyllum spicatum* L. for the removal of Co, Ni, Cu and Zn from industrial effluents was studied. The removal kinetics, tolerance and accumulation capacity of the submerged aquatic plant were assessed. Removal of Cu and Zn was similar and occurred rapidly with time whereas removal of Co and Ni was slower. Plant growth was not adversely affected during the 12 weeks of exposure to the wastewater. Cobalt, Ni, Cu and Zn concentrations of respectively 1,675, 1,529, 766 and 2,883 mg kg^{-1} DM were observed in the biomass. *M. spicatum* is suggested as an efficient plant species for the treatment of metal-contaminated industrial wastewater.

Keywords Cobalt, constructed wetlands, Eurasian water milfoil, plant uptake, sorption

19.1 Introduction

The release of heavy metals into the environment by industrial activities presents a serious environmental threat. Heavy metals can be removed from industrial wastewater by a range of physico-chemical remediation technologies such as precipitation, ion exchange, adsorption, electrochemical processes and membrane processes

[1] Laboratory of Analytical Chemistry and Applied Ecochemistry, Ghent University, Coupure Links 653, 9000 Ghent, Belgium

[2] Department of Environmental Resources, UNESCO-IHE, P.O.Box 3015, 2601 DA Delft, The Netherlands

[3] Laboratory of Environmental Toxicology and Aquatic Ecology, Ghent University, J. Plateaustraat 22, 9000 Ghent, Belgium

(✉) Corresponding author: e-mail: Filip.Tack@UGent.be

J. Vymazal (ed.) *Wastewater Treatment, Plant Dynamics and Management in Constructed and Natural Wetlands*,
© Springer Science + Business Media B.V. 2008

(Kurniawan *et al.*, 2006). However, regulatory standards are not always met. Moreover, these technologies are expensive and energy-intensive, driving towards a search of cheaper alternatives or tertiary treatment steps in both developing and developed countries (Kivaisi, 2001). Qian *et al.* (1999) stated that constructed wetlands (CWs) are highly efficient in removing low levels of metals from large volumes of wastewater.

The use of aquatic plants for the removal of heavy metals from wastewater has been investigated by many authors. Floating plants that offer promising results are *Lemna minor* (Zayed *et al.*, 1998), *Eichhornia crassipes* (Zhu *et al.*, 1999), *Pistia stratiotes* and *Salvinia herzogii* (Maine *et al.*, 2001, 2004). As submerged plants are completely inundated and have the ability to take up metals directly from the water, they are suggested as useful species in reducing metal concentrations from wastewaters. However, there is scarce study about heavy metal uptake by submerged aquatic plants (Guilizzoni, 1991; Keskinkan, 2005). The use of submerged macrophytes for wastewater treatment is therefore still at an experimental stage, with species like *Potamogeton natans* (Fritioff & Greger, 2006) and *M. spicatum* (Keskinkan, 2005) being tested.

M. spicatum L. (Eurasian water milfoil, Haloragaceae) is a rooted, perennial plant which reproduces primarily by vegetative fragmentation. The plant is tolerant to a wide range of water quality conditions and can be found in fresh and brackish water (APIRS, n.d.). The uptake of Cd, Cu, Fe, Hg and Zn by *M. spicatum* has been investigated by Kamal *et al.* (2004), Sivaci *et al.* (2004) and Keskinkan (2005). However, to our knowledge, the combined study of the removal kinetics and accumulation capacity of Co and Ni by *M. spicatum* L. has not been reported in literature. The objectives of this study were (1) to assess the tolerance of *M. spicatum* L. for industrial effluents contaminated with Co, Cu, Ni and Zn and (2) to assess the removal kinetics and uptake levels in the biomass of the studied metals.

19.2   Material and Methods

M. spicatum L. was collected from a local plant nursery and grown for 1 month in a 25% Hoagland nutrient solution (Qian *et al.*, 1999). After the adaptation period, four replicates of about 10 g fresh biomass were analysed for dry matter (DM) and metal contents. Metals were analysed by means of inductively coupled plasma optical emission spectrometry (ICP-OES) (Varian Vista MPX, Varian, Palo Alto, CA) after digestion of the dry plant material with 65% HNO_3 and 20% H_2O_2 (Du Laing *et al.*, 2003). About 70 g fresh biomass was planted into each of four holding tanks (40 cm height, 20 cm diameter) with a 5 cm layer of gravel at the bottom as a rooting medium. Two unplanted tanks were set up as controls.

Two of the planted and the unplanted microcosms were exposed to 9 l synthetic industrial wastewater. A new batch of synthetic wastewater was added every 2 weeks, covering a total experimental period of 12 weeks. The following composition of synthetic wastewater was used (in mg l⁻¹): 1 Co, 1 Ni, 0.5 Cu, 1 Zn, 0.01 Fe, 0.01 Mn, 70 Ca, 15 K, 20 Mg, 1,500 Na, 2,000 SO_4^{2-}, 1,000 Cl^-, 100 HCO_3^-, 1 PO_4^{3-}, 10 NO_3^- and 7 NH_4^+. The synthetic wastewater is representative for an industrial effluent from a metal-processing industry in Flanders, Belgium. The two other planted microcosms were exposed to a 25% Hoagland nutrient solution throughout the experiment to study tolerance properties. Fresh nutrient solution was made from a 20-times concentrated solution that was diluted with tap water and was added every 2 weeks, as for the microcosms exposed to wastewater.

Samples of 10 ml wastewater were collected by syringes after 6, 24, 48, 72, 96, 144, 240 and 336 h of addition of the synthetic wastewater during the first and the last batch, to study the removal kinetics. The initial wastewater added and the samples collected after different times, were analysed for SO_4^{2-} and Cl^- by ion chromatography (IC) (Metrohm 764 Compact IC, Herisau, Switzerland) and total levels of Co, Ni, Cu and Zn by means of ICP-OES. The wastewater was analysed for pH (electrode, HI 1230B, Temse, Belgium) and electrical conductivity (EC) (electrode, WTW LF 537, Weilheim, Germany) at the start and end of each batch.

Plants were exposed to a 14 h light/10 h dark regime. The ambient air temperature varied between 10°C and 19°C during the first batch that started in November 2005. During the last batch starting in February 2006, temperature varied between 9°C and 23°C. Values of evapotranspiration were corrected by daily addition of deionised water. Daily evapotranspiration had a constant value of 1.5 mm day⁻¹ in all microcosms, indicating homogeneous ambient conditions throughout the experiment.

After 12 weeks, *M. spicatum* was harvested from the microcosms and total fresh biomass was determined. The biomass was divided into two parts. One part was rinsed three times with deionised water for analysis of the total metal content. The other part was rinsed three times with 0.1 M HCl to desorb metals from the surface of the biomass. This was based on the cleaning of *M. spicatum* with a 3% HCl solution (0.8 M) as a preparation step in the sorption experiment of Sivaci *et al.* (2004). Plants were dried at 50°C until constant dry weight and analysed for metals by means of ICP-OES after digestion (Du Laing *et al.*, 2003).

A mass balance was performed in order to assess removal pathways of the metals. The masses of metals removed from the wastewater in the unplanted ($M_{unplanted}$) and planted microcosms ($M_{planted}$) were calculated according to the following formula:

$$M_{(unplanted)} = (C_0 - C_{336}) \times V \times N \qquad (19.1)$$

With C_0 the mean initial metal concentration in the wastewater of both batches (mg l⁻¹), C_{336} the mean metal concentration in the wastewater after 336 h of both batches (mg l⁻¹), V the volume of wastewater (9 l) and N the number of batches (6).

The mass of metals removed by the *Myriophyllum* biomass ($M_{Myriophyllum}$) was calculated as:

$$M_{Myriophyllum} = (C_e \times DM_e) - (C_i \times DM_i) \qquad (19.2)$$

with C_i and DM_i the initial metal concentration (mg kg^{-1} DM) in the *Myriophyllum* biomass and the biomass (kg DM), and C_e and DM_e the metal concentration (mg kg^{-1} DM) in the *Myriophyllum* biomass and the biomass (kg DM) after 12 weeks.

Statistical analysis was performed using the S-plus 6.1 software package (Insightful Corp., Seattle, WA). The significance of differences between groups was assessed by means of a Wilcoxon rank sum test for comparison of two groups and a Kruskal–Wallis test for comparison of more than two groups ($\alpha = 0.05$).

19.3   Results

19.3.1   Removal Kinetics of Cobalt, Nickel, Copper and Zinc

A significant decrease of the metal concentrations in the wastewater with time was observed in all microcosms in both batches ($p < 0.001$) (Fig 19.1). The removal of Co and Ni was slower than for Cu and Zn, for which a very rapid decrease was observed. During the first batch, Co and Ni levels decreased to 0.2 mg l^{-1} after 2 weeks, representing a decrease of about 80% of the initial levels. Removal of Co

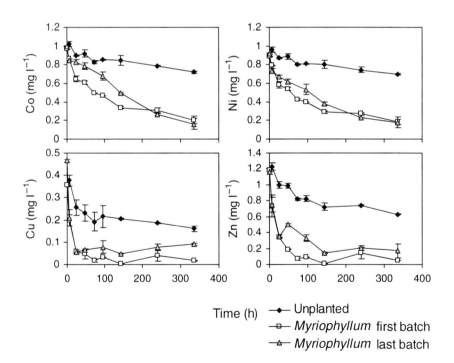

Fig. 19.1 Co, Ni, Cu and Zn concentrations in the wastewater (mg l^{-1}) as a function of time (h) of the unplanted controls and microcosms planted with *Myriophyllum spicatum* during the first and last batch

and Ni was slower in the last batch than in the first batch. Cobalt and Ni levels decreased to values similar to those observed at the end of the first batch.

Contrary to the removal of Co and Ni, removal of Cu and Zn occurred much more rapidly. After 1 day, Cu and Zn concentrations were 0.057 ± 0.009 mg l^{-1} and 0.33 ± 0.00 mg l^{-1}, respectively, representing 16% and 28% of the initial levels. From day 2 onward, Cu and Zn levels more or less reached equilibrium and varied between <0.002 and 0.05 mg l^{-1} for Cu and between 0.007 and 0.19 mg l^{-1} for Zn. A similar rapid removal was observed during the last batch as well, although the levels at equilibrium were higher than during the first batch. Cu levels varied between 0.05 and 0.09 mg l^{-1} from day 2 on, whereas Zn levels reached equilibrium after 144 h and varied between 0.14 and 0.21 mg l^{-1} (Fig. 19.1).

The pH of the wastewater during the first batch decreased only slightly from 8.0 ± 0.0 to 7.3 ± 0.1 in microcosms planted with *M. spicatum* and to 7.5 ± 0.1 in the unplanted controls ($p < 0.01$). However, an important increase to pH 10 ± 0 was observed in the planted microcosms during the last batch ($p = 0.0001$). The EC of the wastewater increased from 7.7 ± 0.0 to 9.0 ± 0.2 mS cm^{-1} in the microcosms planted with *M. spicatum* and to 8.3 ± 0.0 mS cm^{-1} in the unplanted controls during the first batch ($p < 0.01$). At the end of the last batch an EC of 7.4 ± 0.1 mS cm^{-1} was observed in the planted microcosms, which was similar to the initial EC ($p = 0.18$). Chloride and SO_4^{2-} concentrations in the wastewater varied around 1,200 and 2,200 mg l^{-1}, respectively with deviations less than 5%.

19.3.2 Growth of Myriophyllum spicatum

The initial dry mass to fresh mass ratio (DM:FM) was 0.10. The absolute growth of 26 ± 7 g on a fresh biomass basis in the synthetic wastewater was similar to that in the 25% Hoagland nutrient solution ($p > 0.05$). About 40% fresh biomass was formed in both treatments. However, a higher DM:FM of 0.12 in plants harvested from the wastewater than a ratio of 0.095 in those collected from the nutrient solution, was responsible for a higher growth on a dry mass basis in the wastewater. Whereas 30% ± 1% new dry biomass was formed in the nutrient solution (2.2 ± 0.1 g), 67% ± 16% dry mass was formed in the wastewater (4.3 ± 0.8 g). Mean growth rates of 0.05 ± 0.01 g DM day^{-1} and 0.31 ± 0.08 g FM day^{-1} in the wastewater were derived under the assumption of linear growth during the experimental period.

19.3.3 Accumulation of Cobalt, Nickel, Copper and Zinc
         by Myriophyllum spicatum

Table 19.1 presents the initial total metal levels and the total and sorbed metal levels at the end of the experiment. The sorbed metal fraction was considered as the fraction that was rinsed off the *Myriophyllum* surface by 0.1 M HCl. Total Co, Ni,

Cu and Zn levels in *M. spicatum* after 12 weeks of exposure to the wastewater were respectively 3,000, 330, 510 and 5 times higher than initial levels ($p < 0.001$) (Table 19.1). These levels were elevated compared to those in plants grown in the 25% Hoagland nutrient solution ($p < 0.001$).

The microcosms that were exposed to the 25% Hoagland nutrient solution were included in the experimental set-up for tolerance assessment. However, significant Cu and Zn accumulation was observed in plants grown in the nutrient solution as well (Table 19.1). Dilution of the concentrated 25% Hoagland solution was performed with tap water. In order to assess the metal levels in tap water, weekly samples of tap water were collected during a period of 6 weeks. The Cu and Zn concentrations in tap water were in the range of, respectively, 0.04–0.14 and 0.52–1.10 mg l^{-1}. Cobalt concentrations in tap water were lower than 0.002 mg l^{-1}, whereas Ni concentrations varied between <0.002 and 0.24 mg l^{-1}.

Moreover, 74–81% of the total metal concentration in the *Myriophyllum* biomass grown in the wastewater was desorbed by 0.1 M HCl. For plants grown in the nutrient solution, the major part of the total metal level in the biomass was also sorbed on the surface, except for Ni (Table 19.1).

In order to assess the removal pathways of the metals, a mass balance was performed (Table 19.2). The difference between the masses of Cu and Zn removed from the water in the planted ($M_{planted}$) and unplanted microcosms ($M_{unplanted}$) was

Table 19.1 Total initial metal concentrations and total and sorbed metal concentrations in *Myriophyllum spicatum* grown in the wastewater and in the 25% Hoagland nutrient solution at the end of the experiment (mg kg^{-1} DM) together with the percentage of the total metal concentration that was sorbed

	Co (mg kg^{-1} DM)		%	Ni (mg kg^{-1} DM)		%	Cu (mg kg^{-1} DM)		%	Zn (mg kg^{-1} DM)		%
Initial	0.5	± 0.1	0	4.6	± 0.6		1.5	± 0.1		612	± 254	
Wastewater												
Total	1,675	± 195		1,529	± 374		766	± 206		2,883	± 433	
Sorbed	1,243	± 202	74	1,151	± 382	75	564	± 209	74	2,339	± 436	81
25% Hoagland												
Total	12	± 2		12	± 3		111	± 15		1,928	± 340	
Sorbed	8.0	± 2.2	66	2.2	± 3.5	19	69	± 16	62	1,576	± 343	82

Table 19.2 Metal masses removed from the wastewater in the unplanted ($M_{unplanted}$) and planted ($M_{planted}$) microcosms, together with the difference between these masses Δ ($M_{planted} - M_{unplanted}$), and the total metal masses recovered in the *Myriophyllum spicatum* biomass after harvest ($M_{Myriophyllum}$) (mg)

Mass fraction	Co			Ni			Cu			Zn		
$M_{unplanted}$	13	±	1	10	±	1	11	±	1	30	±	1
$M_{planted}$	41	±	2	38	±	1	18	±	0	61	±	1
Δ ($M_{planted} -$ $M_{unplanted}$)	28	±	2	27	±	1	7.7	±	0.9	31	±	2
$M_{Myriophyllum}$	18	±	3	17	±	4	8.3	±	2.4	27	±	5

similar to their masses recovered in the *Myriophyllum* biomass after harvest ($M_{\text{Myriophyllum}}$). However, 23% and 28%, respectively of the Co and Ni masses removed from the wastewater remained unexplained (Table 19.2).

19.4 Discussion

M. spicatum was able to remove Co, Cu, Ni and Zn from the studied industrial wastewater, although differences in the removal kinetics were observed. A relatively long residence time of 2 weeks was necessary to achieve Co and Ni levels of 0.2 mg l^{-1}. On the contrary, a shorter residence time of 48 h was sufficient to achieve 0.05 mg l^{-1} Cu and 0.2 mg l^{-1} Zn. Our findings correspond with those of Keskinkan (2005) who reported rapid removal of Cu, Zn and Cd within the first 1–5 days and a state of equilibrium in systems with *M. spicatum* grown in mixed metal solutions of Cd, Cu and Zn covering a concentration range of 1–16 mg l^{-1}. However, Kamal *et al.* (2004) observed slower removal of Zn and Cu by *M. aquaticum*, probably because of higher initial concentrations of 28 and 5.6 mg l^{-1} respectively.

Non-linear removal kinetics were observed, demonstrating the prevalence of different removal mechanisms (Salt *et al.*, 1995). Sorption onto *M. spicatum* was responsible for the rapid removal of Cu and Zn during the first hours of the experiment. After sorption onto the surface of *M. spicatum*, accumulation and translocation within the biomass were responsible for the slower part of the metal removal. Significant removal of metals was observed in the unplanted microcosms as well, either by sorption onto the recipients and gravel or by precipitation.

Growth of *M. spicatum* was not adversely affected and 67% new dry biomass had developed in the synthetic wastewater. Guilizzoni (1991) stated that concentration limits for toxicity of submerged plants are less commonly found in literature than for terrestrial and emergent aquatic plants and algae. Stanley (1974), cited in Guilizzoni (1991) derived the following levels at which 50% inhibition of root growth of *M. spicatum* occurred (in mg l^{-1}): 2.5 for Al(III), 2.9 for As(III), 7.4 for Cd(II), 0.25 for Cu(II), 0.46 for Cr(VI), 9.9 for Cr(III), 3.4 for Hg(II), 363 for Pb(II) and 21.6 for Zn(II). Copper appeared to be the most toxic metal, although tolerance to Cu levels of 0.5 mg l^{-1} was observed in our experiment. Kamal *et al.* (2004) showed that *M. aquaticum* had the highest tolerance of three aquatic plant species grown in a solution with 28, 5.6, 104 and 0.5 mg l^{-1} of Zn, Cu, Fe and Hg, respectively. *M. spicatum* tolerated a mixed metal solution of 16 mg l^{-1} of Cd, Cu and Zn in the experiment of Keskinkan (2005).

Metal concentrations in aquatic macrophytes are difficult to compare because of differences in wastewater quality, plant species, environmental conditions, analytical procedures, etc. (Guilizzoni, 1991). Moreover, the harvestable parts of *M. spicatum* were analysed for total metal concentrations without distinguishing between leaves, stems and roots. However, keeping this in mind, a comparison of metal concentrations in *M. spicatum* with literature reports is presented in order to obtain a general idea about its accumulation capacity. Submerged aquatic macrophytes generally accumulate higher metal concentrations in their belowground biomass than in their shoots (Guilizzoni, 1991).

Total Co and Ni levels in *M. spicatum* after 12 weeks of exposure to the waste-water were respectively 1,675 and 1,529 mg kg^{-1} DM. These levels largely exceeded Co and Ni levels of respectively 33–57 and 18–19 mg kg^{-1} DM reported by Samecka-Cymerman and Kempers (2004) in shoots of *M. spicatum* collected from a Cu-smelter-impacted surface water with Co and Ni levels of respectively 0.03–0.07 and 0.06–0.07 mg l^{-1}. Qian *et al.* (1999) reported Ni concentrations of 1,077 and 50 mg kg^{-1} DM in respectively roots and shoots of *M. brasiliense* that grew for 10 days in a 25% Hoagland nutrient solution spiked with 1 mg l^{-1} Ni. Zhu *et al.* (1999) reported maximum Ni concentrations of 120 and 1,200 mg kg^{-1} DM in respectively shoots and roots of *E. crassipes* grown for 14 days in a 25% Hoagland nutrient solution spiked with 10 mg l^{-1} Ni. A similar Ni concentration of 1,790 mg kg^{-1} DM was observed in *L. minor* grown for 14 days in a 25% Hoagland nutrient solution spiked with 10 mg l^{-1} Ni (Zayed *et al.*, 1998). However, at a Ni concentration of 1 mg l^{-1} as used in our experiment, accumulation levels in shoots and roots of *E. crassipes* were respectively 20 and 200 mg kg^{-1} DM, being much lower than the reported concentrations for *M. spicatum* in this study. *L. minor* also showed very low accumulation of Ni at a concentration of 1 mg l^{-1}. *M. spicatum* demonstrated to be a better accumulator of Ni than the floating aquatic macrophytes *E. crassipes* and *L. minor*.

The total Cu concentration in *M. spicatum* after 12 weeks of exposure to the wastewater was 766 ± 206 mg kg^{-1} DM. This finding was comparable with levels of 720–1,040 mg kg^{-1} DM reported by Samecka-Cymerman and Kempers (2004) in shoots of *M. spicatum* collected from a surface water with 0.02–0.03 mg l^{-1} Cu. Kamal *et al.* (2004) reported a similar concentration of 549 mg kg^{-1} DM in the shoots of *M. aquaticum* grown for 3 weeks in synthetic wastewater with 5.6 mg l^{-1} Cu. Fritioff and Greger (2006) reported Cu levels of 461 and 357 mg kg^{-1} DM in leaves and stems of the submerged plant *Potamogeton natans* grown in the presence of 0.2 mg l^{-1} Cu. Zhu *et al.* (1999) and Zayed *et al.* (1998) observed much higher Cu concentrations of 2,800 and 3,360 mg kg^{-1} DM in respectively roots of *E. crassipes* and in *L. minor*, grown in 25% Hoagland nutrient solutions spiked with 10 mg l^{-1} Cu. However, at a similar Cu concentration of 1 mg l^{-1}, accumulation levels in roots of *E. crassipes* (600 mg kg^{-1} DM) were in the same range as those reported for *M. spicatum*. *L. minor* showed a much lower Cu accumulation of 300 mg kg^{-1} DM at this lower dose.

The total Zn level of 2,883 ± 433 mg kg^{-1} DM reported in this study largely exceeded levels of 313–515 mg kg^{-1} DM in shoots of *M. spicatum* collected from a surface water with 0.16–0.20 mg l^{-1} Zn (Samecka-Cymerman & Kempers, 2004) and 304 mg kg^{-1} DM in shoots of *M. aquaticum* grown for 3 weeks in synthetic wastewater with 28 mg l^{-1} Zn (Kamal *et al.*, 2004). The initial Zn levels in our study (612 ± 254 mg kg^{-1} DM) were already remarkably higher than these values reported in literature. The total Zn level in *M. spicatum* after harvest were more in line with levels in leaves and stems of *Potamogeton natans* grown in diluted Hoagland's medium spiked with 1.3 mg l^{-1} Zn (Fritioff & Greger, 2006).

The mass balance showed that the difference between the amounts of Co and Ni removed from the wastewater in the planted and unplanted microcosms were higher than those recovered in the *Myriophyllum* biomass, whereas a correct mass balance was set up for Cu and Zn. Sivaci *et al.* (2004) stated that washing the biomass with distilled water may remove electrostatically bound Cd ions from the plant surface. This might have occurred with Co and Ni that was present on the *Myriophyllum* surface. However, it seems unlikely that Zn and part of the Cu would not have been rinsed of the surface as well. It was therefore assumed that another removal pathway not involving uptake by the plants, but induced by the presence of the *Myriophyllum* plants, was responsible for a significant amount of the Co and Ni removal.

Sulphate levels remained constant with time, thereby excluding the possibility of metal sulphide precipitation as the 'missing removal process'. A high pH of 10 was observed in the planted microcosms at the end of the last batch. This increase in pH was attributed to the photosynthesis of the plants that occurred at a greater level in the last batch than at the start of the experiment. It was suggested that Co and Ni were partly removed by precipitation with CO_3^{2-} and OH^- as they were removed slowly from the wastewater by uptake of the plants and were thus more susceptible to this third removal pathway than Cu and Zn.

A simple design calculation was performed under the assumption that a daily flow of 5,000 m^3 of wastewater is to be treated, representative for a metal-processing industry in Flanders. A depth of 2 m was chosen based on the observed depth range of 0.5–3.5 m that can be colonised by *M. spicatum* (APIRS, n.d.). If Cu and Zn are the target pollutants, a residence time of 48 h is sufficient to reach 0.05 and 0.2 mg l^{-1} of respectively Cu and Zn. Hence, a CW with a minimum treatment area of 0.5 ha would be needed. A sevenfold higher residence time and surface area would be required to provide an additional 80% reduction of Co and Ni. Based on the observed growth rates in our experiment, a monthly biomass production of 3.4 ± 0.6 t DM ha^{-1} could be achieved and the following masses of Co, Ni, Cu and Zn could be extracted from the wastewater by harvest of *Myriophyllum* (in kg ha^{-1} $month^{-1}$): 6 ± 1, 5 ± 2, 3 ± 1 and 10 ± 2. The metals in the biomass could be recovered by ashing the biomass (Zhu *et al.*, 1999). Moreover, the combustion of the biomass could produce useful energy. This has been suggested by Maine *et al.* (2001, 2004) as a management strategy for harvested biomass of *Pistia stratiotes*.

Constructed wetlands planted with *M. spicatum* offer potential for the polishing of industrial effluents but their full-scale use is limited. Some factors described by Kivaisi (2001) that have countered the full-scale use of *E. crassipes* in developed countries also apply for *M. spicatum*: (i) the poor performance in northern-hemisphere winters, (ii) the purchase of land, (iii) the intensive harvesting schemes and (iv) the construction and operation of systems for bioenergy recuperation. However, CWs with floating or submerged plants could offer great advantages in developing countries. First of all, year-round growth and thus treatment efficiency could be provided in warmer climates. Secondly, availability of land is greater. At last it is mentioned that maintenance and harvesting schemes could provide employment.

19.5   Conclusions

Removal of Co and Ni by *M. spicatum* was slower than of Cu and Zn. A residence time of 2 weeks was necessary to achieve Co and Ni levels of $0.2\,mg\ l^{-1}$, whereas a shorter residence time of 48 h was sufficient to achieve Cu and Zn levels of respectively 0.05 and $0.2\,mg\ l^{-1}$. A removal pathway associated with the presence of the plant that involved precipitation with CO_3^{2-} and OH^- was probably responsible for about one third of the Co and Ni removal. Growth of *M. spicatum* was not adversely affected by the wastewater. Total Co, Ni, Cu and Zn levels in *M. spicatum* after 12 weeks of exposure accounted for respectively 0.17%, 0.15%, 0.08% and 0.29% on a DM basis.

References

APIRS. (n.d.). Aquatic, Wetland and Invasive Plant Information Retrieval System from the University of Florida. Retrieved in August, 2006, from http://plants.ifas.ufl.edu.

Du Laing, G., Tack, F.M.G., & Verloo, M.G. (2003). Performance of selected destruction methods for the determination of heavy metals in reed plants (*Phragmites australis*). *Analytica Chimica Acta, 497*(1–2), 191–198.

Fritioff, A., & Greger, M. (2006). Uptake and distribution of Zn, Cu, Cd, and Pb in an aquatic plant *Potamogeton natans*. *Chemosphere, 63*, 220–227.

Guilizzoni, P. (1991). The role of heavy metals and toxic materials in the physiological ecology of submersed macrophytes. *Aquatic Botany, 41*(1–3), 87–109.

Kamal, M., Ghaly, A.E., Mahmoud, N., & Côté, R. (2004). Phytoaccumulation of heavy metals by aquatic plants. *Environment International, 29*, 1029–1039.

Keskinkan, O. (2005). Investigation of heavy metal removal by a submerged aquatic plant (*Myriophyllum spicatum*) in a batch system. *Asian Journal of Chemistry, 17*(3), 1507–1517.

Kivaisi, A. (2001). The potential of constructed wetlands for wastewater treatment and reuse in developing countries: A review. *Ecological Engineering, 16*, 545–560.

Kurniawan, T.A., Chan, G.Y.S., Lo, W.-H., & Babel, S. (2006). Physico-chemical treatment techniques for wastewater laden with heavy metals. *Chemical Engineering Journal, 118*, 83–98.

Maine, M.A., Duarte, M.V., & Suñé, N.L. (2001). Cadmium uptake by floating macrophytes. *Water Research, 35*(11), 2629–2634.

Maine, M.A., Suñé, N.L., & Lagger, S.C. (2004). Chromium bioaccumulation: Comparison of the capacity of two floating aquatic macrophytes. *Water Research, 38*, 1494–1501.

Qian J.H., Zayed A., Zhu Y.L., Yu, M., & Terry, N. (1999). Phytoaccumulation of trace elements by wetland plants: III. Uptake and accumulation of ten trace elements by twelve plant species. *Journal of Environmental Quality, 28*, 1448–1455.

Salt, D.E., Blaylock, M., Kumar, N.P.B.A., Dushenkov, V., Ensley, B.D., Chet, I., & Raskin, I. (1995). Phytoremediation: A novel strategy for the removal of toxic metals from the environment using plants. *Biotechnology, 13*, 468–475.

Samecka-Cymerman, A., & Kempers, A.J. (2004). Toxic metals in aquatic plants surviving in surface water polluted by copper mining industry. *Ecotoxicology and Environmental Safety, 59*, 64–69.

Sivaci, E.R., Sivaci, A., & Sökmen, M. (2004). Biosorption of cadmium by *Myriophyllum spicatum* L. and *Myriophyllum triphyllum* orchard. *Chemosphere, 56*, 1043–1048.

Zayed, A., Gowthaman, S., & Terry, N. (1998). Phytoaccumulation of trace elements by wetland plants: I. Duckweed. *Journal of Environmental Quality, 27*, 715–721.

Zhu, Y.L., Zayed, A.M., Qian, J.-H., de Souza, M., & Terry, N. (1999). Phytoaccumulation of trace elements by wetland plants: II. Water hyacinth. *Journal of Environmental Quality, 28*, 339–344.

Chapter 20
Cold Season Nitrogen Removal in a High Loaded Free Water Surface Wetland with Emergent Vegetation

Christer Svedin, Sofia Kallner Bastviken, and Karin S. Tonderski(⊠)

Abstract The aim of this study was to quantify nitrogen removal in high loaded free water surface wetlands dominated by emergent vegetation. It was undertaken in two subsystems of the full-scale wetland Alhagen in Nynäshamn, Sweden. Time proportional samples were taken at the inlets and outlets for 2 weeks in September and November 2005, respectively, and the water flow was monitored. The samples were analysed for ammonium-N (NH_4^+-N), nitrate-N (NO_3^--N), nitrite-N (NO_2^--N) and total-N, and the mass nitrogen removal was calculated. Sediment cores were randomly collected to measure potential denitrification, and the result was related to the actual mass nitrogen removal. Zero total-N removal could be detected in the subsystem with 6 hours hydraulic retention time (HRT). In the one with 3–4 days HRT, the total-N removal rates were 0.6 g N m^{-2} day^{-1} in September and 0.2 g N m^{-2} day^{-1} in November. The potential denitrification rate was 8 times higher than the observed removal in September and 48 times higher in November. This deviation was likely related both to relatively high oxygen levels and to the amount of available organic carbon.

Keywords Emergent plants, hydraulic load, nitrogen removal, wastewater treatment wetland

20.1 Introduction

Constructed wetlands are increasingly used to remove nitrogen (N) from highly pretreated wastewater from larger settlement areas. When constructing such wetlands, several design aspects have to be considered, such as hydraulic efficiency

IFM-Biology, Linköping University, SE-581 83 Linköping, Sweden

(⊠) Corresponding author: e-mail: karsu@ifm.liu.se

and the vegetation that is desirable. It is accepted that the dominant nitrogen-removal processes in wetlands are nitrification/denitrification and plant uptake (Ingersoll & Baker, 1998). Vegetation has an important role for nitrogen removal, and there seems to be several mechanisms. Results by Eriksson and Weisner (1996, 1999) suggested that vegetation increases surfaces for bacterial growth (epiphytic microbial communities) and that denitrification takes place in the biofilms formed. Although epiphytic communities on vegetation might have significant importance for denitrification, Bastviken *et al.* (2003) found lower denitrification capacity on surfaces in the water column than in the sediment.

The vegetation also affects the water flow pattern within wetlands (Jadhav & Buchberger, 1995), and this in turn affects nitrification, denitrification and plant uptake in a nontrivial way (Eriksson & Weisner, 1996, 1999; Eriksson, 2001; Wilcock *et al.*, 2002). For example, Eriksson (2001) showed that the denitrification was sensitive to oxygen levels in microcosms with a high water flow, whereas when the water flow was slow, or stagnant, denitrification could occur in almost oxygen-saturated water. Plants can contribute oxygen to the water as well as to the sediments, as oxygen is transported from the aerial green parts to the roots and rhizomes. Some of this oxygen can leak to the rhizosphere, which can stimulate nitrification and denitrification (Brix, 1997).

There are several studies supporting that denitrification in wetlands is commonly limited by available organic matter (Ingersoll & Baker, 1998; Bachand & Horne, 2000b; Lin *et al.*, 2002). Ingersoll and Baker (1998) found that an organic matter with a C:N ratio of 5:1 caused a nearly 100% removal of the nitrate, independent of the hydraulic loading rate (HLR). This ratio may not be attained in a full-scale wetland (Lin *et al.*, 2002) but it shows the crucial importance of available organic carbon for denitrification. Vegetation can serve as an internal source of organic carbon in wetland systems, and supply the denitrifying bacteria with carbon and energy. In an experiment with addition of sodium acetate to wetland soil samples, Kozub and Liehr (1999) found that the samples with readily degradable carbon sources had a higher denitrification rate than the rate measured prior to the carbon addition. It implies that the form of carbon supplied by the vegetation is important and that there might be interspecies differences that influence the denitrification rates (Ansola *et al.*, 1995; Tanner, 1996).

Several laboratory and pilot-scale studies have shown that in stands of emergent vegetation the conditions for denitrification are more favourable than in stands of submerged vegetation or open water areas (Bachand & Horne, 2000a, b; Lin *et al.*, 2002; Ingersoll & Baker, 1998). However, when scaling up such results, other effects of the vegetation will also play a role. Dense stands of plants may, for example, decrease the hydraulic efficiency and cause increased water velocity in parts of the wetland. The growth pattern of the plants may also contribute to seasonal variations in the denitrification capacity through, e.g. variations in available organic carbon from plant litter and detritus. Hence, it is not trivial to use results from microcosm studies to extrapolate to full-scale wetlands because the environment in a full-scale wetland can be significantly different from, and more complex than, a microcosm.

20.2   Aims of the Study

In the present study, the aims were (i) to quantify nitrogen removal in a free water surface wetland operating to treat nitrified municipal wastewater, and dominated by emergent vegetation, (ii) to analyse how large part of the potential for denitrification was actually realized during operation of the wetland, and (iii) try to identify possible limiting factors for the actual nitrogen removal.

20.3   Materials and Methods

20.3.1   Site Description

This study was performed at Alhagen, a free water surface wetland system in Nynäshamn, south Sweden. The wetland was built on an area previously used for agriculture and pasture, and is in operation since 1997. It receives effluent from the sewage treatment plant (STP), which treats domestic sewage (approximately 19,000 P.E.) and some septic tank sludge. A sequential batch reactor ensures that the effluent is almost completely nitrified during the warm months of the year. The effluent is pumped to the wetland, but at flows > 340 m^3 h^{-1} the excess is discharged into the Baltic Sea. About 84% of the annual flow is treated in the wetland.

The wetland consists of nine ponds and an overland flow area and covers about 26 ha. At the inlet, there are two continuously fed rectangular (100 × 15 m) ponds in series (the Inlet ponds), followed by two parallel ponds, the East pond and the West pond that are loaded intermittently (Fig. 20.1). This study was limited to the Inlet ponds and the West pond due to their relatively high coverage of emergent vegetation. To allow filling of the West pond, the gate at the outlet is closed and the water is directed to the pond from the outlet of the second Inlet pond. This is repeated each Monday and Thursday morning, and the filling period lasts for approximately 24 h. To allow emptying of the West pond, the gate at the inlet is closed to direct the water to the East pond, and the gate at the outlet is opened. This is done in the morning on Tuesdays and Fridays. The gates are then left in those positions until the next filling starts, but after approximately 24 h there is no outflow from the West pond.

Fig. 20.1 Flow scheme of the studied ponds at wetland Alhagen, Nynäshamn. P1–P3 and M1–M6 are sampling points

The Inlet ponds are 0.2–0.3 m deep except for the first 30 m in pond 1, which is approximately 1.5 m, and covered with dense stands of *Phragmites australis* Trin. ex Steud., except for in the deep zone (total coverage is about 85%). A U-shaped pipe transfers the water by gravity from the first to the second pond. After the two Inlet ponds, the water flows to the West pond that is approximately 320 × 30 m (total area 8,400 m²). The first third is open water with mostly algae. Next third is sparsely populated with *Typha latifolia* L., and the last third is densely populated with the same species. There are also filamentous algae, *Elodea canadensis* (Michx.) Rchb. and *Potamogeton natans* L. At the lowest water level, the first meters of the pond have an average depth of 0.1 m and the last meters an average depth of 0.5 m. The increase in water level at maximum filling is about 0.34 m and the volume of water at each filling is 4,000–5,000 m³.

20.3.2   Sampling

Three sampling points were used for mass flow estimations: P1 at the inlet to the wetland system, P2 at the outlet of the second Inlet pond, and P3 at the outlet of the West pond. Six more sampling points (M1–M6) were used to estimate the mass of nitrogen in the residual volume of water at minimum water level in the West pond (Fig. 20.1). Three of these (M1, M3 and M6) were also used to analyse the daily variations in pH, oxygen and temperature.

Samples were collected using three automatic samplers, set to take a sample every 15–30 min to obtain one mixed sample for a 4-hour period. The two samplers located at P1 and P2 were operating on Mondays, Tuesdays, Thursdays and Fridays, and the one at P3 was operating on Tuesdays and Fridays. They were emptied daily and water was frozen until analysis. Samples to estimate residual nitrogen, pH, and temperature were collected manually with a cup fastened to an adjustable handle.

20.3.3   Water Flow, Residual Volume and Nitrogen

The volume of water pumped to the wetland was continuously measured at the sewage treatment plant, and these data were used as the water flow at P1. The theoretical retention time in the two Inlet ponds was approximately 6 h, and the water flow at P2 was set equal to P1 with a time delay of 4 hours. At the outlet of the West pond, the flow was estimated using a fixed millimetre scale. The water-level changes were measured for one filling and emptying cycle (2 days), analysed and compared with the flow data from the sewage treatment plant. This was done once in September and once in November, and the result applied to the other dates, when only the minimum and maximum water levels were measured.

A Matlab program was written to determine the residual water volume (i.e. the volume remaining in the pond at the start of a filling cycle), using the measured water depth at 88 evenly distributed points in the West pond. To estimate the residual nitrogen, water samples were collected in three sampling points on two dates in September and six points (M1–M6; Fig. 20.1) on three dates in November. The data were inserted in the Matlab program for calculation of the amount of nitrogen remaining in the pond at the start of a filling cycle.

20.3.4 Water Analyses

pH was measured in randomly selected samples in bottles from the automatic samplers. The diurnal variation of in situ pH, temperature and oxygen in the West pond was measured during 2 days (one filling and emptying cycle) in M1, M3 and M6 (Fig. 20.1), once in September and once in November. Concentrations of ammonium-N (NH_4^+-N), nitrate-N (NO_3^--N), nitrite-N (NO_2^--N) and total-N were determined using an AutoAnalyzer 3 (BRAN + LUEBBE GmbH SPX Process Equipment, Norderstedt, Germany). Samples for total-N were oxidized with potassium peroxodisulphate according to SS-EN ISO 11905-1 and analysed as NO_2^--N after cadmium reduction as described in the Bran + Luebbe AutoAnalyzer application Method No. G-172-96 Rev. 9, which was also used for NO_3^--N and NO_2^--N analyses. NH_4^+-N was analysed according to Bran + Luebbe AutoAnalyzer application Method No. G-171-96 Rev.1.

20.3.5 Calculations of Mass Nitrogen Removal

The mass nitrogen removal in the two Inlet ponds was calculated as the difference between the mass input and mass output of nitrogen (Equation 20.1).

$$\sum_{t=0}^{12} C_{in}(t)\cdot Q_{in}(t) - \sum_{t=0}^{12} C_{out}(t)\cdot Q_{out}(t) = N_{removal} \qquad (20.1)$$

However, in the West pond, the residual nitrogen had to be included and the nitrogen removal was calculated according to the following equation:

$$\sum_{t=0}^{6} Q_{in}(t)\cdot C_{in}(t) - \sum_{t=0}^{6} Q_{out}(t)\cdot C_{out}(t) - \sum_{t=-3}^{0} Q_{ovflw}(t)\cdot C_{ovflw}(t) + M_N^{(i)} - M_N^{(f)} = N_{removal}$$

$$(20.2)$$

Q_{in}, Q_{out} and Q_{ovflw} are the inflow, outflow and overflow, respectively, and analogous for the concentrations. Q_{ovflw} and C_{ovflw} are correction terms for a leakage at the outlet (Section 20.3.1). $M_N^{(i)}$ is the residual nitrogen prior to the filling of the pond, $M_N^{(f)}$ is the residual nitrogen prior to the next filling, and both are variables estimated by the Matlab program. In order to present the removal in g N m^{-2} day^{-1}, the $N_{removal}$ was divided by the area of the pond and number of days, i.e. for the Inlet ponds 2 days and for the West pond 3 or 4 days.

In order to compare the removal rate in Alhagen with the rates in other treatment wetlands, the first-order rate constant (k) and the temperature corrected rate constant (k_{20}) were calculated for the systems according to the "tanks in series" model (Equation 20.3).

$$\frac{C_{out}}{C_{in}} = \left(1 + \frac{k}{N \cdot q}\right)^{-N} \qquad (20.3)$$

C_{out} and C_{in} are the concentrations in g m^{-3}, q is the hydraulic loading rate in m year^{-1}, and N is the hydraulic efficiency parameter (set to 4.5 in the calculations (Kadlec, 2005)). To calculate the rate constant at 20°C, k_{20}, the Arrhenius equation was used:

$$k = k_{20} \cdot \theta^{(T-20)} \qquad (20.4)$$

T is the water temperature and θ is the temperature factor (set to 1.088 in the calculations (Kadlec, 2005)).

20.3.6   Potential Denitrification

Eight sediment cores were randomly collected from the two Inlet ponds and eight from the West pond to estimate the potential denitrification. This was measured with the acetylene inhibition technique (Balderston *et al.*, 1976), in which acetylene is added to the sediment cores to inhibit the last stage in the denitrification, and the accumulation of N$_2$O is measured. All N$_2$O was assumed to come from the denitrification and hence the production of N$_2$O was a direct measurement of the denitrification rate.

Sixteen sediment cores and four controls were incubated under anaerobic conditions with a nutrient buffer with excess nitrate-N (15 mg NO$_3^-$-N l^{-1}, 42 mg K l^{-1}, 72 mg Ca l^{-1}, 10 mg Mg l^{-1}, 27 mg Na l^{-1} and 2.3 mg PO$_4^{3-}$-P l^{-1}) following the procedure given by Bastviken *et al.* (2005). Prior to the experiment, the buffer was bubbled with nitrogen gas until the dissolved oxygen level was below 2 mg l^{-1} in all cores, and at the start of the experiment, 10% of the core volume of each sample was replaced with buffer saturated with acetylene. The cores were incubated for 4 h in darkness at 20°C. After 2 h, sodium acetate was added to half of the cores to an end concentration of 15 mg C l^{-1}. Water samples for analysis of nitrous oxide (10 ml) and nitrate-N (11 ml) were taken initially, after 2 and 4 h. Nitrous oxide was measured using a gas chromatograph, CHROMPACK CP-9001 (Chrompack, Middleburg, the Netherlands). Henry's law and the ideal gas law was used to calculate the N$_2$O content in the buffer phase of the core.

20.3.7 Statistics

All statistical analyses were performed using Minitab for Windows 13.32 (Minitab Inc., State College, Pennsylvania, USA). Data from the incubations were analysed with the Mann–Whitney's test. Data from the sample points were analysed using a two-factor factorial additive model, the logarithm of the nitrogen concentration (NH_4^+ –N, (NO_2^- + NO_3^-)–N or Tot–N) versus the sample point and day (Montgomery, 2001). In September, a two-way ANOVA was used but in November a general linear model (GLM) had to be used due to an incomplete data set. The data were further processed according to Tukey's method to construct confidence intervals of the difference between the sample points. In all analyses, the level of significance was set to 0.05.

20.4 Results

20.4.1 Water Quality and Water Flow

There were small differences in the quality of the sewage treatment plant effluent between the two periods except for NH_4^+-N, which was much higher in November than in September (Table 20.1). The concentration of phosphorous was low in the influent water and a large fraction of the organic carbon was not directly available for the micro-organisms (low BOD_7 versus COD). Although the flow-weighted mean concentrations of all nitrogen fractions were descending throughout the system studied (Table 20.2), there were no statistical differences in any nitrogen fractions

Table 20.1 Quality of the effluent from Nynäshamn sewage treatment plant during two study periods in 2005. $n = 7$ for total-P, NH_4-N and total-N, $n = 4$ for BOD and COD in September and $n = 3$ in November. Values in mg l^{-1}

Period	Total-P	BOD_7	COD	NH_4-N	Total-N
September	0.2	4.5	35.3	3.5	21.3
November	0.3	6.5	41.3	10.0	20.4

Table 20.2 Flow weighted mean concentrations and standard error of NH_4^+-N (NO_2^- + NO_3^-)-N and total-N at the different sampling points from Inlet (P1) to outlet of the West pond (P3) in wetland Alhagen, Nynäshamn

	$(NO_2^- + NO_3^-)$-N	NH_4^+-N	Total-N
September P1	18.6 ± 0.5	2.0 ± 0.2	19.8 ± 0.6
P2	18.5 ± 0.4	1.9 ± 0.2	20.5 ± 0.4
P3	17.4 ± 0.6	0.8 ± 0.1	19.1 ± 1.0
November P1	9.4 ± 0.3	8.8 ± 1.1	20.0 ± 0.9
P2	9.7 ± 0.2	8.7 ± 1.2	20.4 ± 0.8
P3	9.3 ± 0.7	6.6 ± 2.3	18.9 ± 1.6

between P1 and P2, neither in September nor in November. In contrast, the concentrations of all fractions dropped significantly from P2 to P3 in September (Table 20.2), though in November the decrease was not statistically significant for any of the parameters. In November, the proportion of nitrate was less than half of that in September, and in addition the proportions of ammonium and nitrate varied much more with time (Table 20.2). This may have contributed to the lack of statistical significance in November.

The mean pH varied between 6.9 and 7.2, with no significant difference between the sampling points P1–P3. In both periods, the mean water temperature was higher in P1 than in P3, 17°C versus 15°C in September and 11°C versus 5°C in November. The oxygen levels were highest in P1 and lowest in P2, 3.7–6.5 and 2.0–4.8 mg l^{-1}, respectively, in September. In the middle of the West pond, the dissolved oxygen had a typical diurnal variation, i.e. rising levels of oxygen from early morning to a peak in early afternoon and then falling to a low level in the dark period (Fig. 20.2). This suggests that there was a significant influence of photosynthesis by submerged vegetation and algae on the oxygen levels. This is also supported by the fact that in November the dissolved oxygen fluctuated less.

The HLRs were high in both systems and were ten times higher in the Inlet ponds (1.3–1.5 m day^{-1} in September and 1.4–2.0 m day^{-1} in November) than in the West pond (0.12–0.17 and 0.14–0.19 m day^{-1}, respectively). The high HLRs resulted in a theoretical hydraulic retention time (HRT) of only 6 to 7 h in the Inlet ponds and 3 or 4 days in the West pond. During both periods, the influence of precipitation was negligible (0.1 mm day^{-1} in September and 0.5 mm day^{-1} in November).

The residual volume of water was about 45% of the total volume in the West pond (Fig. 20.3). The sudden change in slope during the filling was caused by an overflow

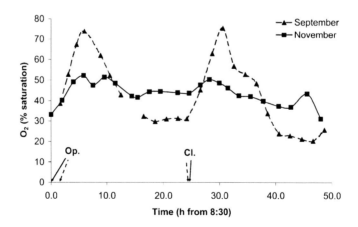

Fig. 20.2 Variation in dissolved oxygen in the middle of the West pond, Alhagen, Nynäshamn. The Op. marks the beginning of filling and the Cl. marks the ending of filling, the dashed arrows represent September and the solid arrows November

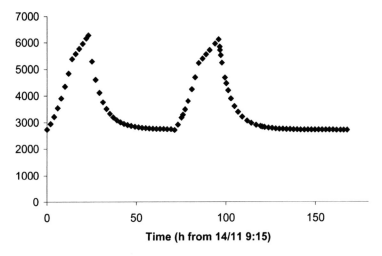

Fig. 20.3 Volume changes in the West pond, Alhagen, Nynäshamn for 1 week in November 2005. The values are estimated using data on times for maximum and minimum water-level height, the relation between water level and volume of water added to the pond, and the calculated residual volume

at the outlet, which started after about 14 h (depending on the filling rate) and continued for about 10 h. According to the MATLAB model, the residual volume of water was 2,730 m^3 at the reference water level. In September, the volume fluctuated from 2,217 to 2,656 m^3 due to the mentioned leakage at the outlet. Correspondingly, the amount of residual nitrogen also varied considerably, from 34 to 51 kg. Generally, the nitrogen concentrations in the West pond were slightly lower on Mondays than on Thursdays in both September and November (data not shown).

20.4.2 Nitrogen Mass Removal

When summarizing data on inflow, residual and outflow nitrogen for each period, it is obvious that the West pond was more efficient than the Inlet ponds in both September and November (Figs. 20.4 and 20.5). The load was almost ten times higher in the Inlet ponds than in the West pond, but there were almost no, or even negative, removal in the former. In the West pond, a significant removal was observed in almost all measured periods.

Because of the relatively large variation of the residual water volume in September, the first order rate constants were calculated only for the November data set. The mean k-value for $(NO_2^- + NO_3^-)$-N removal was 18.9 m year^{-1}, and the $k_{20} = 57.7$ m year^{-1}. The corresponding values for removal of total-N was $k = 5.6$ m year^{-1} and $k_{20} = 16.6$ m year^{-1}.

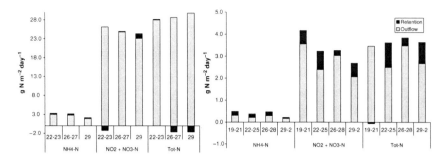

Fig. 20.4 Nitrogen load and removal in the Inlet ponds (left) and the West pond (right) during the September study period. The sum of removal and out represents the load of nitrogen. Note that the scales are different between the Inlet ponds and the West pond

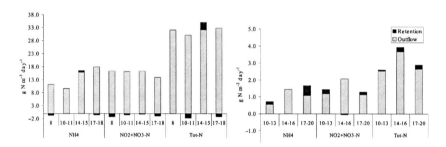

Fig. 20.5 Nitrogen load and removal in the Inlet ponds (left) and the West pond (right) during the November study period. The sum of removal and out represents the load of nitrogen. Note that the scales are different between the Inlet ponds and the West pond

20.4.3 Potential Denitrification

The potential denitrification rates when no acetate was added were almost equal in both systems. After acetate addition, the cores with *Phragmites australis* had about twice as high rate as those with *T. latifolia* (Fig. 20.6). The variance was large, and hence this difference was not statistically significant. For the same reason, the effect of acetate addition was only significant in the cores from the *Phragmites australis* stands ($p = 0.015$, $n = 4$), though it is likely that the denitrification in the *T. latifolia* cores was also limited by available organic matter. With no acetate addition, the potential denitrification was eight times higher than the observed (NO_3^- + NO_2^- + NO^-)-N removal in September (4.8 versus 0.6 g NO_3^--N m^{-2} day^{-1}, respectively). In November the difference was larger, the observed N removal (0.1 g NO_3^--N m^{-2} day^{-1}) was 48 times lower than the potential denitrification.

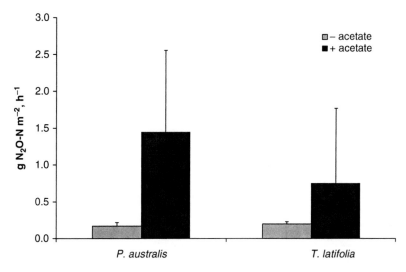

Fig. 20.6 Potential denitrification in intact sediment cores from the Inlet ponds with *P. australis* and from the West pond with *T. latifolia* in October 2005.The cores were incubated in a nutrient buffer without acetate (grey bars) and with addition of acetate (black bars). Error bars indicate standard deviation

20.5   Discussion

Data clearly shows that the water from the sewage treatment plant was more nitrified in September than in November (Table 20.1) and this was probably due to a more efficient nitrification in the sequencing batch reactors (SBR). Other studies have shown that such seasonal variations in the concentrations of NH_4^+-N and NO_3^--N in the effluent from a STP affect the nitrogen removal in subsequent wetlands (Kadlec, 1999). It is, however, likely that the difference in nitrogen removal rates in the West pond between September and November was caused by the drop in temperature rather than the different inflow water quality. The mean temperature was about 10°C lower in November than in September (Table 20.2) and the nitrogen removal rate decreased by more than 60% (from a mean 0.6 to 0.2 g m^{-2} day^{-1} total-N; Figs. 20.4 and 20.5). This agrees with common observations of the temperature influence on microbiological activities. For example, Kadlec (2005) reported a 50% reduction of the wetland nitrogen removal rate constant at a temperature decrease of 8°C. In addition, the (NO_2^--NO_3^-)-N levels at the outlet of the West pond were higher than levels known to limit denitrification. It should also be noted that the removal estimates for September were more uncertain than those in November, when a more accurate sampling system was used (see Section 20.3.3).

The *k*-values calculated for the West pond in November were used to compare the performance of the ponds with literature values. The k_{20}-value for (NO_2^- +

NO_3^-)-N (58 m year^{-1}) was higher than the mean value (34 ± 3 m year^{-1}) reported by Kadlec (2005). The k_{20}-value for total-N removal (equal to 17 m year^{-1}) was in the same range as for other large wetlands in Sweden, 20 m year^{-1} in Magle and 30 m year^{-1} in Ekeby (Andersson *et al.*, 2005).

No removal of total-N could be detected in the Inlet ponds with the high HLR (Fig. 20.4 and 20.5). As shown by the core experiment, the potential denitrification was as high in the Inlet ponds (with *Phragmites australis*; Fig. 20.6) as in the West pond (with *T. latifolia*). This suggests that the difference in removal rates were due to other factors than the plant species per se. The mean potential denitrification rate (0.20 g N_2O–N m^{-2} h^{-1}) was close to rates reported from other micro-and macrocosm studies, e.g. 0.6–1.3 (Lin *et al.*, 2002), 2.8 and 0.3–0.8 (Bachand & Horne, 2000a, b), 0.7–2.4 (Tanner *et al.*, 1999) and 0.1–4.4 g N m^{-2} day^{-1} (Ingersoll & Baker, 1998). It should, however, be noted that the HLR was two to four times higher than in the cited studies. Arheimer and Wittgren (2002) and Braskerud (2002) reported similar HLR for full-scale wetlands receiving nonpoint source runoff.

A short retention time could affect the levels of dissolved oxygen and might also cause resuspension of nitrogenous compounds. Arheimer and Wittgren (2002) also suggested the latter after an analysis of time-series from wetland monitoring programmes, when they could not deduce if wetlands with a retention time less than 2 days had a net release or a net removal of nitrogen. The oxygen levels in the Inlet ponds, with HLR of 1–2 m day^{-1}, were above anoxic conditions in both September and November with a minimum value of 2 mg l^{-1}, though they dropped from the inlet to the outlet. In contrast, a significant removal was always detected in the West pond, with a HLR of 0.12–0.19 m day^{-1}. The removal was larger in the periods 22–25/9, 29/9–2/10, 10–13/11, and 17–20/11 than in the other periods (Figs. 20.4 and 20.5). This was probably due to better nitrogen removal during the weekends when the water was left stagnant for 2 days, i.e. there was less residual nitrogen in the pond on Monday mornings than on Thursday mornings. Probably this favoured the development of anoxic conditions in the sediment, and thus promoted a higher denitrification. This could not be detected from the measurements of oxygen in the water, as those levels were relatively high (Fig. 20.2), but in stagnant water oxygen gradients easily develop, e.g. in the upper sediment layer. This is also supported by other studies (Eriksson, 2001).

The large deviation between observed nitrogen removal and the potential denitrification was probably also related to the oxygen conditions in the pond. In addition, one third of the West pond was open water, which did not promote effective denitrification (Kadlec, 2005). As indicated by the results from the core experiment, the wetland sediment was limited by a lack of available organic matter, and this limitation would have been even more severe in areas without emergent plants. It is also well known that hydraulic aspects might contribute to a lower removal efficiency (Persson, 2005). Any such effects should have been reduced, since the West pond was loaded intermittently.

20.6   Conclusions

This study showed that the choice of plant species was of minor importance for wetland nitrogen removal. Observed field differences in nitrogen removal between the system dominated by *Phragmites australis* and the system dominated by *T. latifolia* were probably due to a much higher HLR in the previous system, which in turn maintained high oxygen levels and might have caused resuspension of nitrogen.

No nitrogen removal was observed in the system with HLR of $1-2$ m day^{-1} despite the high potential denitrification. In the wetland with HLR of $0.12-0.19$ m day^{-1}, the total-N removal rates were relatively stable, 0.6 g N m^{-2} day^{-1} in September and 0.2 g N m^{-2} day^{-1} in November. The difference between months was at least partly explained by a decrease in temperature. This observed removal was much lower than the measured potential denitrification (8 times lower in September and 48 times in November), and this was probably mainly due to high dissolved oxygen levels. In both systems, the denitrification was limited by available organic material as shown by the sediment core study.

The total nitrogen removal rates in the system dominated by *T. latifolia* (West pond) were in the same magnitude as in other large wetland systems in Sweden, despite the predominance of open water surfaces and submergent vegetation in the latter. This further confirms the hypothesis that high oxygen levels in the present wetland systems limited denitrification. The performance could probably be more efficient if more emergent vegetation is established and submerged vegetation removed.

Acknowledgements The authors want to thank Jorgen Mahlgren, Ingrid Rehnlund, Amparo Franquiz-Medina and Lena Lundman for all help with practical issues. Financial support was provided from the Swedish Water Management Research Program (VASTRA), which was financed by the Swedish Foundation for Strategic Environmental Research (Mistra).

References

Andersson, J.L., Bastviken, S.K., & Tonderski, K.S. (2005). Free water surface wetlands for wastewater treatment in Sweden: Nitrogen and phosphorus removal. *Water Science and Technology, 51*(9), 39–46.

Ansola, G., Fernandez, C., & de Luis, E. (1995). Removal of organic matter and nutrient from urban wastewater by an experimental emergent aquatic macrophyte system. *Ecological Engineering, 5*, 13–19.

Arheimer, B., & Wittgren, H.B. (2002). Modelling nitrogen removal in potential wetlands at the catchment scale. *Ecological Engineering, 19*, 63–80.

Bachand, P.A.M., & Horne, A.J. (2000a). Denitrification in constructed free-water surface wetlands: I: Very high nitrate removal rates in a macrocosm study. *Ecological Engineering, 14*, 9–15.

Bachand, P.A.M., & Horne, A.J. (2000b). Denitrification in constructed free-water surface wetlands: II. Effects of vegetation and temperature. *Ecological Engineering, 14*, 17–32.

Balderston, W.L., Sheer, B., & Payne, W.J. (1976). Blockage by acetylene of nitrous oxide reduction in Pseudomonas perfectomarinus. *Applied and Environmental Microbiology, 3*, 504–508.

Bastviken, S.K., Eriksson, P.G., Martins, I., Neto, J.M., Leonardson, L., & Tonderski, K. (2003). Potential nitrification and denitrification on different surfaces in a constructed wetland. *Journal of Environmental Quality, 32*, 2414–2420.

Bastviken, S.K., Eriksson, P.G., Premrov, A., & Tonderski, K. (2005). Potential denitrification in wetland sediments with different plant species detritus. *Ecological Engineering, 25*, 183–190.

Braskerud, B.C. (2002). Factors affecting nitrogen retention in small constructed wetlands treating agricultural non-point source pollution. *Ecological Engineering, 18*, 351–370.

Brix, H. (1997). Do macrophytes play a role in constructed treatment wetlands?. *Water Science and Technology, 35*(5), 11–17.

Eriksson, P.G. (2001). Interaction effects of flow velocity and oxygen metabolism on nitrification and denitrification in biofilms on submerged macrophytes. *Biogeochemistry, 55*, 29–41.

Eriksson, P.G., & Weisner, S.E.B. (1996). Functional differences in epiphytic microbial communities in nutrient-rich freshwater ecosystem: an assay of denitrifying capacity. *Freshwater Biology, 36*, 555–562.

Eriksson, P.G., & Weisner, S.E.B. (1999). An experimental study on effects of submerged macrophytes on nitrification and denitrification in ammonium-rich aquatic systems. *Limnology and Oceanography, 44*, 1993–1999.

Ingersoll, T.L., & Baker, L.A. (1998). Nitrate removal in wetland microcosms. *Water Research, 32*, 677–684.

Jadhav, R.S., & Buchberger, S.G. (1995). Effects of vegetation on flow through free water surface wetlands. *Ecological Engineering, 5*, 481–496.

Kadlec, R.H. (1999). Chemical, physical and biological cycles in treatment wetlands. *Water Science and Technolog, 40*(3), 37–44.

Kadlec, R.H. (2005). Nitrogen farming for pollution control. *Journal of Environmental Science and Health, 40*, 1307–1330.

Kozub, D.D., & Liehr, S.K. (1999). Assessing denitrification rate limiting factors in a constructed wetland receiving landfill leachate. *Water Science and Technology, 40*(3), 75–82.

Lin, Y.-F., Jing, S.-R., Wang, T.-W., & Lee, D.-Y. (2002). Effects of macrophytes and external carbon sources on nitrate removal from ground water in constructed wetlands. *Environmental Pollution, 119*, 413–420.

Montgomery, D.C. (2001). *Design and analysis of experiments*. Hoboken, NJ: Wiley.

Persson, J. (2005). The use of design elements in wetlands. *Nordic Hydrology, 36*(2), 113–120.

Tanner, C.C. (1996). Plants for constructed wetland treatment systems a comparison of the growth and nutrient uptake of eight emergent species. *Ecological Engineering, 7*, 59–83.

Tanner, C.C., D'Eugenio, J., McBride, G.B., Sukias, J.P.S., & Thompson, K. (1999). Effect of water level fluctuation on nitrogen removal from constructed wetland mesocosms. *Ecological Engineering, 12*, 67–92.

Wilcock, R.J., Scarsbrook, M.R., Costley, K.J., & Nagels, J.W. (2002). Controlled release experiments to determine the effects of shade and plants on nutrient retention in a lowland stream. *Hydrobiologia, 485*, 153–162.

Chapter 21
The Role of Vegetation in Phosphorus Removal by Cold Climate Constructed Wetland: The Effects of Aeration and Growing Season

Aleksandra Drizo[1]($\boxtimes$), Eric Seitz[1], Eamon Twohig[1], David Weber[2], Simon Bird[1], and Donald Ross[1]

Abstract The objective of this study was to evaluate the effectiveness and contribution of *Schoenoplectus fluviatilis* (Torr.) (river bulrush) to phosphorus (P) removal from dairy-farm effluent in a cold climate constructed wetland. After 3 years of operation (1,073 days), both nonaerated wetland cell 3 (C3) and aerated cell 4 (C4) exhibited a sharp decline in dissolved reactive phosphorus (DRP) storage, indicating wetlands saturation. The quantities of DRP stored during the three growing seasons (433 days) represented only 10.0%(C3) and 17.7%(C4) of the total amount of DRP (435.06 ± 2.3 g m^{-2}, 97.02 kg) added to each cell (C3 and C4) over the entire 3-year period. However, of the total DRP retained by both wetland cells during 1,073 days of operation, the quantities stored during the three growing seasons (433 days) represented 50.3%(C3) and 36.50%(C4) of the total DRP retention. This indicated that vegetation had an important role in the overall DRP storage regardless of supplemental aeration. Overall, nonaerated C3 DRP mass removal efficiency during the 3-year period of investigation was low, averaging 19.9%. Aerated C4 DRP mass removal efficiency was 2.4-fold higher, averaging 48.4%. Belowground (BG) biomass had significantly higher ($p < 0.001$) P content than aboveground (AG) biomass, throughout the 3-year period of investigation.

Keywords Aeration, agricultural effluent, cold climate, phosphorus removal, *Schoenoplectus fluviatilis*

---

[1] University of Vermont, Department of Plant and Soil Science, Hills Agricultural Building, 105 Carrigan Drive, Burlington, VT 05405, USA

[2] Vermont Agency of Agriculture Food & Markets. 116 State Street, Drawer 20 Montpelier, VT 05620–2901, USA

($\boxtimes$) Corresponding author: e-mail: adrizo@uvm.edu

J. Vymazal (ed.) *Wastewater Treatment, Plant Dynamics and Management in Constructed and Natural Wetlands,*
© Springer Science + Business Media B.V. 2008

21.1   Introduction

Constructed wetlands (CW) have been used to treat a variety of wastewaters for over 2 decades (Reddy & Smith, 1987; Hammer, 1989; Kadlec & Knight, 1996; Vymazal et al., 1998; Vymazal, 2006; Cooper, 2006). Vymazal (2006) recently provided a comprehensive review of CW technology evolution over time. It has been well documented that CW provide sustainable removal of organic material and suspended solids (Kadlec et al., 2000; Brix et al., 2006). However, phosphorus (P) removal by CW has been generally low regardless of the type of wastewater treated, especially in cold climates (Kadlec, 1999; Schaafsma et al., 2000; Luderitz & Gerlach, 2002; Picard et al., 2005; Weber et al., 2006). According to Kadlec (1999) and Kadlec and Knight (1996), P removal via wetland substrate represents only short-term storage, lasting up to a few months. Kadlec (1999) estimated that P removal through vegetation uptake can last for 1–6 years, and the only long-term, sustainable P storage occurs via the biogeochemical cycle and accretion.

The contribution of wetland vegetation to pollutant removal through filtration and sedimentation, stabilization of bed surface, light attenuation, and additional surface area for the attachment of microorganisms has long been established (Brix, 1997). The role of vegetation in providing insulation and thermal protection against ice formations in cold climate CW has also been demonstrated (Allen et al., 2002; Munoz et al., 2006). Oxygen release from plant roots, nutrient content and plant growth in CW were well researched (Brix, 1997; Tanner, 2001; Allen et al., 2002). Despite a relatively large pool of data, there is some discrepancy about the quantities of nutrient uptake and exact contribution of emergent macrophytes to P removal, especially in cold climates (Serodes & Normand, 1999; Luderitz & Gerlach, 2002; Gottschall et al., 2007). In their review of CW for animal wastewater treatment, Hunt and Poach (2001) noted that Scirpus, Typha and Juncus are the most common plant genera used in the USA. However, there is not enough information about the exact contribution of various plant species to P removal from high strength wastewaters in full-scale systems (Tanner, 2001). Agricultural effluents have higher nutrient concentrations, making it difficult to achieve efficient P removal, especially in cold climate wetlands. In addition, vegetation-growing season in these wetlands (average 150 days) is much shorter than in warmer climates.

In this chapter, we evaluated the effectiveness of Schoenoplectus fluviatilis (Torr.) (river bulrush) and its contribution to P removal from dairy-farm effluent in a cold climate CW. The effects of plant-growing seasons (first, second and third growing season) and supplemental aeration on P storage and mass removal were also assessed.

21.2   Site Location and Methods

21.2.1   Site Location

The Constructed Wetlands Research Center (CWRC) was established at the University of Vermont (UVM) Paul Miller Dairy Farm in the fall of 2003. A CW system, consisting of four horizontal subsurface flow wetland cells (1–4) designed to treat dairy effluent (barnyard runoff and milking operations wastewater) represents the main feature of the CWRC. The system has been recently described (Munoz *et al.*, 2006; Drizo *et al.*, 2006; Weber *et al.*, 2006). Each cell is 12 m wide, 18.5 m long and 0.6 m deep with total surface area and total volume of 223 m^2 and 134 m^3, respectively. The first third of each cell from the influent side has been filled with coarse gravel while the remaining two thirds were filled with smaller particle size gravel (2 and 1 cm diameter, respectively) resulting in a total porosity of approximately 38.7%. Munoz *et al.* (2006) presented a detailed schematic of the UVM CW, including the pretreatment settling pit and a series of holding tanks. Each CW cell has been equipped with an aeration system patented by Wallace (2001) and designed by North American Wetland Engineering LLC, (Forest Lake, MN). However, aeration was activated for only two CW cells at a time, in order to test the effect of oxygenation on pollutant treatment performance (Munoz *et al.*, 2006; Drizo *et al.*, 2006). All wetland cells were vegetated with river bulrush (*Schoenoplectus fluviatilis* (Torr.)).

The design of cell 1 and 2 changed several times over a 3-year period of system operation (Drizo *et al.*, 2006); therefore, this data represents a focused study on only two wetland cells: nonaerated C3 and aerated C4.

21.2.2   Methods

Sample locations were selected to be representative of the wetland inlet and outlet areas. Three quadrants were positioned within each of the inlet and outlet areas of C3 and C4. Quadrants covered an area of 0.23 m^2. Plant aboveground (AG) and belowground (BG) biomass were collected from six quadrants per cell during each sampling period (September 2004, 2005 and 2006). For each sampling event, plant samples were taken from a different location within designated cell inlet or outlet sampling areas.

Samples were separated by AG and BG biomass and cut into shorter lengths (0.20 m). They were then brought to the CWRC Laboratory, where they were weighed, dried at 55°C and ground in a Wiley mill to pass a 1 mm screen. Root samples were milled at the Geology Laboratory (UVM). Total nutrient concentration was determined by microwave-assisted nitric acid digestion (CEM Corp. Matthews, NC) following EPA

Method SW-846–3051, with analysis by an inductively coupled plasma atomic emission spectrophotometer (ICP-AES; Perkin-Elmer Corp., Norwalk, CT).

Monitoring of the CW performance began in January 2004 and sampling continues on a weekly basis (Drizo *et al.*, 2006). Dissolved reactive phosphorus (DRP) concentration is determined using the molybdate-reactive P method (APHA *et al.*, 1998).

21.3   Results and Discussion

21.3.1   The Effects of Wetland Age, Plants-Growing Season and Aeration

21.3.1.1   Phosphorus Storage

Nonaerated wetland cell 3 (C3) received 86.23 ± 2.2 g DRP m^{-2} (total 19.23 kg) in the first year (346 days) of operation (2003/2004). Of this, 64.14 ± 1.91 g DRP m^{-2} (14.30 kg) was added during a 147 days growing-season period (Fig. 21.1A). The wetland retained a total of 39.59 ± 0.82 g DRP m^{-2} (8.83 kg), over the entire year, of which 30.01 ± 0.83 g m^{-2} (6.69 kg) was stored during the growing season (147 days).

In the second year (354 days) of operation (2004/2005), C3 received a 2.5-fold increase of DRP (212.60 ± 0.8 g m^{-2}, total 47.41 kg) over the previous year (Fig. 21.1B, A) due to serious malfunctioning of the inlet flow distribution system which occurred in February 2005 and lasted for 97 days (Drizo *et al.*, 2006). Consequently, there was no increase in DRP storage by the wetland in this period: C3 stored 41.55 ± 1.22 g m^{-2} (9.3 kg), which was similar to the amount stored in the first year (Fig. 21.1B, A). Although the amount of DRP added during the second growing season (76.80 ± 1.37 g DRP m^{-2}, 17.13 kg) was similar to the total added over the entire previous season (14.30 kg), the wetland stored 2.5-fold lower amount of DRP (11.77 ± 0.7 g DRP m^{-2}, 2.62 kg) due to hydraulic and organic overloading (Fig. 21.1B, A).

In the third year (372 days) of operation (2005/2006), C3 received 136.23 ± 2.1 g DRP m^{-2} (30.38 kg). The amount of DRP retained during this period was 7.8-fold lower (5.33 ± 0.3 g m^{-2}, 1.19 kg) than the DRP storage documented in the previous year (Fig. 21.1C, B). The decrease in C3 DRP storage during the third growing season followed a similar pattern of declining efficiency as observed between the first two growing seasons. Although the amount of DRP added during this period (147 days) was similar (74.80 ± 1.7 g m^{-2}, 16.68 kg) to the previous growing season, the amount of DRP stored decreased 6.84-fold, from 11.77 ± 0.7 to only 1.72 ± 0.3 g m^{-2} (0.38 kg) (Fig. 21.1B, C).

An exponential decrease in DRP storage occurred after only 3 years of operation, indicating that the wetlands saturated in a shorter period than suggested by Kadlec (1999), who stated 5–6 years as the approximate time it takes for wetland vegetation to reach maturation. We attribute this rapid decrease in DRP storage to the hydraulic and organic overloading of the system (Drizo *et al.*, 2006).

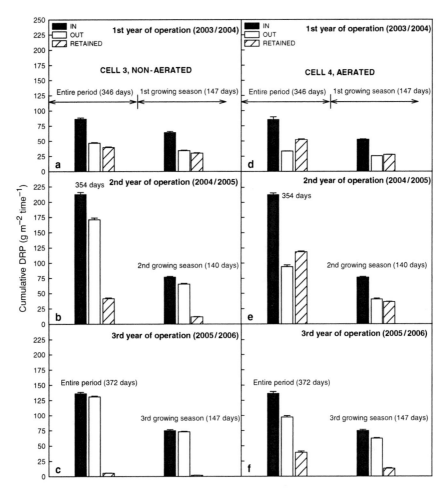

Fig. 21.1 Cumulative DRP added and retained (g m^{-2} time^{-1}) by nonaerated cell 3 and aerated cell 4, during each year of operation and each growing season. Nonaerated cell 3, first year of operation (A), second year of operation (B) and third year of operation (C); Aerated cell 4 first year of operation (D), second year (E) and third year (F)

The amount of DRP load to the aerated cell 4 (C4) was almost equal to the load into nonaerated C3 during the investigated periods (Fig. 21.1D, A). In the entire first year (346 days) of operation (2003/2004), C4 received 85.42 ± 4.39 g DRP m^{-2} (total 19.05 kg) and retained 1.3-fold more DRP (52.27 ± 1.33 g m^{-2}, 11.66 kg) than the amount retained by the C3 during the same period. In the first growing season, C4 received 52.78 ± 0.79 g DRP m^{-2} (11.78 kg) and stored 27.56 ± 0.29 g m^{-2} (6.14 kg), which was nearly equal to the amount of DRP stored by C3 during the same period (30.01 ± 0.8 g m^{-2}, 6.69 kg). These results indicate that the presence of aeration did not have a significant effect on the amount of DRP stored by plant biomass during this first year of investigation.

The amount of DRP added to aerated C4 during the second year (354 days) of operation (2004/2005) was equal to the amount added to nonaerated C3 (212.60 ± 3.04 g m⁻², total 47.41 kg) during the same time period. The amount of DRP stored in C4 during this period (118.63 ± 1.13 g m⁻², 20.96 kg) was 2.3-fold higher than the amount stored in the previous year of operation (Fig. 21.1E, D) and 2.9-fold higher than the amount stored by C3 in the same year of operation. Contrary to the decline in DRP retention of nonaerated C3 observed between growing seasons 1 and 2, the amount of DRP stored by aerated C4 during the second growing season (36.35 ± 0.46 g m⁻², 8.11 kg) was 1.3-fold higher than the amount stored during the first growing season (6.14 kg). The amount of DRP stored by C4 during this 147-day period was 3.0-fold higher than the amount stored by a nonaerated C3 during the same growing season (Fig. 21.1E, B).

In the third year (372 days) of operation (2005/2006), C4 received the same amount of DRP (136.23 ± 2.1 g DRP m⁻², 30.38 kg) as nonaerated C3 (Fig. 21.1F). C4 stored 3.0-fold less DRP than in the previous year (39.12 ± 2.19 g m⁻², total 8.72 kg), showing the same pattern of exponential decrease as observed in C3. However, the amount of DRP stored (8.72 kg) during this entire 372-day period was still 7.3-fold higher than the amount stored by the nonaerated C3 (1.19 kg). During the third growing season, C4 stored 2.8-fold less DRP (12.76 ± 1.10 g DRP m⁻², 2.8 kg) than in the previous growing season, exhibiting a similar pattern of DRP storage decline as observed in C3. However, the amount of DRP stored during this 147-day growing season (2.8 kg) was 7.4-fold higher than the amount stored by nonaerated C3 (0.38 kg), indicating the beneficial effect of aeration on the wetlands' overall phosphorus storage.

In total, nonaerated and aerated wetland cells 3 and 4 received 435.06 ± 2.3 g DRP m⁻² (97.02 kg) each, during the entire 3-year period (1,073 days) of investigation (Fig. 21.2). The amount of DRP added during three growing seasons (433 days) was 2.0-fold lower (215.74 ± 1.66 in C3 and 204.33 ± 3.86 g DRP m⁻² in C4, or 48.11 and 45.6 kg, respectively) than the total DRP load over 3 years, corresponding to a 2.5-fold shorter period of loading (Fig. 21.2). Of the total amount of DRP stored by nonaerated C3 (86.47 ± 2.1 g DRP m⁻², 19.28 kg) over the 1,073-day period, 43.51 ± 1.2 g DRP m⁻² (9.7 kg) was stored during the three growing seasons (Fig. 21.2A). Supplemental aeration in C4 resulted in a 2.4-fold increase of DRP storage (210 ± 1.55 g DRP m⁻², 48.83 kg) during the entire 3-year period, as compared to the DRP retention of the nonaerated C3 (Fig. 21.2B). Aeration also had positive effect on C4 DRP storage during the three growing seasons, which was 1.8-fold higher (76.67 ± 3.83 g m⁻², 17.10 kg) than the amount stored by C3.

The quantities of DRP stored during the three growing seasons within cells 3 and 4 represented 10.0 and 17.7%, respectively, of the total amount of DRP added to these cells over the entire 1,072 days. Our results are similar to those reported by Tanner (2001), who found that after three growth seasons, P accumulation by *Schoenoplectus tabernaemontani* (C.C. Gmelin) Palla (soft-steam bulrush) contributed to only 6–13% of the total P removal by a dairy-treatment wetland in New Zealand. However, DRP storage by *Schoenoplectus fluviatilis* (Torr.) (river bulrush) found in nonaerated cell 3 is 2.0-fold higher than P storage observed in *Typha* (5%)

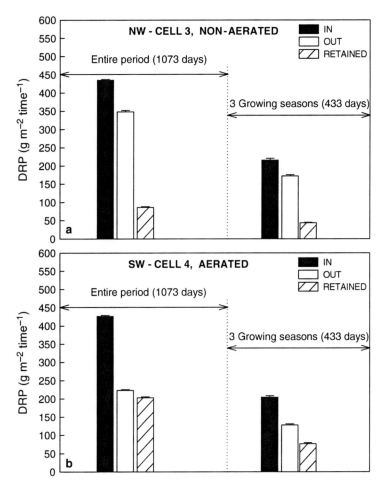

Fig. 21.2 Cumulative DRP added and retained by cell 3 (A) and cell 4 (B) during the entire 3-year period of investigation and the three growing seasons

after 8 years of operation, recently reported for a free surface flow wetland treating dairy-effluent wastewater (Gottschall *et al.*, 2007).

When looking at the total DRP retained by wetland cells 3 and 4 during the entire 3 years of operation (1,073 days), the quantities stored during the three growing seasons (433 days) represented 50.3% and 36.50%, respectively, indicating a potentially important role of vegetation in the overall DRP storage, regardless of supplemental aeration. Our results are similar to the observations made by DiPietro (2004), who investigated the effects of forced aeration on root architecture, plant biomass, and wastewater treatment of three wetland species (*Scirpus atrovirens* Willd., *Iris versicolor* L., and *Schoenoplectus fluviatilis* (Torr.) M.T. Strong. DiPietro (2004) did not observe any statistically significant difference in *Schoenoplectus fluviatilis* root architecture, plant biomass or nutrients storage between the aerated and nonaerated treatment. However, the positive effects of supplemental aeration on organic matter degradation, total suspended solids, nitrogen removal and plant growth development in cold climates have been reported by other authors (Jamieson *et al.*, 2003; Ouellet-Plamondon *et al.*, 2006). Munoz *et al.* (2006) showed evidence that supplemental aeration prevents clogging and preferential flows. Higher DRP storage in aerated cell 4 could have been attributed to the increased organic matter decomposition, suspended solids removal and better contact time between wastewater and the wetland substrate.

21.3.1.2 Phosphorus Mass Removal

DRP mass removal efficiency decreased exponentially with wetlands age in both nonaerated cell 3 and aerated cell 4 (Fig. 21.3). Overall, nonaerated C3 DRP mass removal efficiency during the 3-year period of investigation was low, averaging 19.9%. In the first year of operation (2003/2004), DRP mass removal efficiency was 45.9% ± 1.0%; it decreased 2.3-fold during the second year of operation (2004/2005), falling from 45.91% to 19.54% ± 0.4%. C3 DRP treatment efficiency in the third year of operation (2005/2006) was only 3.9% ± 0.2%, 5.0-fold lower than in the second year and 11.8-fold lower than the efficiency achieved in the first year of operation. Nonaerated C3 DRP mass removal efficiencies during the first, second, and third growing seasons followed the same trend of exponential decrease, representing 34.8% ± 1.1%, 5.5% ± 0.7% and 1.3% ± 1.2% of the total DRP load added each year, respectively. However, it is worth noting that 75.8%, 28.3% and 32.3% of the overall DRP retained during the first, second and third years of operation, respectively, was retained during the growing season, indicating positive effects of vegetation to the overall DRP mass removal efficiency (Fig. 21.3A).

Aerated C4 DRP mass removal efficiency averaged 48.4% over the 3 years, 2.4-fold higher than the efficiency of nonaerated C3 during the same period. In the first year of operation (2003/2004) C4 overall DRP mass removal efficiency was 61.2% ± 1.8%; the efficiency achieved in the second year of operation (2004/2005) was only slightly lower (55.8% ± 1.8%) than during the previous year of operation. However, in the third year of operation (2005/2006) DRP efficiency fell to 28.7% ± 1.2%, 2.0-fold lower than the efficiency achieved in the second year and 1.9-fold lower than the efficiency achieved in the first year of operation. DRP removal efficiency between the first and second growing seasons decreased 1.9-fold (from 32.3% ± 1.2% to 17.1% ± 0.4%), followed by a 1.7-fold decline to 9.8% ± 0.8% during the third growing season (Fig. 21.3B). The positive effect of vegetation on the overall DRP mass removal was also observed in the aerated cell 4: the quantities of DRP retained during the first, second and third growing seasons represented 52.7%, 30.6% and 32.6% of the overall DRP retained during the entire 1,072 days of operation.

The decline in overall DRP reduction performance was most prominent during the third year of operation, in both nonaerated and aerated wetland cells. Although a decrease in phosphorus treatment performance is expected to occur as the wetland ages (Kern & Idler, 1999; Kadlec, 1999; Hunt & Poach, 2001; Schaafsma *et al.*, 2000, Gottschall *et al.*, 2007), in the case of our system, it occurred at an early stage

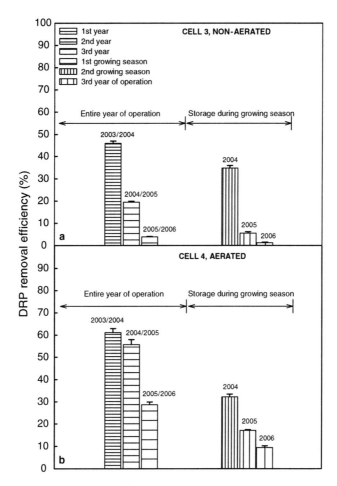

Fig. 21.3 DRP mass removal during each of the 3 years of operation and each of the three growing seasons. (A) nonaerated cell 3; (B) aerated cell 4

of operation (within the first 3 years). Such rapid decrease in treatment performance has been attributed to the high hydraulic loading rate ($10.44 \pm 0.001\,\mathrm{cm\ day^{-1}}$) which was nearly twice as high as those recommended for dairy-wastewater treatment wetlands in North America ($5.5\,\mathrm{cm\ day^{-1}}$) (Drizo et al., 2006).

21.3.2   The Effects of Wetland Age and Aeration on Phosphorus Uptake and Biomass Development

At the end of the first growing season (2004), the P content in the nonaerated wetland C3 AG biomass inlet and outlet areas was 4.05 ± 0.48 and $4.95 \pm 1.05\,\mathrm{g\ P\ kg^{-1}}$, respectively, averaging $4.5 \pm 0.64\,\mathrm{g\ P\ kg^{-1}}$; P content in the BG biomass inlet and outlet wetland areas was 5.0 ± 0.84 and $7.3 \pm 1.27\,g$ P kg^{-1}, respectively, averaging $6.15 \pm 0.92\,g$ P kg^{-1} (Fig. 21.4A).

Phosphorus content in both AG and BG biomass decreased between the first and second growing seasons. At the end of the second growing season (2005), P content in the AG biomass wetland inlet and outlet areas was 1.36 ± 0.26 and $1.42 \pm 0.38\,g$ P kg^{-1}, respectively, being in average 3.0-fold lower than at the end of the first season ($1.39 \pm 0.04\,g$ P kg^{-1}). P contents in the BG biomass inlet and outlet wetland areas were 6.42 ± 0.97 and $6.36 \pm 1.16\,g$ P kg^{-1}, respectively, averaging $6.40 \pm 0.40\,g$ P kg^{-1} and being slightly higher than at the end of the first season (Fig. 21.4B).

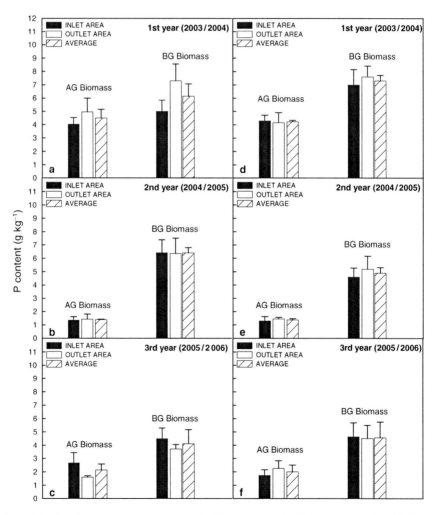

Fig. 21.4 Phosphorus content in nonaerated cell 3 and aerated cell 4 aboveground and belowground biomass at the end of the first (A, D), second (B, E) and third (C, F) growing seasons

When compared to the second growing season, P content of C3 AG biomass during the third growing season showed a slight increase; however, third growing season biomass P content was still significantly lower ($p < 0.01$) than that determined at the end of the first season. Average P content of the AG biomass wetland inlet and outlet areas at the end of the third growing season (2006) was 2.0-fold higher (2.67 ± 0.77 g P kg^{-1}) than the P content in the second growing season (Fig. 21.4B, A); P content in the AG biomass outlet wetland area was 1.60 ± 0.11 g P kg^{-1}, averaging 2.14 ± 0.44 g P kg^{-1} between inlet and outlet areas (Fig. 21.4C). P contents in the BG biomass inlet and outlet wetland areas were 4.51 ± 0.78 and 3.71 ± 0.35 g P kg^{-1}, respectively, averaging 4.11 ± 1.06 g P kg^{-1}; average BG biomass P content at the end of this season was 1.6-fold and 1.5-fold lower than the BG biomass P contents found at the end of the first and second growing seasons (Fig. 21.4).

At the end of the first growing season, P content in the AG biomass of aerated wetland C4 inlet and outlet areas was 4.3 ± 0.42 and 4.15 ± 0.76 g P kg^{-1}, respectively, averaging 4.22 ± 0.11 g P kg^{-1}. P content in the BG biomass inlet and outlet wetland areas was 7.0 ± 1.15 and 7.6 ± 0.82 g P kg^{-1}, averaging 7.3 ± 0.42 g P kg^{-1} (Fig. 21.4D). The AG and BG biomass P contents in this wetland were not significantly different from that found in the nonaerated C3, indicating that supplemental aeration did not have any significant effect on the P content plant biomass during this first growing season.

The P content in the AG biomass wetland inlet and outlet areas at the end of the second growing season (2005) was 1.30 ± 0.32 and 1.44 ± 0.12 g P kg^{-1}, averaging 1.37 ± 0.10 g P kg^{-1}. Second season AG biomass P content was significantly lower ($p < 0.01$) than the AG biomass P content measured at the end of the first season. Similarly, second season P contents in the BG biomass inlet and outlet wetland areas was 4.60 ± 0.67 and 5.18 ± 0.98 g P kg^{-1}, averaging 4.90 ± 0.41 g P kg^{-1}, which was 1.5-fold lower than at the end of the first season (Fig. 21.4E, D) and 1.3-fold lower than in the nonaerated cell 3 during the same time period (Fig. 21.4E, B).

Phosphorus content of the AG biomass wetland inlet and outlet areas at the end of the third growing season (2006) was 1.74 ± 0.42 and 2.25 ± 0.58 g P kg^{-1}, respectively, averaging 2.0 ± 0.50 g P kg^{-1}. P content in the BG biomass inlet and outlet wetland areas was 4.65 ± 1.04 and 4.51 ± 0.98 g P kg^{-1}, respectively, and averaged 4.58 ± 1.16 g P kg^{-1}. Average BG biomass P content at the end of this season was 1.6-fold lower than the P content measured at the end of the first growing season but nearly equal to the P content observed at the end of the second season (Fig. 21.4).

Throughout the 3-year period of investigation, BG biomass P content was significantly higher ($p < 0.001$) than in the AG biomass (Fig. 21.4). These results confirm findings by Tanner (2001) and Gottschall et al. (2007), who observed nutrients allocation from the AG to BG biomass towards the end of a growing season (Tanner, 2001; Gottschall et al., 2007). A significant drop in AG plant biomass was observed during the 3-year study period, indicating a potential decline in plant health and nutrients uptake.

21.4   Conclusions

Supplemental aeration had a positive effect on the overall DRP mass removal efficiency, contributing to a 2.4-fold increase (from 19.9% in nonaerated C3 to 48.4% in aerated C4). However, after 3 years of operation, both C3 and C4 wetland cells exhibited a sharp decline in DRP storage, indicating wetlands saturation regardless of supplemental aeration. Hydraulic and organic overloading greatly impacted DRP mass removal and may have adversely affected plant contribution to wetland treatment.

The quantities of DRP stored during the three growing seasons occurring within this study represented only 10.0% C3 and 17.7% C4 of the total amount of DRP added throughout the entire 3-year study period, initially suggesting that plant contribution to P removal was minimal. However, when considering that the DRP quantities stored only during the three growing seasons represented 50.3% C3 and 36.50% C4 of the total DRP retention measured over the entire study period, such results indicate that vegetation had an important role in the overall DRP storage, regardless of supplemental aeration.

There was no significant difference in the AG or BG P content between C3 and C4 wetland cells. BG biomass had a significantly higher ($p < 0.001$) P content than AG biomass, throughout the 3-year period of investigation. A significant drop in AG plant biomass was observed during the 3-year study period indicating a potential decline in plant health, and therefore, effectiveness.

Future research should focus on in-series multistage CW designs that can improve the systems' DRP retention of nutrient-rich agriculture-wastewater runoff.

Acknowledgements The financial support from Senator Jeffords Office, USDA-CSREES Special Research Grant is greatly appreciated. We also wish to thank Dr. Donald Foss for his efforts in establishing the CWRC and Joel Tilley for conducting plant mineral analyses.

References

APHA, AWWA, and WEF (1998). In A.D. Eaton, L.S. Clesceri, & A.E. Greenberg (Eds.), *Standard methods for the examination of water and waste water*, 19th ed. Washington DC: American Public Health Association.

Allen, W.C., Hook, P.B., Biederman, J.A., & Stein, O.R. (2002). Temperature and wetland plant species effects on wastewater treatment and root zone oxidation. *Journal of Environmental Quality, 31*, 1010–1016.

Brix, H. (1997). Do macrophytes play a role in constructed treatment wetlands?. *Water Science and Technology, 35*(5), 11–17.

Brix, H., Schierup, H.-H., & Arias, C. (2006). Twenty years experience with constructed wetland systems in Denmark – what did we learn? In *Proceedings Of the 10th International Conference Constructed Wetlands for Water Pollution Control* (pp. 1633–1641). MAOTDR: Lisbon, Portugal.

Cooper, P. (2007). The Constructed Wetland Association UK database of constructed wetland systems. *Water Science & Technology 56*(3), 1–6.

DiPietro, T. (2004). Wastewater strength and forced aeration impacts on root architecture, plant biomass, and wastewater treatment of three plant species using laboratory-scale wetland cells. MS Thesis, University of Vermont, Burlington.

Drizo, A., Twohig, E., Weber, D., Bird, S., & Ross, D. (2006). Constructed wetlands for dairy effluent treatment in Vermont: 36 months of operation. In *Proceedings of the 10th International Conference Constructed Wetlands for Water Pollution Control* (pp. 1611–1621). MAOTDR: Lisbon, Portugal.

Gottschall, N., Boutin, C., Crolla, A., Kinsley, C., & Champagne, P. (2007). The role of plants in the removal of nutrients at a constructed wetland treating agricultural (dairy) wastewater, Ontario, Canada. *Ecological Engineering, 29*, 154–163.

Hammer, D.A. (1989) (ed). *Constructed Wetlands for Waste Water Treatment: Municipal, Industrial and Agricultural.* Lewis Publishers. Boca Raton, Florida.

Hunt, P.G., & Poach, M.E. (2001). State of the art for animal wastewater treatment in constructed wetlands. *Water Science and Technology, 44 (11–12)*, 19–25.

Jamieson, T.S., Stratton, G.W., Gordon, R., & Madani, A. (2003). The use of aeration to enhance ammonia nitrogen removal in constructed wetlands. *Canadian Biosystems Engineering, 45*, 9–14.

Kadlec, R.H. (1999). The limits of phosphorus removal in wetlands. *Wetlands Ecology and Management, 7*, 165–175.

Kadlec, R.H., & Knight, R. (1996). *Treatment Wetlands.* Lewis Publishers, Chelsea, MI.

Kadlec, R.H., Knight, R., Vymazal, J., Brix, H., Cooper, P., & Haberl, R. (2000). *Constructed Wetlands for Pollution Control.* Published by International Water Association, London, UK.

Kern, J., & Idler, C. (1999). Treatment of domestic and agricultural wastewater by reed bed systems. *Ecological Engineering, 12*, 13–25.

Knight, R.L., Payne, V.W.E., Borer, R.E., Clarke, R.A., & Pries, J.H. (2000). Constructed wetlands for livestock wastewater management. *Ecological Engineering, 15*, 41–55.

Luderitz, V., & Gerlach, F. (2002). Phosphorus removal in different constructed wetlands. *Acta Biotechnologica, 22*, 91–99.

Munoz, P., Drizo, A., & Hessoin, C. (2006). Flow patterns of dairy farm wastewater constructed wetlands in a cold climate. Water Research, in press.

Ouellet-Plamondon, C., Chazarenc, C., Comeau, Y., & Brisson, J. (2006). Artificial aeration to increase pollutant removal efficiency of constructed wetlands in cold climate. *Ecological Engineering, 27*, 258–264.

Picard, C.R., Lauchlan, F.H., & Steer, D. (2005). The interacting effects of temperature and plant community type on nutrient removal in wetland microcosms. *Bioresource Technology, 96*, 1039–1047.

Reddy, K.R., & Smith, W.H. (1987). *Aquatic plants for water treatment and resource recovery.* Orlando, FL: Magnolia.

Schaafsma, J.A., Baldwin, A.H., & Streb, C.A. (2000). An evaluation of a constructed wetland to treat wastewater from a dairy farm in Maryland, USA. *Ecological Engineering, 14*, 199–206.

Serodes, J.B., & Normand, D. (1999). Phosphorus removal in agricultural wastewater by a recently constructed wetland. *Canadian Journal of Civil Engineering, 26*, 305–312.

Smith, E., Gordon, R., Madani, A., & Stratton, G. (2005). Cold climate hydrological flow characteristics of constructed wetlands. *Canadian Biosystems Engineering, 47*, 1–7.

Tanner, C.C., Clayton, J.S., & Upsell, M.P. (1995). Effect of loading rate and planting on treatment of dairy farm wastewaters in constructed wetlands-II. Removal of nitrogen and phosphorus. *Water Research, 29*, 27–34.

Tanner, C.C. (2001) Growth and nutrient dynamics of soft-stem bulrush in constructed wetlands treating nutrient-rich wastewaters. *Wetlands Ecology and Management 9*, 49–73.

Vymazal, J. (2006). Constructed wetlands with emergent macrophytes: from experiments to a high quality treatment technology. In *Proceedings of the 10th International Conference on Constructed Wetlands for Water Pollution Control* (pp. 3–27). MAOTDR: Lisbon, Portugal.

Vymazal, J., Brix, H., Cooper, P.F., Green, M.B., & Haberl, R. (1998). *Constructed wetlands for wastewater treatment in Europe.* Leiden, The Netherlands: Backhuys.

Wallace, S., Parkin, G., & Cross, C. (2001). Cold Climate Wetlands: Design & Performance. *Water Sci Tech. 44*(11–12), 259–265.

Weber, D., Drizo, A., Twohig, E., Bird, S., & Ross, D. (2007). Upgrading Constructed Wetlands Phosphorus Reduction from a Dairy Effluent using EAF Steel Slag Filters. *Water Science and Technology 56*(3), 135–143

Chapter 22
Performance of Reed Beds Supplied with Municipal Landfill Leachate

Ewa Wojciechowska and Hanna Obarska-Pempkowiak(✉)

Abstract Constructed wetlands can provide an effective and economical solution to landfill leachate treatment. In this chapter, performance of two constructed wetlands (CWs) for municipal leachate treatment, in Gdansk-Szadolki and in Gatka near Miastko (northern Poland) is discussed. The plant in Gdansk-Szadolki, consisting of two horizontal flow (HF) beds, has been in operation since 2001. The performance of the plant in the period 2004–2005 is discussed. Clogging problems and surface flow occurred in the system because of too-low hydraulic conductivity of the filter beds material. Neither the treatment effectiveness nor effluent pollutant concentrations meet the design criteria, though quite high efficiencies of polycyclic aromatic hydrocarbon (PAH) (over 70% at both beds), total nitrogen (N_{tot}) and ammonia nitrogen ($N-NH_4$) (61.7% for the bed I and 51.6% for the bed II) removal are obtained. The willow plantation in Gatka has been in operation since 1996. The leachate was transpired to the atmosphere and there was no outflow from the plant. Prior to the investigations of the plant performance (in the period December 2005–November 2006), the polyethylene wells were installed in order to collect pore water samples. Good removal efficiency of N_{tot} and $N-NH_4$ (70–78% and 78–88%, respectively) was observed.

Keywords Constructed wetlands, hydraulics, landfill leachate, reed, removal of pollutants, willow

22.1 Introduction

In the last years, several on-site constructed wetland systems for landfill leachate treatment were built in Europe and North America. According to the literature reports (Mæhlum, 1995; Peverly *et al.*, 1995; Kowalik *et al.*, 1996; Bulc *et al.*,

Gdansk University of Technology, Faculty of Civil and Environmental Engineering, Narutowicza 11/12, 80-952 Gdansk, Poland

(✉) Corresponding author: e-mail: hoba@pg.gda.pl

1997; Martin *et al.*, 1999; Kadlec, 2003; Johansson Westholm, 2003 Kinsley *et al.*, 2006), constructed wetlands can provide an effective treatment for landfill leachate and have potential to remove micropollutants present in the leachate as well as heavy metals. The most common plant species are *Phragmites australis* and *Typha latifolia*. In comparison to the highly effective treatment methods (microbiological and chemical oxidation processes and membrane techniques), applied for leachate treatment, the wetland systems are by far less expensive, both during construction and operation.

Roots of emergent macrophytes, like *P. australis*, grow in anaerobic environment and thus are capable of tolerating low oxygen concentrations, elevated concentrations of iron and manganese, as well as organic acids and other breakdown products of organic matter (Peverly *et al.*, 1995). Many authors also report the capability of emergent macrophytes to uptake heavy metals (Hawkins *et al.*, 1997; Crites *et al.*, 1997; Vymazal, 2003; Karathanansis & Johnson, 2003; Weis *et al.*, 2004; Obarska-Pempkowiak *et al.*, 2005). According to Peverly *et al.* (1995) *P. australis* grows well in the presence of high concentrations of $N-NH_4^+$ ($300 \, mg \, l^{-1}$), BOD_5 ($300 \, mg \, l^{-1}$), Fe ($30 \, mg \, l^{-1}$), Mn ($1.5 \, mg \, l^{-1}$) and K ($500 \, mg \, l^{-1}$).

The willow (*Salix viminalis*) is another hydrophyte species that can be applied in the wetland systems treating landfill leachate. Willow plantations are not so popular as reed bed systems, however they are successfully used for instance at the Örebro site (southern Sweden). *Salix* is a pioneer plant, easily growing at devastated and degraded sites, with a high potential of heavy metals accumulation (Hasselgren, 1989; Kowalik *et al.*, 1996).

In this chapter, the treatment performance of two constructed wetlands (CWs) for landfill leachate treatment is discussed. One of the facilities is a reed bed system at Szadolki near Gdansk and another one is a the willow plantation at Gatka near Miastko (northern Poland).

22.2   Study Facilities and Methods

A constructed wetland for leachate treatment was built in a municipal landfill in Gdansk-Szadolki in 2001. It consists of two parallel horizontal flow (HF)-CW beds (subsurface, horizontal flow of sewage) (Figs. 22.1 and 22.2). The area of each bed is $50 \times 50 \, m$ and the depth is $0.6 \, m$. The beds were planted with common reed (*P. australis*). According to the design project, the leachate treated in the beds ought to be discharged to Kozacki stream flowing through the landfill area. However, since the effectiveness of treatment is below expectations, the leachate is recirculated to one of the landfill's compartments.

The CW plant in Szadolki was designed according to the guidelines of Kickuth (1981) who recommended fine-grained soils as filter-bed materials. Kickuth was convinced that the initial low hydraulic conductivity of the medium would increase due to root penetration. Unfortunately, in many CWs built in 1980s, where this

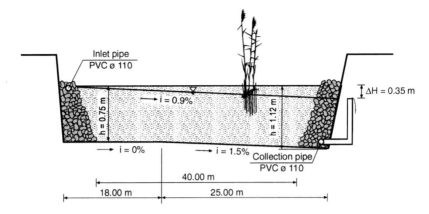

Fig. 22.1 Cross-section of constructed wetland in Szadolki

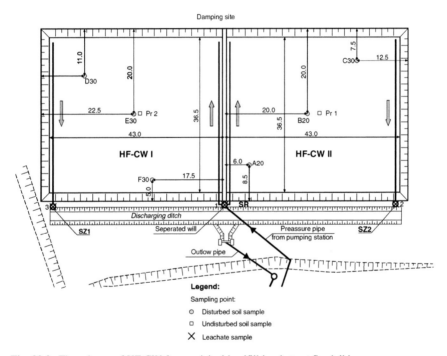

Fig. 22.2 The scheme of HF-CW for municipal landfill leachate at Szadolki

approach was applied, clogging of the beds occurred. This resulted in surface flow of sewage and deterioration of treatment effectiveness (Brix & Schierup, 1989). Hence medium and coarse sands are considered to be better filter bed materials till date (Peverly *et al.*, 1995).

In the CW Szadolki also, clogging took place at both beds, resulting in flooding and surface flow of leachate. Since the design criteria for the surface flow beds are different than for subsurface flow ones, expected treatment effectiveness was not achieved. Moreover, *P. australis* started to die off due to permanent flooding (another reason was probably very high water salinity – average Cl^- concentration was $821.3\,mg\,l^{-1}$), so the chance for hydraulic conductivity improvement due to root penetration vanished. The determined quantity of leachate water from landfill area varied from 6 to $240\,m^3\,day^{-1}$.

In 2004 and 2005, seven series of analyses of pollutants concentrations in the leachate were performed. The samples of leachate were collected at the inflow to the CW system and at the outflow of each of two beds and the efficiencies of pollutant removal were calculated. The concentrations of the following pollutants were measured: TSS, BOD_5, COD_{Cr}, $N-NH_4$, N_{tot}, TOC, polycyclic aromatic hydrocarbon (PAH), total alkalinity, Fe Mn and Cl^-. The analyses were performed according to Polish Standard Methods. Based on the pump capacity and pump working period, a quantity of leachate waters from landfill area in period 2002–2005 was evaluated.

In November 2004, the granulometric analyses of filtration beds media and permeability coefficients were performed. Two types of soils samples were collected from randomly chosen sampling points at the beds I and II: the disturbed-structure samples for grain size analyses (from the depth below 30 cm) and the undisturbed-structure samples for the measurement of permeability coefficient (from the depth 30–40 cm). The analyses were carried out according to Polish Norm PN-88/B0481.

Another analyzed facility for landfill leachate treatment was a willow plantation in Gatka, near Miastko (northern Poland). The facility has been constructed in 1996 to treat the leachate from municipal landfill. The average flow of leachate was approximately $5\,m^3\,day^{-1}$. The leachate is collected by drainage and outflows to a retention tank (30 × 30 m), where averaging of the leachate volume and composition takes place. The leachate outflow from the tank is directed to the soil filter (43 × 31 m) planted with willow *Salix* (Fig. 22.3). The depth of the filter is 0.8 m. *Salix* was growing only around point III, while the remaining area of the bed was covered by orchard grass (*Dactylis glomerata* L.). The leachate percolating through the soil filter is transpirated to the atmosphere by the plants and there is no outflow from the filter.

In the period December 2005–November 2006 the analyses of pollutants concentration in the leachate from CW Gatka were performed. The samples of inflowing leachate were collected at the inflow to the retention tank vegetated filter (point II, Fig. 22.3). Since the total amount of leachate applied to the bed got transpirated to the atmosphere, in order to evaluate the treatment processes taking place at the filter, two polyethylene wells were installed at the filter to assemble pore water samples collection (Fig. 22.3). Four polyethylene pipes 150 mm diameter were placed in the bed (Fig. 22.3, sampling points III, IV). The leachate percolating through the filter gathered at the bottom of the pipes and could be collected using a syphon.

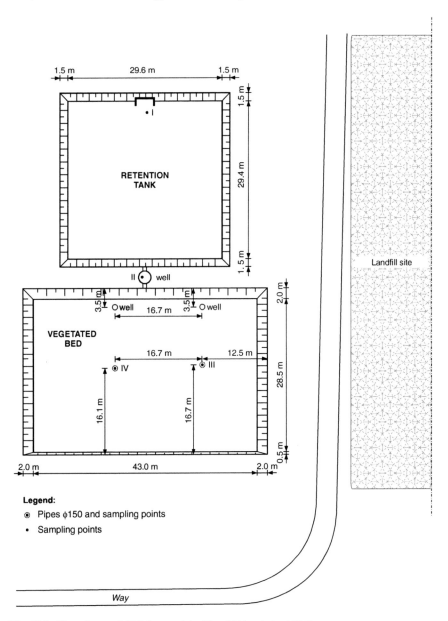

Fig. 22.3 The scheme of CW for municipal landfill leachate at Gatka

In the period from December 2005 to November 2006, five series of analyses were performed. The following parameters were measured: pH, TSS, BOD_5, COD_{Cr}, $N-NH_4^+$, N_{tot}, $N-NO_3^-$, organic N, total P, chlorides, Fe and total alkalinity. The analyses were performed according to Polish Standard Methods.

22.3 Results and Discussion

22.3.1 Reed Beds Hydraulics at Szadolki

In the design project of the CW system in Szadolki neither the grain size nor permeability coefficient were defined. Since the grain size distribution was considered to be the main reason for the beds flooding, the granulometric analyses and permeability coefficient measurement was undertaken. The results of soil analyses are presented in Table 22.1.

Permeability coefficients of both beds were very low in comparison to the recommended values (Kadlec et al., 2000; Reed et al., 1995).

Based on the results of filtration medium analyses, the hydraulic conductivity of the beds I and II in Szadolki was calculated from the law of mass conservation and the Darcy law:

$$Q_S = F \cdot v \tag{22.1}$$

$$v = k \cdot I \tag{22.2}$$

where:

Q_S – flow rate ($m^3 \, s^{-1}$)
F – the cross-section area of the bed perpendicular to the flow direction (m^2)
v – flow velocity ($m \, s^{-1}$)
k – permeability coefficient ($m \, s^{-1}$)
$I = \Delta H / L$, hydraulic gradient (–).

$$F = h \cdot W \tag{22.3}$$

where:
h – bed depth (m)
W – bed width (m)

Table 22.1 The results of filtration medium analyses from CW Szadolki

Parameter	Unit	Bed I Sample number			Bed II Sample number		
		A20	B20	C30	D30	E30	F30
The depth of sample collection	cm	20–30	20–30	30–40	30–40	30–40	30–40
Effective grain size	mm	0.0011	0.0007	0.0024	0.0016	0.001	0.0012
Coefficient of graining nonuniformity $U = d_{60}/d_{10}$	–	449	583	174	252	694	356
Permeability coefficient	$m \, s^{-1}$ $m \, day^{-1}$	$k_{10} = 5.77 \cdot 10^{-5}$ $k_{10} = 4.98$			$k_{10} = 2.55 \cdot 10^{-5}$ $k_{10} = 2.20$		

Table 22.2 The minimal permeability coefficients enabling the subsurface flow of leachate and the corresponding minimal effective grain size d_{10} calculated for different hydraulic loadings

Hydraulic loading Q (m³ day⁻¹)	Minimal permeability coefficient k (m s⁻¹)	Minimal effective grain size d_{10} (mm)
50	1.2×10^{-3}	0.35
100	2.4×10^{-3}	0.50
200	4.8×10^{-3}	0.70

Then the flow rate for each of the beds is equal to:

$$Q_S = F \cdot v = h \cdot W \cdot k \cdot I = h \cdot W \cdot k \cdot \Delta H / L \qquad (22.4)$$

The geometrical parameters of the beds (depth, width, length, etc.) are given in the Fig. 22.2. The values of permeability coefficients for the beds I and II are given in Table 22.1.

The calculated flow rates are 0.527 m³ day⁻¹ for bed I and 1.193 m³ day s⁻¹ for bed II. The total hydraulic capacity of the CW system (the sum of flow rates of both beds) is equal to 1.72 m³ day s⁻¹. In contrast, the hydraulic loading of the beds, evaluated based on the pump capacity and pump working period for 2002–2004, was varying from 6 to 240 m³ day s⁻¹. The hydraulic capacity of the beds is very small in comparison to the hydraulic loading of the beds, which resulted in surface flow of sewage and flooding of the beds.

The minimal permeability coefficient enabling the subsurface flow of leachate was calculated from the formula:

$$k = \frac{Q_S \cdot L}{h \cdot W \cdot \Delta h} \qquad (22.5)$$

The minimal effective grain size assuring the required permeability coefficient was calculated from the Hazen equation:

$$k = \frac{\left(d_{10}\right)^2}{100} \qquad (22.6)$$

where: k (m s⁻¹), d_{10} (mm)

on the condition that coefficient of grain nonuniformity: $U = \dfrac{d_{60}}{d_{10}} \leq 5$

The minimal permeability coefficients and corresponding effective grain sizes calculated for three different hydraulic loadings Q_s are given in the Table 22.2.

22.4 The Raw Leachate Quality

The quality of leachate discharged to the CW at Szadolki is presented in Table 22.3. The leachate composition at Szadolki was very unstable, which is indicated by very high standard deviation values and large differences between the minimal and maximal values. The most unstable parameters were BOD_5, COD, TOC, TSS, $N-NH_4^+$ and PAH. The additional tank at the beginning of treatment process would enable qualitative and quantitative averaging of the unstable discharge of the leachate.

The quality of raw leachate from the CW at Gatka is presented in Table 22.4. The leachate composition was more stable than at Szadolki, though high standard deviation values were also observed.

Table 22.3 The quality of leachate discharged to the CW treatment plant at Szadolki

Parameter	Mean ± standard deviation		Maximum	Minimum
pH	7.5	±0.06	7.4	7.5
TSS (mg l⁻¹)	150	±134	380	63.2
BOD_5 (mg l⁻¹)	792	±969	2,119	41
COD_{Cr} (mg l⁻¹)	1,616	±1,645	4,380	503
TOC (mg l⁻¹)	411	±221	622	181
$N-NH_4^+$ (mg l⁻¹)	302	±206	523	40.3
N_{tot} (mg l⁻¹)	433	±93	551	342
$P-PO_4^{3-}$ (mg l⁻¹)	1.8	±1.5	2.73	0.11
PAH (ng l⁻¹)	1,127	±1,349	2,680	238
Chlorides (mg l⁻¹)	749	±163	922	532
Total alkalinity (mval l⁻¹)	57.9	±13.5	71.7	39.5
SO_4^{2-} (mg l⁻¹)	34.6	±34.05	72	5.4
Fe (mg l⁻¹)	22.6	±13.95	38.12	11.17
Mn (mg l⁻¹)	0.98	±0.69	1.76	0.42
Cu (mg l⁻¹)	0.2	±0	0.2	0.2
Pb (mg l⁻¹)	0.3	±0	0.3	0.3
Cd (mg l⁻¹)	0.1	±0	0.1	0.1
Hg (mg l⁻¹)	0.001	±0	0.001	0.001
$BOD_5:COD_{Cr}$	0.49		0.56	0.18

Table 22.4 The quality of leachate discharged to the CW treatment plant at Gatka (point I)

	Inflow to the retention tank			
Parameter	Mean ± standard deviation		Maximum	Minimum
pH	8.43	±0.31	8.7	62.3
TSS (mg l⁻¹)	2,714	±139	2,830	2,561
COD_{Cr} (mg l⁻¹)	804	±271	111	600
BOD_5 (mg l⁻¹)	76.5	±20.6	101	51.4
N_{tot} (mg l⁻¹)	182	±74	270	101
$N-NH_4^+$ (mg l⁻¹)	85	±42	148	55
$N-NO_3^-$ (mg l⁻¹)	4.4	±2.8	8.2	1.7
N_{org} (mg l⁻¹)	93	±53	153	28.1
P_{tot} (mg l⁻¹)	5.0	±0.5	5.4	4.5
Chlorides (mg l⁻¹)	446	±107	574	333
Fe (mg l⁻¹)	16.1	±22.9	50.4	3.2
Total alkalinity (mval l⁻¹)	36.3	±8.0	48.2	31.5
$BOD_5:COD_{Cr}$	0.10	–	0.11	0.07

In the literature, considerable variation in the quality of leachate from different landfills has been reported. In older studies, the authors concluded that leachate from young landfills is characterized by high COD, even several thousands of milligrams per litre, while in leachate from old sites the COD concentrations decrease below a few hundreds of milligrams per litre (Henry et al., 1987). There are, however, newer reports suggesting that sometimes 'methanogenic' conditions are established relatively fast. Hence, even the leachate from a relatively young landfill can contain low COD (Lo, 1996).

The measure of organics bioavailability in the leachate is the BOD_5 to COD ratio. According to Surmacz-Górska (2000), at young landfill sites (the age of landfilled wastes not higher than 3–5 years) the BOD_5 to COD ratio is high, reaching even 0.7, indicating high biodegradability of organics in the leachate. Both COD and BOD_5 concentrations are very high (over 4,000 mg l^{-1} and over 6,000 mg l^{-1}, respectively). The pH is acidic (<6.5) indicating that the acidogenic fermentation phase products (volatile fatty acids) are present. At the mature landfills (5–10 years) the BOD_5 to COD ratio decreases to 0.3–0.5, since the easily biodegradable organics (BOD_5) are consumed. The pH increases to 6.5–7.5. The leachate from old landfills (over 10 years) is characterized by the low BOD_5 to COD ratio (<0.1) and pH over 7.5.

The average COD concentration in the leachate from Szadolki was 1,616 mg l^{-1} (with the minimum of 503 mg l^{-1} and the maximum of 4,380 mg l^{-1}), which is relatively high. The COD of the leachate from Gatka was lower (the average value 804 mg l^{-1}, minimum 600 mg l^{-1}, maximum 1,111 mg l^{-1}), indicating that 'methanogenic' conditions are already established at the landfill. The BOD_5 to COD ratio at Szadolki was 0.49, indicating only partial stabilization of wastes and relatively high bioavailability of the organics present in the leachate, while in Gatka it was 0.1, showing that the biodegradation of wastes is already accomplished and the bioavailability of the leachate organics is low. These observations are confirmed by the pH values, that are lower at Szadolki (7.5) than in Gatka (8.43).

Apart from organics, $N-NH_4$ is the principal pollutant in the leachate. Ammonia nitrogen is present in the leachate from young landfills owing to the deamination of amino acids during destruction of organic compounds (Tatsi & Zoubolis, 2002 after Crawford & Smith, 1985). Leachate from older landfills is rich in $N-NH_4$ due to hydrolysis and fermentation of the nitrogenous fractions of biodegradable substrates. Ammonia concentrations in leachate from different landfills may vary from tens or hundreds of milligrams of $N-NH_4^+$ l^{-1} (Surmacz-Górska, 1999) to even over 10,000 mg of $N-NH_4^+$ l^{-1} (Tatsi & Zoubolis, 2002). The average $N-NH_4$ concentrations in the analysed leachates were as follows: 301.8 mg l^{-1} and 85.0 mg l^{-1} at Szadolki and Gatka, respectively. These values are in agreement with the findings of Chu et al. (1994), who report that after a period of 3–8 years the $N-NH_4$ concentration reaches mean value of 500–1,500 mg l^{-1} and it will remain at this level for at least 50 years.

The concentrations of toxic heavy metals (Cu, Pb, Cd and Cr) were quite low – slightly above the detection level, so the potential environmental hazard is minimal. Among the analysed metals only the Fe concentrations were high. High concentration of Fe probably contributed to clogging of the beds at Szadolki.

Chlorides concentration in the leachate from Szadolki was high (748 mg l⁻¹), although from the literature even higher values are known – (Kulikowska & Klimiuk, 2007, mean chlorides concentration in the leachate from Wysieka landfill, Poland: 954 mg l⁻¹). The chlorides concentration in the leachate from Gatka was lower: 446 mg l⁻¹.

22.5   Treatment Efficiency

In case of the CW Szadolki, the reed beds I and II applied for landfill leachate treatment significantly differed in treatment efficiency (Table 22.5, Fig. 22.4). The bed I more effectively removed organic substances and nitrogen compounds, while the bed II achieved higher efficiency of iron and manganese removal.

Table 22.5 Composition of landfill leachate after treatment at Szadolki

	Bed I		Bed II	
	Mean	SD	Mean	SD
TSS (mg l⁻¹)	84.8	17.2	124	36.8
BOD₅ (mg l⁻¹)	303	68.9	576	116
COD$_{Cr}$ (mg l⁻¹)	1,045	281	1,422	426
TOC (mg l⁻¹)	272	29	356	48
N-NH₄⁺ (mg l⁻¹)	98.4	11.2	146	27.1
N$_{tot}$ (mg l⁻¹)	148	23.1	209	45.1
PAH (ng l⁻¹)	330	12.4	302	22.3
Total alkalinity (mval l⁻¹)	38.5	6.9	37.8	8.9
Fe (mg l⁻¹)	15.1	8.8	13.0	4.6
Mn (mg l⁻¹)	0.5	0.13	0.4	0.22
BOD₅:COD	0.29		0.40	

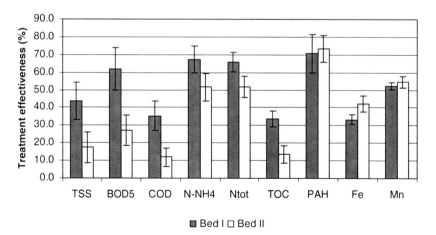

Fig. 22.4 Average treatment efficiency with standard deviation – bed I and bed II at Szadolki

High efficiency of PAH removal was observed – over 70% at both beds (Fig. 22.4). Also the N_{tot} and $N-NH_4^+$ removal efficiency at the bed I was high, though at the bed II it was lower. The decrease of N_{tot} and $N-NH_4^+$ concentrations as well as the decrease of total alkalinity and pH in the effluent indicate nitrification process took place. The BOD_5 to COD ratios after treatment at the beds I and II were lower than in the raw leachate (0.49). The decrease of the ratio to 0.29 (bed I) indicated that biodegradation of available organic substance took place at the bed. In case of the bed II, the BOD_5 to COD ratio decrease was slight (from 0.49 to 0.40), confirming poor efficiency of treatment processes.

Efficiency of organic substance removal (especially COD) in CW Szadolki was low in comparison to other CWs for leachate treatment. Mæhlum (1995) reported the BOD_5- and COD-removal efficiencies in the plant in Esval (Norway) of 91% and 88%, respectively. Also TOC-removal effectiveness was high (71%). Iron was removed in 88%. The CW in Esval had similar construction to CW Szadolki (two HF-CW beds working in parallel), but in Esval, the leachate was pretreated in an aeration lagoon and the effluent of HF-CW beds was polished in a surface flow bed. The major difference between Esval and Szadolki was the filter bed material – in Esval gravel (10–20 mm diameter) was used. The treatment plant in Esval works in a cold climate (average February temperature $-12°C$); thus, similar treatment efficiencies are possible to achieve in Poland, if the CW is properly designed.

Other literature reports confirm that pollutants-removal efficiencies (especially organic substances) can be significantly higher than in CW Szadolki. In Dragonja (Slovenia) removal efficiencies of COD, BOD_5, $N-NH_4$ and Fe were as follows: 68%, 46%, 81% and 80%, respectively (Bulc et al., 1997). Kinsley et al. (2006) reported 93–99% BOD_5 and 97–99% $N-NH_4^+$ removal efficiencies.

The composition of landfill leachate from the sampling points II–IV, after retention tank and at the vegetation filter at Gatka is presented in the Table 22.6. Figure 22.5 presents average treatment efficiencies of selected pollutants.

Table 22.6 Composition of landfill leachate after treatment at Gatka (II – after retention tank; III – the vegetation filter – spot with willow; IV – the vegetation filter – spot with orchard grass)

Parameter	II		III		IV	
	Mean	SD	Mean	SD	Mean	SD
pH	8.4	0.1	7.2	0.2	7.4	0.1
TSS (mg l^{-1})	2,596	171	312	142	413	369
COD_{Cr} (mg l^{-1})	648	55.2	788	2.4	576	14.1
BOD_5 (mg l^{-1})	60.3	16.9	43.7	9.9	54.0	11.7
N_{tot} (mg l^{-1})	116	17.4	34.2	6.2	25.9	8.1
$N-NH_4^+$ (mg l^{-1})	75.0	7.4	16.5	0.6	8.8	0.5
$N-NO_3^-$ (mg l^{-1})	2.0	1.1	1.4	0.3	1.3	0.3
N_{org} (mg l^{-1})	38.8	20.3	12.9	3.5	15.8	7.8
P_{tot} (mg l^{-1})	5.2	0.5	1.1	0.6	1.7	1.7
Chlorides (mg l^{-1})	321	44.8	582	17.8	574	211.4
Fe (mg l^{-1})	3.1	0.2	5.1	0.7	8.2	3.7
BOD_5:COD	0.09		0.06		0.09	

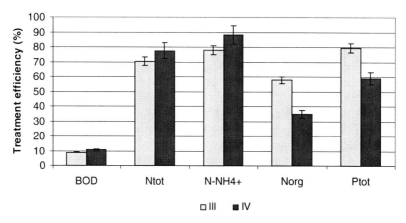

Fig. 22.5 Average treatment efficiency at willow plantation in Gatka with standard deviation

Comparison of the pollutants concentrations before and after the retention tank in Gatka (Tables 22.4 and 22.6) leads to the conclusion that the retention tank fails to play its role. The pollutants concentrations in leachate outflowing from the retention tank (point II) are only slightly lower than in the point I (before the tank). Especially the COD and BOD_5 concentrations are high at the point II (648 mg l^{-1} and 60.3 mg l^{-1}, respectively). The TSS concentration at the inflow to retention tank (point I) is equal to 2,714 mg l^{-1}, while after the tank (point II) it is 2,596 mg l^{-1}, indicating that resuspension of sediments takes place in the retention tank. The thick layer of sediments (approximately 0.8 m) has gathered over the 10 years of tank operation. The sediments were dredged out in December 2006 (no samples were collected after this event).

The organics concentrations (COD, BOD_5) decrease was not significant at the vegetation filter. The mean COD concentration decreased between points II and IV from 648 mg l^{-1} to 576 mg l^{-1}, while at the point III the increase to 788 mg l^{-1} was observed. The average BOD_5 concentrations decreased from 60.3 mg l^{-1} (II) to 43.7 mg l^{-1} (III) and 54.0 mg l^{-1} (IV). The BOD_5 to COD ratio decreased from 0.10 in raw leachate (point I) to 0.09 after the retention tank (point II). The same value of the BOD_5 to COD ratio was observed at point IV (orchard grass), while at the point III (willow) the ratio decreased to 0.06. These results indicate that although the concentration of bioavailable organics was already low in raw leachate, the willow plants were able to uptake some more of the organic substrate, causing further depletion of the BOD_5 to COD ratio.

Generally, the hydrophyte bed in Gatka effectively removed nitrogen. The total nitrogen (N_{tot}) decreased from 115.7 mg l^{-1} (point II) to 34.2 mg l^{-1} (point III) and 25.9 mg l^{-1} (points IV), giving 70% and 78% removal efficiency, respectively. The concentrations of organic nitrogen decreased from 38.8 mg l^{-1} (point II) to 12.9 mg l^{-1} and 15.8 mg l^{-1} at the points III and IV, respectively (removal efficiencies of 78% and 88%, respectively), which was due to ammonification process.

Nitrification and denitrification processes took place in the bed. The $N-NH_4^+$ concentration decreased from 75.0 mg l^{-1} (II) to 16.5 mg/l (III) and 8.8 mg l^{-1} (IV). Denitrification also took place, which is indicated by the decrease of pH and $N-NO_3^-$ concentration from 2.0 (II) to 1.4 mg l^{-1} (III) and 1.3 (IV). The depletion of $N-NO_3^-$ seems small; however, intensive evapotranspiration by *Salix* takes place around point III (which is indicated for instance by chlorides concentration growth), causing concentration of pollutants.

The increase of chlorides concentration from 321 mg l^{-1} (II) to 581 mg l^{-1} (III) and 574 mg l^{-1} (IV) could only be explained by evapotranspiration process. At the point III (willow) high phosphorus-removal efficiency was observed – 79%, while at the point IV only 60% removal efficiency was reached.

22.6 Conclusions

The quality of leachate inflowing to the CW Szadolki was very unstable. An additional tank at the beginning of treatment process would enable qualitative and quantitative averaging of the unstable discharge of the leachate. The leachate contained typical concentrations of $N-NH_4$. The mean BOD_5 to COD ratio of 0.49 indicates that the landfilled wastes are only partly decomposed. Concentrations of Cl and Fe are high, contributing to the operation problems (*P. australis* die off and beds clogging, respectively). The low hydraulic conductivity of the beds was the fundamental reason of the beds clogging. In spite of the operation problems, the CW for leachate treatment in Szadolki provides quite good treatment efficiencies of BOD_5 (bed I), PAH, N_{tot} and $N-NH_4^+$. Treatment efficiencies of other parameters were much lower than reported for other landfill leachate treatment CW systems.

The leachate from the landfill at Gatka also contained typical concentrations of $N-NH_4$. The average BOD_5 to COD ratio was 0.1, indicating low bioavailability of the organics. The retention tank situated before the inflow to the vegetated filter failed to play its role (pollutants concentrations were only slightly decreased). The soil filter was partly vegetated by *Salix*, while the rest of the filter was vegetated by orchard grass. Nitrogen-removal efficiencies at the vegetated filter were high (70–78%). Ammonification as well as nitrification and denitrification processes take place at the filter. High phosphorus-removal efficiency (79%) was observed at the part of the filter covered with *Salix*.

Acknowledgements Funding support from the Committee of Scientific Research in Poland for constructed wetland study (3 T09D 017 27) is gratefully acknowledged.

References

Brix, H., & Schierup, H.H. (1989). Danish experience with sewage treatment in constructed wetlands. In D.A. Hammer (Ed.), *Constructed wetlands for wastewater treatments* (pp. 565–573). Chelsea, MI: Lewis.

Bulc, T., Vrhovsek, D., & Kukanja, V. (1997). The use of constructed wetland for landfill leachate treatment. *Water Science and Technology*, *35*(3), 301–306.

Chu, L.M., Cheung, K.C., & Wong, M.H. (1994). Variations in the chemical properties of landfill leachate. *Environmental Management*, *18*, 105–112.

Crawford, J.F., & Smith, P.G. (1985). Landfill Technology. Butterworths, London, pp. 84–85 (Chapter 4).

Crites, R.W., Dombeck, G.D., Watson, R.C., & Williams, C.R. (1997). Removal of metals and ammonia in constructed wetlands. *Water Environmental Research*, *69*, 132–135.

Hasselgren, K. (1989). Landfill leachate treatment and reuse in soil – plant systems. In *Proceedings of the 2nd International Landfill Symposium Landfill Concepts, Environmental Aspects, Lining Technology, Leachate Management, Industrial Waste and Combustion Residues Disposal* (pp. 1–17). Porto Conte, Alghero, Italy.

Hawkins, W.B., Rodgers, J.H. Jr., Gillespie, W.B., Dunn, A.W., Dorn, P.B., & Cano, M.L. (1997). Design and construction of wetlands for aqueous transfers and transformations of selected metals. *Ecotoxicology and Environmental Safety*, *36*, 238–248.

Henry, J.G., Prasad, D., & Young, H. (1987). Removal of organics from leachates by anaerobic filter. *Water Research*, *21*, 1395–1399.

Johansson Westholm, L. (2003). Leachate treatment with use of SBR-technology combined with a constructed wetland system at the Isätra landfill site, Sweden. In *Proceedings of the Sardinia 2003, the 9th International Waste Management and Landfill Symposium* (pp. 75–81). S. Margherita di Pula, Cagliari, Italy.

Kadlec, R.H., Knight, R.L., Vymazal, J., Brix, H., Cooper, P.F., & Haberl, R. (2000). *Constructed wetlands for pollution control: Processes, performance, design and operation*. Scientific and Technical report No. 8. London: IWA Publishing.

Karathanansis, A.D., & Johnson, C.M. (2003). Metal removal by three aquatic plants in an acid mine drainage wetland. *Mine Water and the Environment*, *22*, 22–30.

Kadlec, R.H. (2003). Integrated natural systems for landfill leachate treatment. In J. Vymazal (Ed.), *Wetlands – nutrients, metals and mass cycling* (pp. 1–33). Leiden, The Netherlands: Backhuys.

Kickuth, R. (1981). Abwasserreinigung in Mosaikmatritzen aus aeroben und anaeroben Teilbezirken. In F. Moser (Ed.), *Grundlagen der Abwasserreinigung* (pp. 639–665). Munchen, Wien: Verlag Oldenburg.

Kinsley, C.B., Crolla, A.M., Kuyucak, N., Zimmer, M., & Lafleche A. (2006). Nitrogen dynamics in a constructed wetland system treating landfill leachate. In *Proceedings of the 10th International Conference on Wetland Systems for Water Pollution Control* (pp. 295–305). Lisbon, Portugal: MAOTDR.

Kowalik, P., Slatyer, F.M., & Randerson, P. (1996). Constructed wetlands for landfill leachate treatment. In W.L. Thofelt & A. Englund (Eds.), *Ecotechnics for a sustainable society* (pp. 189–200). Ostersund, Sweden: Mid Sweden University.

Klimiuk, E., Kulikowska, D., & Koc-Jurczyk, J. (2007). Biological removal of Organics and nitrogen from landfill leachates–A review. In: Pawlowska M. & Pawlowski L. (eds.) Management of pollution emission from landfills and sludge. Taylor & Francis Group, London: 187–204.

Lo, I.M.-C. (1996). Characteristics and treatment of leachates from domestic landfills. *Environment International*, *22*, 433–442.

Mæhlum, T. (1995). Treatment of landfill leachate in on-site lagoons and constructed wetlands. *Water Science & Technology*, *32*(3), 129–135.

Martin, C.D., Johnson, K.N., & Moshiri, G.A. (1999). Performance of constructed wetland leachate treatment system in Chunchula landfill, Mobile County, Alabama. *Water Science & Technology*, *40*(3), 67–74.

Obarska-Pempkowiak, H., Haustein, E., & Wojciechowska, E. (2005). Distribution of heavy metals in vegetation of constructed wetlands in agricultural catchment. In J. Vymazal (Ed.), *Natural and constructed wetlands: Nutrients, metals and management* (pp. 125–134). Leiden, The Netherlands: Backhuys.

Peverly, J.H., Surface, J.M., & Wang, T. (1995). Growth and trace metals absorption by *Phragmites australis* in wetlands constructed for landfill leachate treatment. *Ecological Engineering*, *5*, 21–35.

Reed S.C., Crites R.W., & Middlebrooks E.J. (1995). Natural systems for waste management and treatment. Second edition. McGraw-Hill, Inc, New York: 198–199.

Surmacz-Górska, J. (2000). *Degradation of the landfill leachate organic compounds*. Lublin, Poland: Polish Academy of Science (in Polish).

Tatsi, A.A., & Zoubolis, A.I. (2002). A field investigation of the quantity and quality of leachate from a municipal solid waste landfill in a Mediterranean climate (Thessaloniki, Greece). *Advance in Environmental Research, 6*, 207–219.

Vymazal, J. (2003). Distribution of iron, cadmium, nickel and lead in a constructed wetland receiving municipal sewage. In J. Vymazal (Ed.), *Wetlands – nutrients, metals and mass cycling* (pp. 341–363). Leiden, The Netherlands: Backhuys.

Weis, J.S., Glover, T., & Weis, P. (2004). Interactions of metals affect their distribution in tissues of *Phragmites australis. Environmental Pollution, 131*, 409–404.

Chapter 23
Enhanced Denitrification by a Hybrid HF-FWS Constructed Wetland in a Large-Scale Wastewater Treatment Plant

Fabio Masi ($\boxtimes$)

Abstract The municipal centralized treatment plant (a classic activated sludge technology) receiving the wastewater produced by the municipality of Jesi (Ancona province, Central Italy), was upgraded in 2002. A new "nitrification–denitrification" compartment and a hybrid constructed wetland (CW) were added to the system to enhance denitrification and provide final treatment polishing. The whole system treats about 19,000 m^3 day^{-1} (more than 60,000 PE) and a part of the effluent is reused in a nearby industrial area. All the new sections have been provided with an online monitoring system, in order to reduce as much as possible the energy consumption for the denitrification process, and relying more on the final wetland stage whenever it obtains sufficient treatment levels. The upgraded wastewater treatment plant has been operating since January 2003, and after the first year of operation, good development of the macrophyte communities has been observed. The hybrid CW system consists of an initial sedimentation pond (volume of 2,000 m^3), a 1 ha horizontal subsurface flow system (HF) second stage, and 5 ha free water system (FWS) as a final component. It has taken almost 18 months since the CW start-up for denitrification to occur at considerable levels. Nitrates removal rates ranged from 25% to 98% during the hottest period of the year.

Keywords Constructed wetlands, denitrification, hybrid systems, tertiary treatment

23.1 Introduction

Treatment performances of both free water surface (FWS) and subsurface flow (SSF) constructed wetlands receiving primary or secondary effluents have been reviewed for the USA by Kadlec and Knight (1996) and more recently by the

IRIDRA Srl, via Lorenzo il Magnifico 70, Florence, 50129, Italy

($\boxtimes$) Corresponding author: e-mail address: masi@iridra.com

J. Vymazal (ed.) *Wastewater Treatment, Plant Dynamics and Management in Constructed and Natural Wetlands,*
© Springer Science + Business Media B.V. 2008

Constructed Wetland Association (CWA) for the UK (2006). The two existing data-bases on CWs (CWA, 2006; NADB, 1993) are showing that several hundred CWs are polishing secondary or tertiary wastewaters in both countries. Designing a terti-ary SSF treatment wetland at $0.7–1\,m^2\,PE^{-1}$ is currently widely accepted across Europe (640 of the 677 tertiary CWs in the CWA database are those with horizontal flow – HF CWs), while generally the FWSs need a longer hydraulic retention time (HRT) and consequently a larger surface area. Tertiary CWs showed a high removal rates for all the organic residual substances, achieving effluents with less than $5\,mg$ $BOD_5\,l^{-1}$ and offering an optimum environment for capturing, by adsorption phe-nomena, the bioresistant pollutants. In addition, their high filtration capacity results in less than $10\,mg$ TSS l^{-1} in the final effluent and in many cases good nutrient reductions, in particular nitrogen. In the CW mesocosm, anaerobic conditions often prevail, due to the relatively low re-aeration potential of the units. As a conse-quence, having an appropriate nitrification in the mechanical section of the waste-water treatment plant (the new nitro-denitro section), the tertiary CW can enhance denitrification process fairly efficiently (Kadlec et al., 2000).

Wetlands that receive significant and sustained nitrate loadings have in fact dem-onstrated high rates of nitrate reduction with more than 80% of the externally loaded nitrate lost through denitrification (Moraghan, 1993; Crumpton et al., 1993).

The main objectives of the CW hybrid system implemented to Jesi municipality tertiary treatment plant were:

1. To buffer occasional bad performances of the municipal wastewater treatment plant
2. To enhance denitrification process to enable effluent reuse in a nearby industrial area (cooling in a sugar company) and to minimize the effluent discharge impact on the receiving Esino River

The tertiary CW system in Fig. 23.1 was designed to provide an easy way to either use or turn off the nitro-denitrification system during the secondary treatment. This was done for saving energy by limiting its use in the technical nitro-denitro com-partment for periods when the existing wetland system is not meeting treatment performance standards. This will allow for comparison between the CW system with or without the new nitro-denitro compartment, and examination of the cost-benefit of installing the additional CW stage to the mechanical section. During the warm season, when vegetation and microbial activity are greatest, the existing terti-ary CW may not need the additional secondary treatment compartment to meet the water treatment goals for denitrification.

It is important to understand the exact contribution of the additional nitro-denitro system to the overall treatment performance of the multistage system. In areas where energy consumption is of great concern, the addition of energy-dependent technolo-gies must look at both the ability of the technology to improve treatment as well as the cost-benefit of the improved treatment in relation to the increased cost of energy (WERF, 2005). Apart from a need to decrease energy consumption, there is a growing demand to improve water quality across Italy in order to comply with the EC Directive 2000/60. Several rivers will have to improve their water quality within the next 10

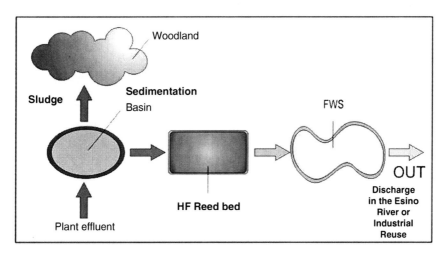

Fig. 23.1 Treatment scheme for the tertiary hybrid CW system in Jesi

years, and consequently, there will be stringent requirement for the large-scale waste-water treatment plants discharges into these rivers. Therefore, new treatment technologies are sought, capable of efficient nutrients removal at affordable costs.

Constructed wetlands have proved to be the more efficient means of nutrient control, despite the seasonability of the involved removal processes and within reasonable economic limits, in the comparison with traditional biological/mechanical treatments (nitro-denitro compartments in common activated sludge plants) (WERF, 2005). In reviewing CW technology for wastewater treatment, Benndorf (2005) recently concluded that they can be accepted as a valid "ecotechnology" for reducing the loading of nutrients in the final receiving water body.

23.2 Material and Methods

The Jesi tertiary CW hybrid system operative parameters are shown in Table 23.1. The hybrid CW system consists of an initial sedimentation pond with a volume of $2,000\,m^3$, 1 ha second stage (two series of five parallel HF beds, each bed 50 m wide and 20 m long, all planted with *Phragmites australis*) and finally a 5 ha FWS part (Fig. 23.1). The accumulated sludge in the sedimentation basin is periodically pumped into a wet woodland planted with *Populous alba*. The final outlet can be further disinfected by a new UV station just before the reuse in a nearby industrial area.

Since the installation of the system, four different monitoring campaigns were performed:

1. The first one took place immediately after the CW system start-up in January 2003 and consisted of a 20-day daily monitoring of the inlet water and the outlets of the HF and FWS stages.

Table 23.1 Operative parameters of the Jesi tertiary CW system

Number of P.E.	60,000
Average temp. – warm season	18°C
Average temp. – winter	9°C
Daily flow	13,000–19,000 m³ day⁻¹
Total HRT	≅ 1.5 day
Organic load	1,650 kg COD day⁻¹
Nitrogen load	198 kg day⁻¹
TSS load	462 kg day⁻¹
Sedimentation pond volume	15,000 m³ (As = 5,000 m²; depth = 3 m)
HF area	10,000 m² (depth = 0.7.m; gravel size 0.5–1 mm)
FWS area	50,000 m² (depth = 0.4 m)

2. The second campaign lasted for 7 months (June to December 2003), with a monthly sampling of the inlet and FWS outlet effluents.
3. During 2004, effluent samples of the final FWS stage were collected once a week.
4. During the last campaign, that took place from March to December 2005, both inlet and outlet of the FWS were monitored once a week.

The following parameters were analyzed: alkalinity, TSS, COD, BOD$_5$, NH$_4^+$, NO$_2^-$, NO$_3^-$, N$_{tot}$, P$_{tot}$. Grade samples were taken for the evaluation of the level of some heavy metals in the final effluent (Cd, Cr, Pb, Ni, Cu, Zn).

Standard analysis methods IRSA/CNR were used for both chemical and microbiological measurements in all cases (IRSA/CNR – Water Research Institute of National Research Centre, which is the Governmental Research Institute that provides the analytical standards for Italy; standards methods are almost the same provided by APHA, 1992).

23.3 Results and Discussion

From the first monitoring campaign, a low nitrification rate was observed throughout the hybrid CW system (Fig. 23.2) as also a small increase in the nitrates concentrations. The system was in that period completely unvegetated and the water temperature varied between 10°C and 13°C. Almost 50% of initial retention was achieved for phosphorous, mainly by the action of the first HF stage.

Even in the following 7 months no substantial reduction of the nitrate concentration in the outlet was observed, even though the outlet values had already met the discharge limit of 10 mg l⁻¹ (nitrate-sensitive areas) for the whole period (Fig. 23.2). Also, in the warm months, July and August, an increase in concentration was detected in the outlet, showing that the bacteria communities which are involved in the denitrification process were not developed yet in the wetland system (Fig. 23.3). The water flow has been continuously monitored since the beginning of operation (Fig. 23.4).

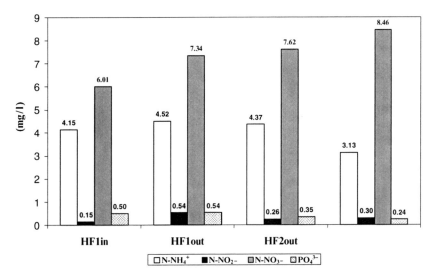

Fig. 23.2 First monitoring campaign from 01/24/2003 to 02/15/2003: mean values on 20 samples for NH_4^+, NO_2^-, NO_3^-, PO_4^{3-} concentrations (start-up of the system: 20th January 2003). HF1 = first series of beds, HF2 = second series of beds

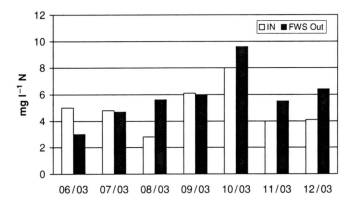

Fig. 23.3 Inlet and outlet nitrate concentrations for the CW system during the first year of operation of CW Jesi

At present, due to the unplanned nature the monitoring, only the outlet concentrations are available for all the operation period, from the start-up to the end of 2005. Only the last monitoring campaign, performed weekly since June 2003 to December 2005, gave us the possibility to correlate inlet and outlet concentration values. The main monitored parameters are reported in Fig. 23.5.

The main aim of this tertiary CW treatment seems to have been reached at the moment. The denitrification process appeared to perform well, considering quite

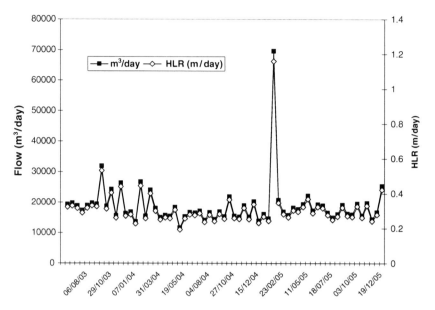

Fig. 23.4 Measured flow and hydraulic loading rate for the Jesi CW between June 2003 and December 2005

temperate climatic conditions of that area in Central Italy. The mean annual removal of nitrate amounted to 77%, maintaining a minimum activity even in the coldest months with about 25% removal and reaching up to 98% in summer (Fig. 23.6).

The average removals during the first 3 years of operation are 76%, 10%, 50%, and 30% for TSS, BOD_5, NO_3^-, and TN, respectively (Table 23.2). No removal has been observed for phosphorous, which seems to be trapped in the system during the cold months but almost completely released during summer. The wetland seems to be a source of reduced nitrogen, forming a small amount of ammonium ion and nitrites which were almost absent in the inlet (at the end of the technological treatment). Also a certain amount of reduced carbon, that permits the high denitrification rate, must be generated by the wetland itself, because the inlet carbon coming from the wastewater treatment plant was not high enough to support nitrate reduction.

Nitrogen removal processes have proved to be highly affected by the temperature and plant density, in agreement with the results obtained by Kim *et al.* (2004). The authors pointed out that a fairly stable nitrogen removal should be anticipated whenever the temperature is higher than 10°C and fresh plant biomass is between 5 and 17 kg m^{-2}.

Considering the Jesi CW design, it can be noticed that the surface area commonly used in Europe for tertiary HF-CWs treatment, about 0.5–1 m^2 PE^{-1}, has not be adopted in this case, where the value is about 0.15 m^2 PE^{-1} for the HF stage and 0.8 m^2 PE^{-1} for the FWS. Similar applications of CWs as tertiary treatment have obtained comparable results with longer HRT (Kadlec & Knight, 1996; Kadlec *et al.*, 2000).

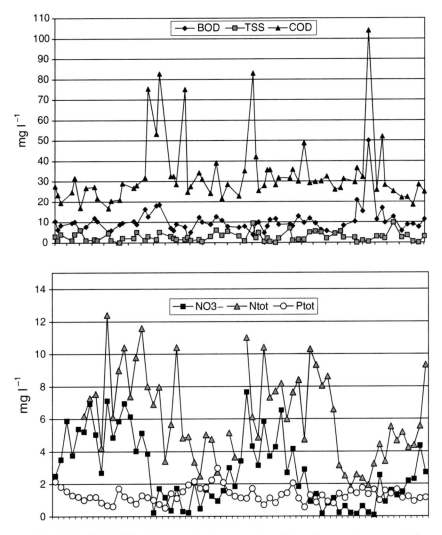

Fig. 23.5 BOD$_5$, TSS, COD, nitrate, total nitrogen and total phosphorus outlet concentrations between June 2003 and December 2005 in CW Jesi

The Italian law for the reuse of treated wastewater (DM 185/2003) imposes the limits reported in Table 23.2. The obtained performances showed that the whole wastewater treatment plant has reached the desired output levels for the discharge in the Esino River for all the considered parameters since its upgrading. Even the appropriate water quality for reuse, based on the stringent Italian legislation, has been reached for almost all the parameters. Only the total surfactant concentrations have been always over the law limit, even though few comments can be made due to the lack of the inlet data for this parameter and the possibility of formation of humic acids in the wetland which could interfere with the analysis.

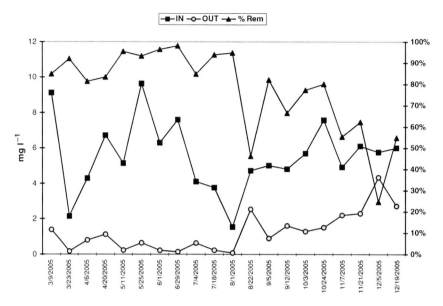

Fig. 23.6 Nitrate inlet and outlet concentrations and percentage removal between March 2005 and December 2005 in CW Jesi

Table 23.2 Mean concentration of inlet and outlet for the tertiary CW system and limits for reuse and for discharge in fresh water following the Italian legislation (D.Lgs. 152/2006)

		Limits for reuse	Limits for discharge	Inlet	St dev.	Outlet	St. dev.
TSS	mg l^{-1}	10	35	11.4	±11.1	2.7	±2.1
COD	mg l^{-1}	100	125	37.7	±22.3	33.5	±16.6
BOD$_5$	mg l^{-1}	20	25	11.6	±9.8	10.1	±6.1
Ammonium	mg l^{-1} NH$_4^+$	2	15	0.07	±0.15	1.6	±2.2
Nitrates	mg l^{-1}N-NO$_3^-$	–	20	5.5	±2.0	2.8	±2.1
Nitrites	mg l^{-1}N-NO$_2^-$	–	0.6	0.05	±0.06	0.11	±0.17
N$_{tot}$	mg l^{-1}	15	15	8.5	±2.5	6.2	±2.7
P$_{tot}$	mg l^{-1}	2	2	1.4	±0.5	1.3	±0.5
Total surfactants	mg l^{-1}	0.5	2	–	–	2.1	±0.8
Chlorides	mg l^{-1} Cl$^-$	250	1,200	80.7	±12.1	82.6	±30.6
Sulphates	mg l^{-1} SO$_4^{3-}$	500	1,000	93.5	±10.3	89.8	±17.1

23.4   Conclusions

The constructed wetland was consistent in meeting the Italian regulatory limits for wastewater reuse for all main parameters. At no period during the investigation did the system fail at achieving the discharge limits with the exception of total surfactants. The desired denitrification rates have been reached continuously since the establishment of the vegetation. There continued to be a small reduction in organics throughout the study. The efficient reduction of the influent TSS levels is credited for facilitating the UV lamp action before water reuse. The study revealed that the periods of highest nitrate removal occurred when the density of vegetative increased. This observation suggested that the vegetation increased the carbon source to the system, and directly correlates with the increase in the denitrification rate of the system.

References

APHA. (1992). *Standard methods for the examination of water and wastewater analysis*. 19th ed. American Public Health Association. Washington DC: AWWA and WPCF.

Benndorf, J. (2005). Ecotechnology: Basis of a new imission concept in water pollution control. *Water Science and Technology*, *52*(5), 17–24.

Crumpton, W.G., Isenhart, T.M., & Fisher, S.W. (1993). The fate of nonpoint source nitrate loads in freshwater wetlands: Results from experimental wetlands mesocosms. In G.A. Moshiri (Ed.), *Constructed wetlands for water quality improvement* (pp. 283–291). Boca Raton, FL: CRC Press/Lewis.

CWA. (2006). Constructed Wetland Association. www.constructedwetland.org

Kadlec, R.H., & Knight, R.L. (1996). *Treatment wetlands*. Boca Raton, FL: CRC Press/Lewis.

Kadlec, R.H., Knight, R.L., Vymazal, J., Brix, H., Cooper, P.F., & Haberl, R. (2000). *Constructed wetlands for water pollution control: Processes, performance, design and operation*. London: IWA Scientific and Technical Report No. 8.

Kim, Y., Lee, D.R., & Giokas, D. (2004). Agricultural reuse of the secondary effluent polished by an algal pond system coupled with a constructed wetland. *Water Science and Technology*, *50*(6), 79–86.

Moraghan, J.T. (1993). Loss and assimilation of ^{15}N-nitrate added to a North-Dakota Cattail Marsh. *Aquati Botany*, *46*, 225–234.

NADB. (1993). North American treatment wetland database. Electronic database.

Water Environment Research Foundation – WERF. (2005). Nutrient farming and traditional removal: An economic comparison. Report 2005. 03-WSM-6CO.

Chapter 24
Growth Dynamics of *Pistia stratiotes* in Temperate Climate

Silvana Perdomo[1](✉), **Masanori Fujita**[2], **Michihiko Ike**[3], and **Masafumi Tateda**[4]

Abstract In order to understand the growth dynamics of the floating macrophyte *Pistia stratiotes* under natural conditions in a temperate climate, and the influence of higher levels of N and P concentrations in water on the specific growth rate, μ, a number of experiments were carried out in two series. Series I: 29 determinations of μ in the stationary growth stage using an oligotrophic medium were carried out. The experimental data was fitted by a function which rules cyclic processes in wetlands ($R = 0.93$), and a maximum μ of 0.0395 day^{-1} over the year was calculated. Series II: several batch experiments were carried out in Osaka 34° northern latitude, and in Montevideo 34°50' southern latitude. The transitory stage of growth dynamics for plants, which were transferred to a medium with higher concentrations of N and P, was considered. In this case, the Double Monod model with double inhibition affected by a constant term adequately explains the variations in μ with $R = 0.85$ and 0.88 at 24°C and 29°C respectively.

Keywords Double Monod model, growth dynamics, *Pistia stratiotes*, temperate climate

[1] Limnosistemas®, Avda. Costanera Mz 205 S2, El Pinar, Canelones, Uruguay

[2] Deanery, Kochi National College of Technology, 200-1 Monobe Otsu, Namgoku, Kochi 783-8508, Japan

[3] Department of Environmental Engineering, Graduate School of Engineering, Osaka University, 2-1 Yamadaoka, Suita, Osaka 565-0871, Japan

[4] Department of Environmental Technology, College of Technology, Toyama Prefectural University, 5180 Kurokawa, Kosugi-machi, Imizu-Gun, Toyama, Japan

(✉) Corresponding author: e-mail: sperdomo@internet.com.uy

J. Vymazal (ed.) *Wastewater Treatment, Plant Dynamics and Management in Constructed and Natural Wetlands*,
© Springer Science + Business Media B.V. 2008

24.1   Introduction

The growth rate of aquatic plants depends on many factors like efficiency of using solar energy, composition and nutrient levels in water, cultural method, and environmental variables (Reddy *et al.*, 1983). The efficiency of the plant to use solar radiation is firstly based on the photosynthetic process to accomplish fixation of the atmospheric CO_2 (Reddy *et al.*, 1983). Such a process will determine the highest potential of every aquatic plant species for biomass production (Reddy *et al.*, 1983). The response of biomass to solar light intensity is an individual characteristic of each species (Reddy *et al.*, 1983). Vymazal (2001) pointed out that there is the lack of information on the critical level of nutrients to obtain the highest growth rate of a species. The growth rate of aquatic plants is also influenced by the cultural practices (Reddy *et al.*, 1983) like the starting density of the population and the frequency of harvesting the plants. Temperature is one of the most important environmental factors for productivity of a given species. Most of the aquatic plants grow well between 20°C and 30°C, and stop growing below 10°C. In 1998, Kadlec proposed a cyclic function for processes that rule the wetlands for wastewater treatment. In the present work, it is proposed that the specific growth rate, μ, under natural light and temperature conditions, follows that cyclic function. The effects of transplanting a plant population already adapted to low concentrations of both N and P to a medium with higher levels of those nutrients are also analyzed in this work. For such cases, the author proposes that a "transitory stage" takes places where the Double Monod model with inhibition dominates until the plants adapt themselves to the new medium.

24.2   Purpose of Study

The work attempts to contribute to a deeper understanding of the growth dynamics of *Pistia stratiotes* under natural conditions in temperate climate and the influence of concentrated N and/or P water. The study also intends to help to understand whether the optimal conditions for growth exist, and what is the influence of the growth dynamics on plant feasibility.

24.3   Materials and Methods

24.3.1   *Growth Dynamics Under Steady State and Natural Conditions*

Pistia plants were cultured unsheltered under the Uruguayan climate, and average specific rates of growth were determined several times during the year. The population density of the plants was kept constant, and the average levels and composition of the nutrient medium were kept almost constant for all calculations. The results were determined for 1995, 1996, 1997, and 2000.

The original *Pistia* plants were taken from an oligotrophic pond (Laguna del Cisne, Depto. de Canelones, Uruguay) and were cultured in a nutrient solution (modified Arnon & Hoagland solution, Arnon & Hoagland, 1940) as shown in Table 24.1. The nutrient solution was renewed every fortnight except in summer when it was done every week. The average concentration of N and P were maintained at values $< 3\,mg\;l^{-1}$ and $< 1\,mg\;l^{-1}$, respectively. A tank of 679 l with $1.13\,m^2$ of water surface was used (Fig. 24.1). This container was divided into four similar sections to cultivate three replicates in sections S_{c1}, S_{c2}, and S_{c3}, leaving section S_{c4} to vary in the population so that total density was kept constant and equal to approximately $3,540\,g_{ww}\,m^{-2}$ (Fig. 24.1). For each experimental run, $1,000\,g_{ww}$ were placed in each section. The biomass was measured every week for 4 weeks keeping total density constant, and returned to initial conditions in the fourth week.

Average daily temperatures were obtained from the Servicio Oceanográfico e Hidrográfico de la Armada, SOHMA, Uruguay.

In a month's time μ may be considered approximately constant for each run of 4 weeks when four points of the graph given by Equation (24.1) were obtained and the slope determined for each section $-S_{c1}$, S_{c2}, and S_{c3} – resulting in μ_1, μ_2, and μ_3,

Table 24.1 Modified Arnon and Hoagland nutrient solution (Based on Arnon and Hoagland, 1940)

Compound	Concentration (mg l⁻¹)	Compound	Concentration (mg l⁻¹)
NH_4NO_3	8.57	H_3BO_3	0.953
$NaH_2PO_4.2H_2O$	5.03	$MnSO_4.H_2O$	0.393
K_2SO_4	58.6	$CuSO_4.5H_2O$	0.027
$MgSO_4.7H_2O$	2.8	$ZnSO_4.5H_2O$	0.073
$CaCl_2.2H_2O$	29.4	$(NH_4)_6Mo_7O_{24}.4H_2O$	0.249
$FeSO_4\;0.5\%$	$0.2\,ml\;l^{-1}$		
Tartaric acid 0.4%	$0.2\,ml\;l^{-1}$		

Fig. 24.1 Experimental tank for μ determination under natural conditions

respectively. Those values were averaged and the μ mean value of the experimental run was then obtained. This μ is related to the final day of the experimental run and also to the daily average atmospheric temperature during the period of experiments. The obtained experimental values of μ are correlated with the day of the year following the function given by Kadlec in 1998 (Equation 24.1), which is as follows:

$$\mu = \mu_m \left(1 + A_{mp} \cos \left(w \left(t_n - t_{max}\right)\right)\right) \tag{24.1}$$

Where,

μ = specific growth rate (day^{-1})
μ_m = average specific growth rate (day^{-1})
t_n = day of the year, in consecutive numbers
t_{max} = day of the year when μ is maximum
$w = 2\pi/365$
A_{mp} = amplitude of function
μ_m, A_{mp} and t_{max} are parameters of the function to be determined for the applied experimental conditions.

24.3.2   Growth Dynamics in Transitory State

In order to determine the transitory behavior of *Pistia* plants, several experimental batches were carried out exposing fresh plants to different levels of nutrients (N and P) concentrations under natural conditions of light and temperature.

The experiments were carried out in runs of six batches each. Plastic sinks with a volume of 28 l each were used to contain the nutrient solution (modified Arnon & Hoagland solution; Arnon & Hoagland, 1940). Three of the runs were carried out in Montevideo, under climatic conditions of Uruguay (between 33° and 38° southern latitude) and two runs were carried out in Osaka, under climatic conditions of Japan (between 30° and 34° northern latitude). Both countries are of temperate climate. The experimental guidelines may be summarized as follows:

1. Fresh plants from oligotrophic ponds were used with similar initial biomass density for each batch and for each same run (4.1–7.5 kg$_{ww}$ m^{-2}).
2. The biomass was weighed approximately every 7 days.
3. The nutrient solution was completely renewed everyday or every other day as long as the experiment lasted – between 7 and 21 days in total.
4. The average daily atmospheric temperature was registered so as to determine the average daily temperature for the whole run.
5. The content levels of N and P were determined at the beginning and at the end of every solution-renewal period of 2 or more days. These values were averaged, respectively, to obtain a mean start and a mean end value for nutrients concentration. The mean concentration for each batch during the time of the experiment was determined by averaging the mean content value in the water at the beginning

Table 24.2 Experimental conditions to determine μ in batch systems. Nutrient levels are expressed as mean concentration in mg l^{-1}

Run	Location	$T_{m,atm}$(°C)	P	N	Run	Location	$T_{m,atm}$(°C)	P	N
1	Montevideo	24	3.0	0.72	4	Osaka	29	0.020	20[b]
			2.6	3.09				0.520	20
			2.5	5.60				1.02	20
			2.1	9.30				2.02	20[b]
			1.5	30.0				3.02	20[b]
			1.6	47.0				5.02	20[b]
2	Montevideo	28	<0.01[a]	0.580	5	Osaka	30	4	1.30
			0.06	2.72				4	2.30
			0.04	4.82				4	6.30
			0.08	9.18[b]				4	11.3
			0.19	30.6[b]				4	31.3
			0.24	44.8[b]				4	51.3
3	Montevideo	24	<0.01[a]	5.40					
			0.24	5.10					
			4.2	5.10					
			12.5	7.30[b]					
			22.7	4.80					
			80.6	5.50					

[a] For calculation the value used was 0.005
[b] Conditions of $\mu < 0$.

and the corresponding one at the end. For the case in which the renewal time was only 1 day the end value was assumed to be approximately equal to start value constant concentration of nutrients.

6. The specific growth rate, μ, was determined by lineal correlation between LnB (Neper's logarithm of B) and time elapsed since day 0 of the experiment.

Table 24.2 summarizes the experimental conditions for the batch trials.

24.4   Results

24.4.1   Growth Dynamics Under Steady State and Natural Conditions

Following the methodology explained in 24.3.1, 29 values for μ were obtained at different times of the year (Table 24.3), and in different years. *P. stratiotes* presented values of μ between a minimal experimental value of -0.037 day^{-1} in March 1995 and a maximal of 0.039 day^{-1} in January 1997. The average daily atmospheric temperature varied between 8°C and 28°C. Figure 24.2 shows μ experimental variations during the runs.

The correlation parameters obtained for eq. 24.1 for temperate climate using *P. stratiotes* plants cultivated in water with oligotrophic characteristics are: $n = 29$; $R = 0.93$; $\mu_m = 0.011$ day^{-1}; $A_{mp} = 2.77$; $t_{max} = 22$; $\mu_{max} = 0.040$ day^{-1}.

Table 24.3 Experimental data obtained in the determination of μ under natural conditions in temperate climate

Date	t_n	T_m (°C)	μ (day^{-1})	Date	t_n	T_m (°C)	μ(day^{-1})
24.01.1997	24	–	0.039	23.06.1995	173	11.5	−0.012
31.01.1997	31	–	0.033	04.07.1996	185	7	−0.009
24.02.2000	54	27.5	0.034	23.07.1996	204	6	−0.006
23.03.2000	82	25	0.027	24.08.1995	235	17.0	−0.016
30.03.2000	89	17	0.025	27.08.1996	239	11	−0.011
13.04.2000	103	16	0.011	31.08.1995	242	14.5	−0.026
27.04.2000	117	18.5	0.019	07.09.1995	249	19.0	−0.017
04.05.1995	123	17.7	−0.005	26.09.1996	269	13	−0.008
12.05.1995	131	15.2	0.003	05.10.1995	277	12	−0.007
20.05.1996	140	11	−0.007	09.10.1996	282	21	0.005
23.05.1996	142	–	−0.000	11.10.1995	283	13	−0.007
29.05.1996	148	12.1	−0.013	01.11.1995	304	20.5	0.033
03.06.1996	154	8	−0.010	14.11.1995	317	18.0	0.028
05.06.1995	155		−0.007	14.11.1996	318	21	0.023
				04.12.1996	338	–	0.029

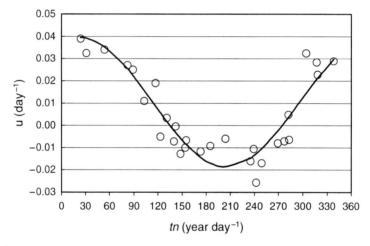

Fig. 24.2 Change of μ with the day of the year for *Pistia stratiotes* in Uruguay

24.4.2 Growth Dynamics in Transitory State

As seen in Table 24.2, 7 out of 30 pairs of concentration levels (N and P) resulted in $\mu < 0$. This implies the plant was in the death phase under such conditions. Those values are of no significance for modeling the effects of nutrients concentrations of N and P in the medium over the growth rate. Therefore, only values with $\mu > 0$ were considered. In turn, the data from runs of similar T_m (±2°C) were correlated together to produce two populations of data: $T_m \sim 24$°C with $n = 12$ and $T_m \sim 29$°C with $n = 11$.

The following function was fitted to the experimental data:

$$\mu = \mu_0 \frac{NP}{(K_{SNc} + N + N^2/K_{INc})(K_{SPc} + P + P^2/K_{IPc})} - C_P \quad (24.2)$$

where,

μ = specific growth rate in transitory phase (day^{-1})
μ_0= kinetic constant (day^{-1})
K_{SNc}, K_{SPc} = kinetic constants of saturation for N and P respectively (mg l^{-1})
K_{INc}, K_{IPc} = inhibition kinetic constants for N and P respectively (mg l^{-1})
C_P= $R_0 E$ for *P. stratiotes* under experimental conditions, day^{-1}.
R_0= plant respiration factor
E = efficiency of biomass conversion from energy
N = nitrogen concentration (mg l^{-1})
P = phosphorus concentration (mg l^{-1})

R_0, the plant respiration factor, is similar (~0.001) for different species such as the water hyacinth and the soybean plant (Lorber *et al.*, 1984), and therefore, no great variations are expected among similar species such as the water hyacinth and the water lettuce.

E, the plant efficiency of biomass conversion from energy, is approximately 0.83 for the case of the water hyacinth. At the present moment no estimated value for C_P in the case of *P. stratiotes* is available. Therefore C_P or ($R_0 E$) is estimated to be 0.008, as is similar to water hyacinth. The adjustment was also made by considering C_P = 0.060, 0.070, and 0.080 in order to detect any increase in the correlation coefficient that might indicate an approach to the real value of C_P for *Pistia* plants. Table 24.4 presents the resulting values of the calculated correlations. Considering C_P = 0.008 as in the case of water hyacinth, the model would

Table 24.4 Kinetic parameters for the Double Monod model with double inhibition, at different T_m and with μ in transitory phase

C_P = approx. 0.008 (water hyacinth)	T_m = 29°C	T_m = 24°C	C_P = approx. 0.008 (water hyacinth)	T_m = 29°C	T_m = 24°C
N	11	12	K_{INc}	72.5	103
R	0.880	0.850	K_{SPc}	2.22	4.5×10^{-4}
μ_0	0.113	0.038	K_{IPc}	4.5×10^5	288
K_{SNc}	0.155	1.03			
C_P = approx. 0.007			C_P= approx. 0.007		
N	11	12	K_{INc}	70.4	56.3
R	0.90	0.90	K_{SPc}	2.64	4.3×10^{-4}
μ_0	0.113	0.040	K_{IPc}	4.8×10^5	805
K_{SNc}	0.165	1.67			
C_P = approx. 0.006			C_P= approx. 0.006		
N	11	12	K_{INc}	68.8	95.7
R	0.91	0.84	K_{SPc}	3.04	4.8×10^{-4}
μ_0	0.125	0.035	K_{IPc}	1.7×10^6	252
K_{SNc}	0.172	1.16			

Table 24.5 Conditions of μ_{max} predicted by the model for R_0E approx. 0.0083

T_m (°C)	N_{opt} (mg l^{-1})	P_{opt} (mg l^{-1})	μ_{max} (day^{-1})
24	10.3	0.360	0.031
29	3.40	999	0.103

introduce conditions where μ is maximal. Such optimal conditions for biomass growth are presented in Table 24.5.

24.5   Discussion

The plants used for the experiments to determine specific rates came from an oligotrophic lake near the experimentation site, and the mean concentrations of N and P in the cultivation media during the tests were maintained at levels below 3 mg l^{-1} and 1 mg l^{-1}, respectively, during experiments. The experiments were carried out for several months, each one always in a culture media with a similar composition, and at the same geographical site. Therefore, it is clear that the populations of plants used in each period were made up of plants adapted to the working conditions. That is to say, the dynamic of cosenoidal growth observed for the variation in the specific rate of growth of *P. stratiotes* corresponds to a population which was stable both with respect to climate and with respect to the culture media. Hence, it is sensible to call this the dynamics of biomass growth under "steady state conditions". Under the above-mentioned conditions, the rate of growth, μ, will depend only on annual cyclical ambient variables.

The equation proposed by Kadlec (1998) for the cyclical variables associated with wetlands has a good fitting with the results obtained in this study, with a correlation coefficient $R = 0.93$ for a population of $n = 29$. The average value of the specific growth rate given by the model is $\mu = 0.011$ day^{-1}, and the maximum annual value predicted is $\mu = 0.040$ day^{-1}, for a range of mean atmospheric temperatures between 8°C and 28°C. The mean value predicted is coherent with those found by Tucker (1983), μ between 0.025 and 0.150 day^{-1}, for populations of *P. stratiotes* in Florida, USA. However, generally lower values are to be expected for Uruguay than for Florida, due to the differences in mean temperature between the two geographical areas.

The level of agreement found for experimental data from different runs carried out in different years (1995, 1996, 1997, and 2000) supports the hypothesis of an annual variation which repeats cyclically. The model explains satisfactorily the variations in the specific rate of growth of *P. stratiotes* in the year, and indicates its dependence on solar radiation and also on mean temperature, which are ecosystem variables that maintain a similar annual dynamic. The cosenoidal variation of μ in the year suggests, for the case of the *P. stratiotes*, that there is a time of year in which solar radiation allows net growth $(\mu > 0)$, or the dynamics of the biomass sustains a static population $(\mu = 0)$, or enters a phase of net decay $(\mu < 0)$.

In Uruguay, *Pistia* does not show vegetal growth between approximately day 132 and day 275 of the year, which is from the middle of May to the middle of September. This behavior becomes important when it comes to designing treatment systems which require a minimum level of efficiency throughout the year. When working in a temperate climate, *Pistia* would be more appropriate for seasonal treatment systems.

Starting with plants in the steady state, the transitory growth phase of *P. stratiotes* was analyzed as the concentrations of the main nutrients (N and P) had been increased. To do this, plants obtained from an oligotrophic lake near the work site were used. They were placed in experimental solutions with concentrations of N and/or P higher than those in the original culture media (Table 24.3). Each run included the monitoring of the biomass for 7 days, which is a period short enough to consider the lack of stabilization of the population, and to have a situation of transitory growth dynamics, a stage prior to the state of equilibrium of the population with the media and with the conditions of the culture. Under the experimental conditions thus established, two groups of experimental values of specific rates were obtained; these were for mean atmospheric temperatures of approximately 24°C and 29°C respectively. In both cases the results yielded by conditions of population decline ($\mu < 0$) were disregarded. These occurred under conditions of $T_m \sim 24$°C when $0.080 < P < 0.240$ mg l^{-1} and $9.18 < N < 44.8$ mg l^{-1}; and in the case of $T_m \sim 29$°C when $2.02 < P < 5.02$ mg l^{-1} and $N =$ approximately 20 mg l^{-1}. In this respect, the framework of experimental conditions of Table 24.3 suggests a relation which is not simple among μ, N, and P, where the value of μ is determined by a function which depends on both concentrations. Once the fitting analysis for both populations had been done, in line with the model of Lorber *et al.* (1984), and modified so as to take into account possible inhibiting effects (Equation 24.2), coefficients of correlation of $R = 0.88$ ($n = 11$) and $R = 0.85$ ($n = 12$) were found. This is when we consider that *Pistia* possesses the same R_0E as *Eichhornia crassipes* (~0.008). Since great variations in this parameter between one species and another are not to be expected, the above merely serves to cover the lack of real data about this parameter for *P. stratiotes*, as there might be significant differences of fitting the model with a slight difference in the value of the R_0E utilized. The proof of this is that for the fittings tested for the same populations of data but with $R_0E \sim 0.007$ (Table 24.4), the coefficients of correlation R, gave 0.90 for both cases. For $R_0E \sim 0.006$, the R obtained was 0.91 and 0.84 for $T_m = 24$°C and 29°C respectively, results which confirm the observation. According to this model it would seem that $R_0E \sim 0.007$. According to the model utilized it would seem that $R_0E \sim 0.007$ could be more appropriate for *Pistia* than that of *E. crassipes*. We consider, therefore, that in fact the model satisfactorily explains the growth dynamic of *Pistia* in the transitory state for the conditions of the tests, although larger populations, and working with a real value of R_0E, would help definitively confirm these results in future research.

In this first approximation it is clear that during the adaptation phase of the population to a higher level of N and P there is a growth dynamic with kinetics in line with the Monod model for double substrate with double inhibition, added to the constant C_p, thus confirming the effect of the concentration of these nutrients on the growth of the biomass. From the behavior predicted by the model it was found that, for these conditions, *Pistia* will have a maximum specific growth rate of 0.031 day^{-1} at 24°C, and 0.103 day^{-1} at 29°C. These values will be obtained when $N \sim 10$ mg l^{-1} and $P = 0.4$ mg l^{-1} for $T_m \sim 24$°C. While at $T_m \sim 29$°C, the concentration of N should be ~ 3 mg l^{-1} and that of P should be 999 mg l^{-1}. These results indicate that, for practical purposes, the maximum predicted for μ by the model will not be reached under normal experimental conditions or in natural culture media at $T_m = 29$°C, but could be observed when $T_m = 24$°C.

24.6 Conclusions

When a population of *Pistia* plants is adapted to the conditions of temperature and to an oligotrophic culture (concentrations of N and P), the biomass develops according to a dynamic of a stationary state, in which the specific rate of growth only depends on the day of the year. With cosenoidal annual cyclical behavior it is possible to determine the period in which the population grows ($\mu > 0$), remains constant ($\mu = 0$), or declines ($\mu < 0$). According to the model used, it is possible to determine the mean value of μ which the population will have in a year in Uruguay, the maximum it will reach, and the day of the year on which it will reach it. This curve, found in the particular case of a temperate subtropical climate for the species *P. stratiotes*, is by nature characteristic of this species and of the geographical location.

On the other hand, during the plants' first stage of adaptation to a media with higher concentrations of nutrients, in particular N and P, the following considerations emerge:

- In the transition stage which occurs while the *Pistia* plants are adapting to a culture media with higher levels of N and P, the specific rate of biomass growth varies with temperature, and with the levels of N and P in the new media.
- The double Monod with double inhibition by substrate model, affected by a constant term associated with the respiration factor and with the efficiency of the utilization of energy which are characteristic of the species, gives a good explanation of the variation in μ in the transitory phase ($R > 0.85$).

Although, within the scope of this study, the fittings found for each case adequately explain the behavior of the data it is important to confirm these results with larger statistical populations. Beside this, it would be interesting to check if the behavior shown by *Pistia* also governs other similar species.

Acknowledgements We thank the Comisión Sectorial de Investigación Científica (CSIC) of Universidad de la República (UDELAR), Uruguay, and Osaka University of Japan for their financial support to the experimental works. We take much pleasure in acknowledging Laboratorio Tecnológico del Uruguay (LATU) for making available their laboratory facilities to conduct the experiments in Uruguay.

Our sincere gratitude to Dr. Jan Vymazal for the review of the manuscript and to Mr. Carlos Rivero for his help and support. The authors also wish to express their sincere acknowledgement to Olga Urbanc-Ber i , the person and the scientist, who will remain in vivid memory among us.

List of symbols

μ_0 = kinetic constant (day^{-1})
μ_m = average specific growth rate (day^{-1})
μ_{max} = maximum specific growth rate (day^{-1})
μ = specific growth rate (day^{-1})
A_{mp} = amplitude of function
C_p = kinetic constant for *P. stratiotes*
E = efficiency of biomass conversion from energy
K_{INc}, K_{IPc} = inhibition kinetic constants for N and P respectively (mg l^{-1})
K_{SNc}, K_{SPc} = saturation kinetic constants for N and P respectively (mg l^{-1})
N = nitrogen concentration (mg l^{-1})
N_{opt}, P_{opt} = nitrogen and phosphorus optimal concentrations for biomass growth, respectively (mg l^{-1})
P = phosphorus concentration (mg l^{-1})
R_0 = plant respiration factor
S_{c1}, S_{c2}, S_{c3}, S_{c4} = sections of the experimental tank
T_m = mean ambient temperature
t_{max} = day of the year when μ is maximum
t_n = day of the year, in consecutive numbers
$w = 2\pi/365$
ww = wet weight, dry weight

References

Arnon, D.J., & Hoagland, D.R. (1940). Crop production in artificial culture solution with special reference to factors influencing yields adsorption of inorganic nutrients. *Soil Science, 60*, 463–485.

Kadlec, R. (1998). Chemical, physical and biological cycles in treatment wetlands. In *Proceedings of the 6th International Conference on Wetland Systems for Water pollution control* (pp. 42–53). Aguas de Sao Pedro, Brasil: IWA.

Lorber, M.N., Mishoe, J.W., & Reddy, K.R. (1984). Modelling and analysis of water hyacinth biomass. *Ecological Modelling, 24*, 61–77.

Reddy, K.R., Sutton, D.L., & Bowes, G. (1983). Freshwater aquatic plant biomass production. *Soil and Crop Science Society of Florida Proceedings, 42*, 28–40.

Tucker, C.S. (1983). Culture density and productivity of *Pistia stratiotes*. *Journal of Aquatic Plant Management, 21*, 40–41.

Vymazal, J. (2001). Types of constructed wetlands for wastewater treatment: Their potential for nutrient removal. In J. Vymazal (Ed.), *Transformations of nutrients in natural and constructed wetlands* (pp. 1–93). Leiden, The Netherlands: Backhuys.

Chapter 25
Fractionation, Biodegradability and Particle-Size Distribution of Organic Matter in Horizontal Subsurface-Flow Constructed Wetlands

Jaume Puigagut[1], Aracelly Caselles-Osorio[1,2], Nuria Vaello[1], and Joan García[1](✉)

Abstract In order to analyse the fate of organic matter and particles along the length of horizontal subsurface-flow constructed wetlands, two pilot wetlands were monitored in terms of BOD_5, COD and their fractions, TOC and particle counts in various sampling campaigns. Anaerobic biodegradability properties of the effluent organic matter were also measured by means of a methane-production test. The most abundant particles in the influent, at intermediate points along the length of the wetlands, and in the effluent were those comprised in the range of 0.7 to 2 µm. Most of the particles (close to 80%) were removed within the first quarter of the wetland length. Most organic matter was also removed within the first quarter (50% and 80% in terms of TOC and COD, respectively). Further organic matter removal along the rest of the length of the wetland only accounted for a 0–10% of the total removal. The organic matter of the effluents is difficult to biodegrade under anaerobic conditions, but it is readily biodegradable under aerobic conditions.

Keywords Biodegradability, clogging, constructed wetlands, horizontal subsurface flow, suspended solids

25.1   Introduction

Chemical and physical characterisation of organic compounds present in wastewater has become a major issue in wastewater technology research due to the great influence of organic matter characteristics on both treatment efficiency and costs

---

[1] Environmental Engineering Division; Hydraulics, Maritime and Environmental Engineering Department; Technical University of Catalonia, Jordi Girona, 1-3, 08034-Barcelona, Spain

[2] Department of Biology, Atlantic University, km 7 Highway Old Colombia Port, Barranquilla, Colombia

(✉) Corresponding author: e-mail: joan.garcia@upc.edu

(Dulekgurgen *et al.*, 2006; Levine *et al.*, 1991). Although subsurface-flow constructed wetlands (SSF CWs) have been shown to efficiently remove organic matter and nutrients from urban wastewater, there are several scientific evidences that indicate that particulate organic matter is one of the major causes for reduction of treatment efficiency because it is related with the gradual clogging of the media (Langergraber *et al.*, 2003). Clogging generally occurs near the inlet zone of the wetlands as a result of solids entrapment and sedimentation, biofilm growth and chemical precipitation (Langergraber *et al.*, 2003). In relation with this, and despite that the effect of characteristics and particle-size distribution of the organic matter has been scarcely investigated in constructed wetlands, organic loading rates below 6 g DBO m^{-2} day^{-1} are recommended in order to avoid a rapid clogging (Caselles-Osorio & García, 2007). A physico-chemical pretreatment is one suitable way to reduce both the organic load and the total amount of suspended solids in the influent of a SSF CW. This pretreatment may at least partially prevent the reduction of the hydraulic conductivity of the wetlands, thereby delaying the clogging of the system (Caselles-Osorio & García, 2007). However, certain requirements of the physico-chemical pretreatment can make this process unsuitable in the context of CW technology: the cost of the coagulants, the energy required for adding and mixing coagulants, and the increase in sludge handling.

This chapter aims to contribute to close the knowledge gap on the fate and characteristics of organic matter in horizontal SSF CWs. We conducted organic-matter fractionation analyses, anaerobic biodegradability tests and particle counts. The obtained results would benefit both the technical and the scientific communities by increasing the available tools for a better process understanding as well as design criteria.

25.2   Material and Methods

25.2.1   Treatment Plant and Reed Beds

Samples were taken from two pilot reed beds (located in a more complex treatment system) during several sampling campaigns. The treatment system was set up in May 2000 and dismantled in late 2005. The system treated part of the urban wastewater generated by the Can Suquet housing development in the town of Les Franqueses de Vallès (Barcelona, north-east Spain). See García *et al.* (2004a, b) for a detailed outline of the treatment system. The climate at the site is Mediterranean, with an average annual temperature of 13.5°C and an accumulated rainfall of 460 mm (data for 2000). In the system, the screened wastewater flowed to an Imhoff tank. From there, the primary effluent was pumped and equally distributed into eight parallel reed beds with a surface area of 54–56 m^2 each. Experiments were conducted in two of these eight reed beds. Table 25.1 shows the design characteristics and operational conditions of these two beds during the experiments.

Table 25.1 Operational conditions (of the three sampling campaigns conducted in this study) and main design characteristics of the reed beds used for the experiments

Name of reed bed	Organic loading rate (g DBO m^{-2} day^{-1})	Hydraulic loading rate (mm day^{-1})	L:W ratio	Porosity (%)	Type of media	D_{60}[a] (mm)	Cu[b]	Gravel depth (cm)	Wetted depth (cm)
C2	5.3[c]–3.8[d]	41[c]–29[d]	2:1	39	Fine gravel	3.5	1.7	55	50
B2	5.8[e]	32[e]	1.5:1	39	Fine gravel	3.5	1.7	55	50

[a] D_{60} is the pore diameter of a sieve at which 60% of the gravel media is retained
[b] Cu is the uniformity coefficient Cu = D_{60}/D_{10}
[c] June 2003 campaign
[d] December 2003 campaign
[e] April–July and December 2005 campaign

In order to analyse the organic matter (TOC, COD and COD fractions) and particle counts, water samples were taken from the influent, at intermediate points along the length, and from the effluent of the two reed beds. Intermediate samples were obtained from perforated tubes inserted in the gravel. These perforated tubes where placed at the centre of the wetlands and at a fractional distance of 0.25, 0.5 and 0.75 in bed C2 and at 0.08, 0.25, 0.5 and 0.75 in bed B2. Each perforated tube consisted of a metal mesh cylinder emptied of gravel. Samples from the perforated tubes were obtained at a mean depth (around 25 cm) using a peristaltic pump working at a very low velocity in order to avoid any water disturbance.

25.2.2 Particle Counts

Particle count samples of the influent, intermediate points, and effluent of the two reed beds were taken for 5 consecutive days in two sampling campaigns (in June and December 2003). These samples were analysed using a Coulter Multisizer II, which measures the number of particles sorted by size. In our case, the Coulter counted the particles with a diameter greater than 0.7 μm and smaller than 21.5 μm. Each sample was analysed at least twice in order to ensure reliable measurements.

25.2.3 Organic Matter Measurements and Fractionation

Total organic carbon (TOC) was analysed, based on the same samples used for particle counts (taken from both beds in June and December 2003). TOC was measured using a Shimadzu TOC 5000 Analyser.

Samples for COD fractionation were taken during a period of 4 months (once every 15 days from April to July 2005) from bed B2. In previous trials, total COD and COD filtered at 121 μm and 15 μm were considered; nevertheless, because most of the COD was found to be less than 15 μm, a new fractionation procedure was considered. Finally, the analyses were: total COD, COD filtered at 1.2 μm and COD filtered at 0.2 μm. Therefore, the fractions considered were: above 1.2 μm (macrocolloidal and settleable fraction), between 0.2 μm and 1.2 μm (colloidal fraction), and below 0.2 μm (soluble fraction).

BOD$_5$ and total COD data were available from a long-term study (influent and effluent grab samples taken three to four times per month during 2002 and 2003) (García *et al.*, 2004b, 2005).

25.2.4   Biodegradability Tests

We studied the anaerobic biodegradability of the organic matter of two effluent samples of bed B2 taken in December 2005 using a procedure similar to the one described in García *et al.* (2007). Specifically, we used airtight 2.2 L glass bottles (two per sample) with a screw top and a microvalve to extract gas samples. The bottles were filled with 3 kg of fresh gravel (obtained from the middle part of the bed), which occupied 90% of the volume of the bottles, and 800 ml of effluent (samples) or bottled water (control), offering a headspace of 200 ml. The remaining air in the headspace was removed by injecting He for 1 min. The bottles were placed in dark chambers at 20°C and gently shaken by hand everyday in order to avoid methane accumulation in the interstitial gravel spaces. A 1 ml gas sample was obtained from the headspace reactor using Hamilton gas syringes every day for 25–30 days. The gas samples were spiked to a gas chromatographer (Thermo Finnigan Trace GC) equipped with a flame ionisation detector for methane measurements.

25.3   Results

25.3.1   Particle Counts

Figure 25.1 shows the number of particles in the more numerous size ranges at the inlet, at intermediate points, and at the effluent of bed C2 in the June 2003 campaign. Figure 25.2 shows the percentage of the total number of particles remaining along the length of bed C2 in the same campaign. In Fig. 25.1 it can be seen that smaller particles (mainly those within the ranges of 0.7 to 1 μm and 1 to 2 μm) are the most abundant at each sampling point. Furthermore, in Fig. 25.2 it is possible to see that most of the particles are removed as they pass through the wetlands.

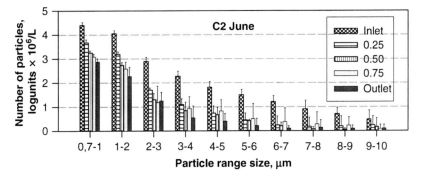

Fig. 25.1 Number of particles grouped by (the most numerous) size ranges at the inlet, at intermediate points (in fractional distance: 0.25, 0.50 and 0.75), and at the outlet in reed bed C2 during the June 2003 sampling campaign. Each bar represents the average results for five samples obtained over 5 consecutive days

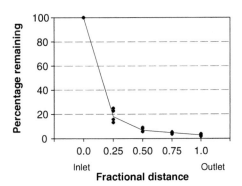

Fig. 25.2 Percentage of remaining particles along the length of bed C2 during the June 2003 sampling campaign. The line connects the average results for five samples obtained over 5 consecutive days

In addition, the greatest removal of particles occurs in the first fourth of the wetland. This trend is observed for particles of all sizes (Fig. 25.1). All the other particles count results—even for a different bed (B2) and sampling campaign (December 2003)—were very similar to those presented in Figs. 25.1 and 25.2.

25.3.2 Organic Matter and Fractionation

TOC concentrations along the length of wetland C2 have a pattern similar to that of the particles (Fig. 25.3). The greatest TOC removal clearly occurs in the first fourth of the bed. This trend (together with the particle counts) suggests that most organic matter removal is due to the retention of particulate organic matter. The other TOC concentration results—even for a different bed (B2) and sampling campaign (December 2003)—were very similar to those presented in Fig. 25.3.

Fig. 25.3 TOC concentrations at the inlet, at intermediate points, and at the outlet of bed C2 during the June 2003 sampling campaign. The line connects the average results for five samples obtained over 5 consecutive days

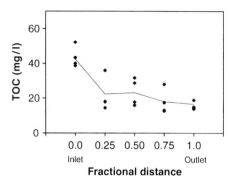

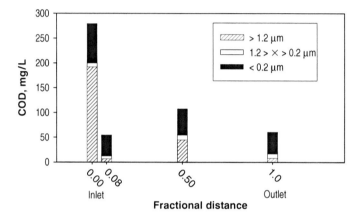

Fig. 25.4 COD fractions analysed at the inlet, at intermediate points, and at the outlet of bed B2. Each section of the bars represents the average result for seven samples obtained in a period of 4 months (April–July 2005)

COD fractionation was studied in bed B2. Figure 25.4 illustrates the distribution of the COD fractions along the length of this wetland. The most important COD fraction in the influent is the macrocolloidal and settleable fraction ($> 1.2\,\mu m$), and the least important is the colloidal fraction (between 0.2 and $1.2\,\mu m$). The removal of the macrocolloidal and settleable COD fraction occurs very close to the inlet (between the inlet and a fractional distance of 0.08), whereas the soluble ($< 0.2\,\mu m$) and colloidal COD fractions ($1.2\,\mu m > x > 0.2\,\mu m$) experience almost no removal along the length of the wetland. There is even a systematic increase in macrocolloidal and settleable COD in the middle of the bed.

Table 25.2 shows the average BOD_5 and COD effluent concentration in the two beds (B2 and C2) during the long-term study. It is clear that most of the remaining organic matter is still biodegradable, at least under aerobic conditions, because BOD_5 usually accounts for 70–95% of the total COD.

Table 25.2 Averages (±SD) COD and BOD$_5$ in the effluents of the two beds (B2 and C2) in a long-term study. Calculations based on 39 and 40 measurements for 2002 and 2003, respectively

	Bed	
Year and parameter	B2	C2
2002, COD, mg l^{-1}	69 ± 12.8	67 ± 36.8
2002, BOD$_5$, mg l^{-1}	59 ± 14.8	65 ± 32.2
2003, COD, mg l^{-1}	66 ± 14.6	65 ± 28.5
2003, BOD$_5$, mg l^{-1}	43 ± 16.6	43 ± 24.8

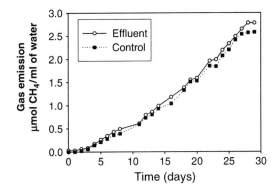

Fig. 25.5 Changes in methane concentration over time in one of the anaerobic biodegradability tests conducted with an effluent sample from bed B2 (and a control)

25.3.3 Biodegradability Tests

Figure 25.5 shows the results of one of the anaerobic biodegradability tests carried out with an effluent sample from bed B2 (the results from the other test were very similar). The figure shows that, although the methane concentration in the effluent sample was always greater than the concentration found in the control, the difference was very small. At the end of the two anaerobic biodegradability tests, the total amount of methane produced was very low (1.0–1.3 mg CH$_4$ l^{-1}), which indicated a very low amount of COD-equivalent reduction (4.0–5.4 mg l^{-1}). The COD of the two tested effluent samples ranged from 54 to 57 mg l^{-1}, and the BOD$_5$ was 54 mg l^{-1} in both cases. Therefore, the organic matter was almost completely readily biodegradable through aerobic pathways.

25.4 Discussion

Organic loading rates and the amount of solids in wastewater are factors closely related to the progressive clogging of SSF CWs (Langergraber *et al.*, 2003). Furthermore, solids mainly accumulate near the inlet of horizontal SSF CWs

(Caselles-Osorio *et al.*, 2007). Our results are consistent with these general statements because the greater removal of the organic matter parameters evaluated in this study (TOC, COD and particle-size distribution) takes place within the first fourth of the horizontal SSF CWs. TOC removal in the first fourth of the wetlands was around 50%, whereas COD removal was close to 80%. From the first fourth of the wetland onwards, organic matter removal was very low, ranging from 0% to 10%, depending on the parameter in question.

The decrease in COD along the length of the horizontal SSF CWs was mainly due to a reduction in the macrocolloidal and settleable fraction. Soluble and colloidal COD fractions experienced very low removal rates and remained quite constant along the length. However, it is likely that part of both the soluble and colloidal COD fractions was removed within the beds, but this removal could be compensated by internal organic matter production. Caselles-Osorio *et al.* (2007) also found that the removal of the macrocolloidal and settleable COD fraction was the main fraction responsible for total COD removal in full-scale horizontal SSF CWs.

The anaerobic biodegradability tests indicated that the organic matter in the effluents of the horizontal SSF CWs was not readily biodegradable under anaerobic conditions. The results of the long-term evaluation, however, indicated that the organic matter of the effluents was readily biodegradable through aerobic pathways since the BOD_5 was only slightly lower than the COD. Therefore, in order to increase the efficiency of horizontal SSF CWs similar to those studied in the present investigation, design criteria and operational conditions that favour more oxidised conditions should be considered. As García *et al.* (2004b, 2005) pointed out, shallow horizontal SSF CWs (wetted depth of 0.3 m) could be a feasible means of further removing the remaining organic matter through aerobic pathways.

25.5   Conclusions

The most abundant particles in the influent, at intermediate points along the length of the horizontal SSF CWs, and in the effluent are those comprised in the range from 0.7 to 2.0 μm out of a studied range that comprised particles between 0.7 and 21.5 μm. Most of the particles (close to 80%) are removed within the first quarter of the wetland length.

Organic matter removal (in terms of TOC and COD) in horizontal SSF CWs occurs mainly between the inlet and the first fourth of the wetland. Within this first fourth, TOC removal was approximately 50% and COD removal was around 80%. Further organic matter removal along the rest of the length of the wetland accounts for just 0–10% of total removal.

The organic matter remaining in the effluents of the horizontal SSF CWs tested in this study is mainly soluble and difficult to biodegrade through anaerobic pathways, but it is readily biodegradable under aerobic conditions.

Acknowledgements This study was possible thanks to the work of many people at the Environmental Engineering Division of the Technical University of Catalonia, in particular the fieldwork and technical support of Paula Aguirre, Virginia Campos, Alessandro Porta and Eduardo Álvarez. We would also like to thank the Consorci per a la Defensa dels Rius de la Conca del Besòs and Les Franqueses del Vallès Town Council for their assistance. This study was financed by grants awarded by the Spanish Ministry of Education and Science (research projects REN2002-04113-C03/TECNO and CTM2005-06457-C05/TECNO).

References

Caselles-Osorio, A., & García, J. (2007). Effect of physico-chemical pretreatment on the removal efficiency of horizontal subsurface-flow constructed wetlands. *Environmental Pollution, 146,* 55–63.

Caselles-Osorio, A., Puigagut, J., Segú, E., Vaello, N., Granés, F., García, D., & García, J. (2007). Solids accumulation in six full-scale subsurface flow constructed wetlands. *Water Research, 41,* 1388–1398.

Dulekgurgen, E., Dogruel, S., Krahan, Ö., & Orhon, D. (2006). Size distribution of wastewater COD fractions as an index for biodegradability. *Water Research, 40,* 273–282.

García, J., Ojeda, E., Sales, E., Chico, F., Aguirre, P., & Mujeriego, R. (2004a). Spatial variations of temperature, redox potential and contaminants in horizontal flow reed beds. *Ecological Engineering, 21,* 129–142.

García, J., Aguirre, P., Mujeriego, R., Huang, R., Ortiz, Y., & Bayona, J.M. (2004b). Initial contaminant removal performance in horizontal flow reed beds treating urban wastewater. *Water Research, 38,* 1669–1678.

García, J., Aguirre, P., Barragán, J., Mujeriego, R., Matamoros, V., & Bayona, J.M. (2005). Effect of key design parameters on the efficiency of horizontal subsurface flow constructed wetlands. *Ecological Engineering, 25,* 405–418.

García, J., Capel, V., Castro, A., Ruiz, I., & Soto, M. (2007). Anaerobic biodegradation tests and gas emissions from subsurface flow constructed wetlands. *Bioresource Technology, 98*(16), 3044–3052.

Langergraber, G., Harberl, R., Laber, J., & Pressl, A. (2003). Evaluation of substrate clogging processes in vertical flow constructed wetlands. *Water Science and Technology, 48*(5), 25–34.

Levine, A.D., Tchobanoglous, G., & Asano, T. (1991). Size distributions of particulate contaminants in wastewater and their impact on treatability. *Water Research, 25,* 911–922.

Chapter 26
Wastewater-fed Aquaculture, Otelfingen, Switzerland: Influence of System Design and Operation Parameters on the Efficiency of Nutrient Incorporation into Plant Biomass

Andreas Graber[1] and Ranka Junge-Berberović[1](✉)

Abstract A wastewater-fed, partly indoor aquaculture plant consisting of 36 basins with a total area of 360 m^2, charged with effluent from a methanization plant processing organic household waste (Kompogas System), started operation in spring 1998 in Otelfingen, Switzerland. Due to high effluent concentrations on one hand (total organic carbon, total nitrogen, and total phosphorus concentrations being 670, 255, and 52 mg l^{-1}, respectively) and stringent wastewater regulations on the other hand, the aquaculture was split into modules and stocked with organisms of different environmental requirements. The modular structure, as opposed to traditional one-pond polyculture, allowed manipulation and better nutrient budgeting of the system. Research focused on nutrient-recycling efficiency of a wide array of modules, producing biomass such as algae, fish, zooplankton, and aquatic macrophytes as ornamental plants (*Eichhornia*, *Pistia*, *Ipomoea*, *Lemna*, *Azolla*) which were suitable for sale on the Swiss market. Different arrangements of modules, of water and wastewater flows and of nutrient concentrations were tested in 3 years, 1998–2000. Our research showed that aquaculture can be set up to either produce treated wastewater ready for discharge into surface waters or to maximize nutrient recycling through biomass production. In the first case, the recommended loadings are below 0.47 g N m^{-2} day^{-1} and 0.084 g P m^{-2} day^{-1}, in the latter 4.3 g N m^{-2} day^{-1} and 0.31 g P m^{-2} day^{-1} still resulted in a 78% removal of nitrogen and 28% removal of phosphorus. Through weekly biomass harvest, up to 0.36 g N m^{-2} day^{-1} and 0.06 g P m^{-2} day^{-1} were converted into plants. With appropriate planning and plant selection and management, it was possible to increase the fraction of nutrients eliminated via primary production up to 40%, which is significantly higher than reported in the literature.

[1] University of Applied Sciences Waedenswil, Institute of Natural Resource Sciences, Section Ecological Engineering, Gruental, CH - 8820 Waedenswil, Switzerland

(✉) Corresponding author: e-mail: jura@zhaw.ch

J. Vymazal (ed.) *Wastewater Treatment, Plant Dynamics and Management in Constructed and Natural Wetlands*,
© Springer Science + Business Media B.V. 2008

Keywords Biogas-effluent reuse, aquatic macrophytes, nutrient recycling, wastewater-fed aquaculture

26.1   Introduction

The basic concept to treat the effluent from a methanization plant was developed in 1997 (Junge-Berberovic & Staudenmann, 1997). The Kompogas methanization plant processes organic household waste and consists of a one-step anaerobic, thermophilic reactor followed by an aerobic polishing unit (Edelmann *et al.*, 1993). Besides biogas and solid residues, a liquid fraction with high organic and inorganic nutrient loads is released. The primary purpose of the wastewater-fed aquaculture at Otelfingen was to treat and reuse the biogas effluent according to the postulates of ecological engineering, in order to meet the requirements for wastewater disposal by Swiss law. The secondary purpose was to convert the nutrients contained in the effluent into usable biomass and thus generate income.

The fundamental idea for designing the pilot plant was to observe the postulates of ecological engineering (Mitsch & Jørgensen, 1989) which call for a solar-based design. This implies that the majority of energy in the system originates from renewable energy resources but also that primary producers play a central role in the system. In most constructed wetlands, plants do not play a central role in the elimination of nutrients, but are facilitators of bacterial degradation. Even in the so-called productive treatment processes, like "Living Machines" (Todd & Josephson, 1996) or Stensund Aquaculture (Guterstam, 1996), the nutrient elimination through biomass harvesting accounts for less than 10% of the total reduction (Maddox & Kingsley, 1989; Gumbricht, 1993).

In tropical and subtropical countries, cyprinid fish and vegetables are traditionally produced for human consumption (Jana, 1998; Yan & Yao, 1989). In temperate climate zones, the demand for such products is rather low, hence new products had to be developed. For example, water hyacinth (*Eichhornia crassipes*) is an annoying weed in subtropical climates, where it is sometimes used as renewable resource (animal feed, fiber). On the other hand, because of their quick growth, these plants have the potential to be the main nutrient recyclers. Individual elimination rates by *Pistia stratiotes* and *Eichhornia crassipes* reported by Staudenmann and Junge-Berberovic (2003) were comparable to maxima reported by Reddy and Smith (1987). Being very decorative, they are increasingly used in Switzerland as indoor and outdoor ornamental plants.

The research at Otelfingen thus focused on screening of suitable aquatic organisms with a market potential and their testing under different environmental conditions. In addition, the qualities of the products (nutritional value, market value) as well as management methods (cultivation, harvest, pest susceptibility) were examined and reported (Mathis, 2000; Moura, 2001). All together 34 plant and 8 animal species were examined. The complete list of publications is available at www.cascadesystems.ch. This chapter concentrates on the operating seasons 1999 and 2000, and on the efficiency of nutrient incorporation into biomass.

26.2 Methods

26.2.1 System Design and Operating Conditions

The wastewater-fed, partly indoor aquaculture plant at Otelfingen, Switzerland was constructed in May 1998 and has been in continuous operation since then. During 3 consecutive years (1998–2000), operation parameters were varied, to define an optimum modus operandi (Table 26.1). The system consisted of 36 basins, arranged in 3 × 12 plots, each basin measuring 2 × 5 m with a depth of either 0.5 m (24 basins at the east end) or 1.5 m (12 basins at the west end), resulting in a total water surface of 360 m². Flexible basin interconnections and freshwater inlets allowed the variation of operational conditions.

During the pilot phase between May and December 1998, the 12 eastern basins were covered by a plastic tunnel greenhouse. The system was designed as a combination of "constructed food chains", where the basins were arranged in seven major steps, each containing between one and four modules of primary producers or consumers (Fig. 26.1). The main goal was to meet the wastewater treatment requirement according to the Swiss law, which is < 0.8 mg P l⁻¹. The details of the technical installations, such as aeration and heating, and the design and results of 1998 were described by Staudenmann and Junge-Berberovic (2000, 2003).

In January 1999, a double-layered tunnel greenhouse was installed (Filclair, 40 × 10 m), which covered all 24 shallow basins and also offered place for hydroponic cultivation of vegetables (Mathis, 2000). The operating conditions

Table 26.1 Priorities of objectives and basic design of the wastewater-fed aquaculture Otelfingen

	1998	1999	2000
Main objectives	1. Wastewater treatment to the levels of Swiss law 2. Introduction of a wide array of modules	1. Wastewater treatment to the level required by Kompogas (internal water recycling) 2. Maximising plant quality for marketing 3. Nutrient recycling	
Treatment steps and modules	Mosaic of primary producer and consumer modules	Experimental plant divided in two distinct parts: mainly primary producers and consumers	
Basins	Interconnected in parallel and in series	Shallow basins form three independent lines with concentration gradient from in- to outflow	
Average nutrient concentrations in basins	4–152 mg N l⁻¹	9.9–64 mg N l⁻¹	25–118 mg N l⁻¹
	<0.05–18 mg P l⁻¹	2.4–15 mg P l⁻¹	7.7–17 mg P l⁻¹

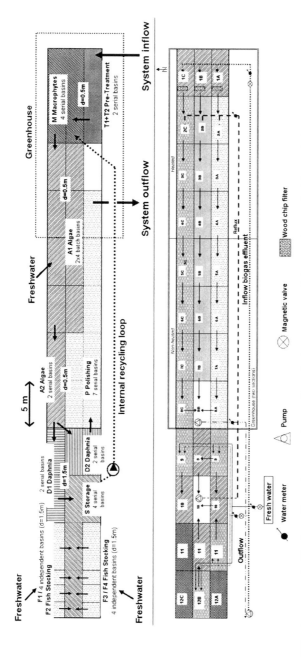

Fig. 26.1 Arrangement of basins with wastewater and water flows in the wastewater-fed aquaculture at Otelfingen during the summers of 1998 (above) and 1999 (below). The independent lines C, B, and A in 1999 and 2000 were arranged from north to south. In 2000, the wood chip filters were removed and both effluent and reflux water entered the system through the basins of row 1

were modified based on the findings of 1998. Instead of striving for very low outflow concentrations, the goal was to maximize growth rates and marketing quality of the plants. The aquaculture effluent still contained high nutrient concentrations but the treatment allowed it to be reused as process water in the Kompogas facility.

Thus, in 1999 and 2000 the plant was divided in consumer modules (deep basins) and production modules (shallow basins). The shallow basins were arranged in three separate production lines and operated at different loading rates (Fig. 26.1). This resulted in dilution rates of the biogas effluent of 42, 15, and 22 for the lines A, B, and C. Plant species were stocked in rows in the same pattern for each line according to the findings of 1998. Additional species were tested (*Ipomoea aquatica*, *Eleocharis dulcis*, *Cyperus esculentes*, rice varieties) and maintenance was reduced to 1 working day per week.

In 2000, much higher wastewater loadings were used to attain higher production rates. Reflux water from basin 10B was added daily to basins 1ABC to dilute the wastewater (Fig. 26.1). Wastewater input into basins 1B and 1C occurred daily, whereas to 1A wastewater was added every 7 days, but the same amount as to 1B, i.e., the weekly total. This resulted in dilution rates of the biogas effluent of 5.6, 4.9, and 2.9 for the lines A, B, and C. This setup was chosen to measure dynamic versus continuous wastewater addition (comparing lines A and B) and the difference in the wastewater dilution factor (comparing lines B and C). Water outflow for each line was calculated using the total water outflow of the plant from basin 8B and input flows into basins 1ABC.

26.2.2   Monitoring of the System

All major waste- and freshwater movements were monitored by mechanical water counters. The flows assessed were system inflow and system outflow, as well as every freshwater addition and flows in the recycling loop. Water temperatures were measured in situ with data-loggers. Water samples were collected weekly at 20 cm below surface and stored for a maximum of 24 h at 8°C before analysis. Total organic carbon (TOC) and total nitrogen content (TN) were determined by combustion and infrared detection (CHNS-Analyzer Vario EL, Elementar GmbH, Germany; HighTOC, Gerber Instruments AG, Germany). Chemical oxygen demand (COD), ammonium-nitrogen (NH_4-N), nitrate-nitrogen (NO_3-N), and total phosphorus (TP) content were determined by photometry (Cadas 30, Dr. Lange AG, Switzerland). Sporadically, biomass composition of macrophytes, daphnia, and fish was determined. Analyses included dry weight, ash, carbon, nitrogen, and phosphorus content. All biomass harvests and transfers between basins were quantified by fresh weight. Standing crops of macrophytes and zooplankton, as well as fish fresh weight, were determined monthly in 1998, and at the beginning and the end of the growing periods in 1999 and 2000.

26.3 Results

26.3.1 Higher Wastewater Loads Improve Nutrient Recycling by Plants

The untreated biogas-effluent was rich in nutrients; TOC, TN, NO_3-N, NH_4-N, and TP concentrations being 1,600, 670, 130, 370, and 61 mg l^{-1}, respectively (Staudenmann & Junge-Berberovic, 2003), and it contained 15–20% suspended solids. Before it entered the aquaculture, suspended solids were reduced to < 1% in a centrifugal decanter, and ammonia was partly oxidized to nitrate in a fixed bed biological reactor. Even thus pretreated, the effluent was highly concentrated and varied more than an order of magnitude over a season (Table 26.2). The atomic N: P ratios for 1998, 1999, and 2000 were 12, 32, and 26, respectively. This was due to different operating conditions of the biogas plant and varying efficiencies of the pretreatment.

The setup in 1998 allowed for an almost complete elimination of nitrogen and phosphorus by the system (92% and 98%, Table 26.3), whereby 25% of nitrogen and 38% of phosphorus removal was achieved by plant uptake.

In 1999 and 2000, hydraulic loading was raised, thus decreasing residence time by more than a factor of 5 compared to 1998 (Table 26.3). This also had consequences for nutrient concentrations in the basins with macrophytes (Fig. 26.2). In 1999, elimination rate of the overall system increased two times for nitrogen but plant uptake remained the same and nutrient levels in the aquaculture effluent were also higher. Phosphorus elimination remained the same for both system and plants.

In 2000, biogas-effluent loading was five times higher than in 1999. Nitrogen elimination increased four times for both system and plant uptake. The system was not able to eliminate more phosphorus than in the previous years but plant uptake doubled. This was probably due to more consistent harvesting and to higher

Table 26.2 Composition of pretreated biogas-effluent that entered the wastewater-fed aquaculture (*n*: number of measurements during each season)

Para-meter	Unit	1998			1999			2000		
		n	mean	range	*n*	mean	range	*n*	mean	range
COD	mg l^{-1}	15	1,672	1,310–2,180	12	7,462	1,918–16,200	4	2,608	1,720–3,440
TOC	mg l^{-1}	60	649	483–858	12	2,576	463–3,345	15	1,458	405–9,160
TN	mg l^{-1}	60	253	154–512	11	1,142	202–3,033	15	495	219–2,460
NH_4-N	mg l^{-1}	14	93.0	71.5–128	12	726	42–1,340	23	162	3.1–634
NO_3-N	mg l^{-1}	15	162	40.3–325	12	66.8	33.7–112	23	134	15.0–359
TP	mg l^{-1}	15	48.4	33.6–126	3	78.5	53.2–97.2	23	42.0	5.6–81.4
TSS	mg l^{-1}	4	5.1	5.1–5.2	7	8.1	5.5–1.3	3	5.2	4.4–5.9

Table 26.3 Seasonal water balances and elimination rates for basins with primary producers

		1998	1999	2000
		18/06–01/10	01/09–11/11	18/05–21/08
General	Duration (days)	103	70	95
	Planted area (m²)	220	210	210
	Total harvest (FW kg)	1,844	1,074	2,204
	Mean water temperature (°C)	21.1	19.0	23.2
Hydraulic loading	Biogas effluent (in)	1.8	1.9	9.9
(mm day⁻¹)	Freshwater (in)	8.3	39.6	30.3
	Outflow (out)	2.2	36.5	33.6
	Hydraulic residence time (day)	68	12.0	12.4
	Losses (evapotranspiration)	7.9ᵃ	5.1ᵇ	6.6ᵇ
N (g m⁻² d⁻¹)	Inflow	0.468	1.519	4.326
	Outflow	0.038	0.713	0.932
	Elimination by the system	0.430	0.806	3.394
	Elimination by plant uptake	0.108	0.097	0.357
N elimination (%)	Total (% of Input)	92	53	78
	Removed by plants (% of Total)	25	12	11
P (g m⁻² d⁻¹)	Inflow	0.084	0.195	0.314
	Outflow	0.002	0.109	0.225
	Elimination by the system	0.082	0.086	0.089
	Elimination by plant uptake	0.032	0.028	0.060
P elimination (%)	Total (% of Input)	98	44	28
	Removed by plants (% of Total)	38	33	67

ᵃ Measured with evaporation boxes
ᵇ Calculated from water balance

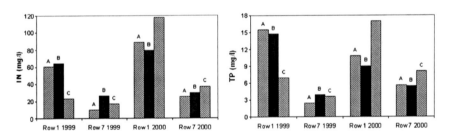

Fig. 26.2 Nitrogen and phosphorus levels in the start and end basins of differently operated lines of wastewater-fed aquaculture in 1999 and 2000

percentage of rooted plants, who could extract phosphorus from decomposing sediments in the substrate. Through weekly harvests, as much as 0.36 g N m⁻² day⁻¹ and 0.06 g P m⁻² day⁻¹ could be converted into plant biomass. Evapotranspiration losses were higher in the first year, probably due to more uncovered area. Dynamic addition of biogas effluent resulted in the highest percentage of nutrient elimination by plant uptake (Fig. 26.3, Line A, 2000).

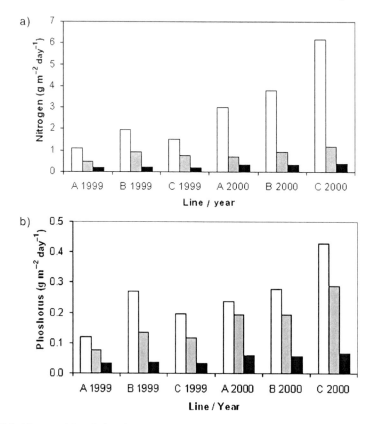

Fig. 26.3 Nitrogen (a) and phosphorus (b) loading and elimination of differently operated lines of wastewater-fed aquaculture. White bars show input, grey bars effluent, and black bars elimination by harvest of plants

26.3.2   Performance of Free-Floating Macrophytes

The best results were achieved by duckweed *(Lemna minor, Spirodela polyrrhiza)*, water fern *(Azolla filiculoides)*, water hyacinth *(Eichhornia crassipes)*, and water lettuce *(P. stratiotes)*. Their growth performance was excellent in high-nutrient conditions and nutrient-recycling rates of nitrogen and phosphorus of these plants were thus high during all years of operation (Table 26.4, Figs. 26.4 and 26.5).

26.4   Discussion

Based on the 3 years of experience, the aquaculture can be set up to either produce treated wastewater ready for discharge into surface waters or to maximize nutrient recycling through biomass production. In the first case, the recommended loading

Table 26.4 Biomass in 1998. Seasonal averages of biomass composition (in % of dry weight) and elimination rates of three aquatic macrophytes in dependence to decreasing nutrient concentrations in modules (M1 pretreated biogas effluent, M2 secondary macrophyte ponds, D Daphnia population, P polishing) (after Staudenmann and Junge-Berberovic, 2003). The increasing atomic ratios of nutrients (C:P and C:N) point to the increasing nutrient limitation, which is especially severe for phosphorus

Species	Module	n	%C	%N	%P	C/P	C/N	N g m^{-2} day^{-1}	P g m^{-2} day^{-1}
Pistia	M1	7	32.8	2.37	0.75	113	16	0.133	0.042
stratiotes	M2	5	33.0	2.14	0.85	100	18	0.069	0.028
	D	7	34.2	2.70	0.83	107	15	0.051	0.015
	P	3	35.9	0.88	0.13	694	47	0.008	0.001
Eichhornia	M1	3	33.1	3.73	0.87	98	10	0.049	0.011
crassipes	M2	5	33.9	3.39	0.90	97	12	0.176	0.047
	D	4	35.1	2.50	0.43	210	16	0.065	0.011
	P	7	37.3	0.94	0.26	376	46	0.013	0.004
Lemna minor	M2	6	35.5	4.27	1.61	57	10	0.027	0.010
	D	3	36.2	2.74	1.39	67	15	–	–
	P	3	38.5	1.42	0.17	593	32	0.004	0.000

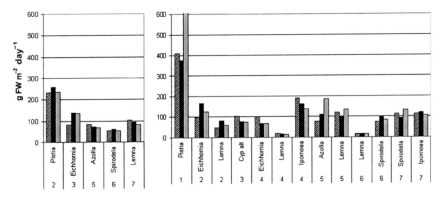

Fig. 26.4 Growth rates of aquatic macrophytes in aquaculture Otelfingen during autumn 1999 (left) and summer 2000 (right) in Lines A (hatched bars), B (black bars), and C (grey bars). The numbers denote the row of the line. The rows further in the line received lower concentration of nutrients. Note that in 1999 the lines differed only in concentration of nutrients, but in 2000 the Line A also had another dynamic of loading (weekly pulses)

rates are below 0.47 g N m^{-2} day^{-1} and 0.084 g P m^{-2} day^{-1}, allowing for a nutrient removal above 90%. In the latter case, the loading rates were 4.3 g N m^{-2} day^{-1} and 0.31 g P m^{-2} day^{-1}, which still resulted in 78% removal of nitrogen and 28% removal of phosphorus. Through weekly harvests, as much as 0.36 g N m^{-2} day^{-1} and 0.06 g P m^{-2} day^{-1} could be converted into plant biomass. Still, there is a trade-off between biomass quality and quantity: plants with high elimination rates may not be suitable for selling and vice versa.

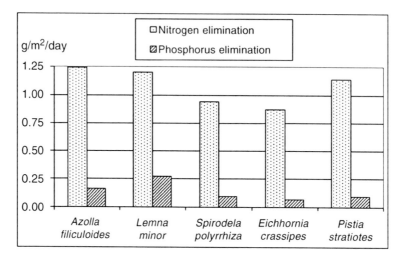

Fig. 26.5 Nutrient elimination by plant uptake in July 2002. The system was operated as in the year 2000, plants were placed in the basins 2A, 3A, 7C, 2B and 2C (from left to right)

Duckweed (*S. polyrrhiza, L. minor*) and water fern (*A. filiculoides*) can be used to feed fish and other animals. Because of their lack of sufficient fats and carbohydrates, the plants are not suitable for total substitution of conventional fish fodder, but could be used for dietary supplement (Marti *et al.*, 2000).

Floating macrophytes proved to be a suitable option for biomass production. Selection of plant species should take into account nutrient concentrations in the wastewater, and thus possible peaks of toxicants. In Otelfingen, plant species were placed where they found their optimum in the nutrient gradient between in- and outflow. *P. stratiotes* tolerated higher nutrient concentrations than *Eichhornia crassipes*. Growth speed and in particular growth forms of *Eichhornia* were dependent on the nutrients offered (Todt *et al.*, 1999). In later experiments, *C. alternifolius* proved to be extremely tolerant to high salinity and nutrient levels (unpublished data). As a consequence, in 2002 all basins in row 1 were stocked with *Cyperus* planted in gravel pots, which were harvested as ornamental plants.

26.5 Conclusions

Continuous development of productive systems for wastewater treatment imposes a paradigm shift. The implementation of a wastewater-fed aquaculture for the sole purpose of effluent purification is too expensive compared to planted soil filters. In order to fully utilize the special merits of this method, the system should be operated as a productive unit, allowing for high growth rates of selected crops and producing effluent water treated to levels necessary for reuse. Thus, multiple requirements of product quality and income generation can be fulfilled. In certain cases, the addition of a complementary fertilizer should be considered to prevent nutrient limitation by phosphorus or potassium (especially if using water hyacinths).

Acknowledgements This research was supported by the Swiss Federal Commission for Technology and Innovation (grants No. CTI-3487.2 and CTI-4516.2). Without contributions of our colleagues, students, and research assistants from the Institute of Natural Resource Sciences of UASW, this project would not have been possible. We especially thank Jean Bernard Baechtiger, Juerg Staudenmann, and Alex Mathis for their continual support and inputs.

References

Edelmann, W., Engeli, H., Glauser, M., Hofer, H., Membrez, Y., Meylan, J., & Schwitzguébel, J. (1993). Vergaerung von haeuslichen Abfaellen und Industrieabwaessern. Bern, Switzerland: Bundesamt fuer Konjunkturfragen, IP Erneuerbare Energien.

Gumbricht, T. (1993). Nutrient removal capacity in submersed macrophyte pond systems in a temperate climate. *Ecological Engineering, 2*, 49–61.

Guterstam, B. (1996). Demonstrating ecological engineering for wastewater treatment in a Nordic climate using aquaculture principles in a greenhouse mesocosm. *Ecological Engineering, 6*, 73–97.

Jana, B.B. (1998). Sewage-fed aquaculture: The Calcutta model. *Ecological Engineering, 11*, 73–85.

Junge-Berberovic, R., & Staudenmann, J. (1997). Wastewater-fed aquaculture in temperate climate: Possibilities and limits of nutrient recycling. In A. Moser (Ed.), *The green book of Eco-Tech. Strategy report from ICEE4* (p. 187). Graz, Austria: 4th International Ecological Engineering Conference, Sept. 22–24.

Maddox, J.J., & Kingsley, J.B. (1989). Waste treatment for confined swine with an integrated artificial wetland and aquaculture system. In D.A. Hammer (Ed.), *Constructed wetlands for wastewater treatment: Municipal, industrial and agricultural* (pp. 191–200). Chelsea, MI: Lewis.

Marti, B., Graber, A., & Junge-Berberovic, R. (2000). Wasserpflanzen als Tierfutter: Optimierung des Proteingehaltes. Diploma thesis at the University of Applied Sciences Waedenswil, Switzerland, 45 pp.

Mathis, A. (2000). Gemueseproduktion mit Ueberschusswasser einer Kompogas-Anlage. *Der Gemuesebau/Le Maraîcher, 9*, 13–14.

Mitsch, W.J., & Jørgensen, S.E. (1989). *Ecological engineering: An introduction to ecotechnology.* New York: Wiley.

Moura, M. (2001). Kulturversuche mit tropischen Gemuesearten in abwassergespiesener Aquakultur. Dissertation, University of Applied Sciences Waedenswil, Switzerland.

Reddy, K.R., & Smith, W.H. (1987). *Aquatic plants for water treatment and resource recovery.* Orlando, FL: Magnolia.

Staudenmann, J., & Junge-Berberovic, R. (2000). Treating biogas plant effluent through aquaculture: First results and experiences from the pilot plant Otelfingen (Switzerland). In B.B. Jana, R.D. Banerjee, B. Guterstam, & J. Heeb (Eds.), *Waste recycling and resource management in the developing world* (pp. 51–57). India: University of Kalyani and Switzerland: International Ecological Engineering Society.

Staudenmann, J., & Junge-Berberovic, R. (2003). Recycling nutrients from industrial wastewater by aquaculture systems in temperate climates (Switzerland). *Journal of Applied Aquaculture, 13*, 67–103.

Todd, J., & Josephson B. (1996). The design of living technologies for waste treatment. *Ecological Engineering*, 6, 109–136.

Todt, D., Junge-Berberovic, R., & Staudenmann, J. (1999). Produktion von tropischen Pflanzen zur Naehrstoffelimination von Prozesswasser. Diploma thesis, University of Applied Sciences Waedenswil, Switzerland.

Yan, J., & Yao, H. (1989). Integrated fish culture management in China. In W.J. Mitsch & S.E. Jørgensen (Eds.), *Ecological engineering* (pp. 375–340). New York: Wiley.

Chapter 27
Is Concentration of Dissolved Oxygen a Good Indicator of Processes in Filtration Beds of Horizontal-Flow Constructed Wetlands?

Jan Vymazal[1,2]($\boxtimes$) and Lenka Kröpfelová[1]

Abstract Filtration beds of subsurface horizontal-flow constructed wetlands (HF CWs) are generally considered as anoxic or anaerobic and, therefore it is assumed that the outflow concentration of dissolved oxygen (DO) is usually very low. During 2004 and 2005, nearly 60 HF CWs in the Czech Republic were visited and it was found that many systems really provided low DO outflow concentrations ($<2\,\mathrm{mg\,l^{-1}}$). However, a substantial number of systems surprisingly provided relatively high concentration of DO ($>5\,\mathrm{mg\,l^{-1}}$). We selected several systems where monitoring program included a broader selection of parameters, especially ammonia-N, nitrate-N, and sulfate, and set up more frequent measurements of dissolved oxygen. In our study, we also used some older data that were obtained during research projects in the past. We focused on nitrification and sulfate-reduction as processes occurring under strictly aerobic and anaerobic conditions, respectively. When evaluating the data, we found that in systems with very low outflow concentrations of DO, nitrification was frequently very limited but in some systems a substantial reduction of ammonia occurred. On the other hand, several systems with relatively high O_2 outflow concentrations provided nearly zero removal of ammonia. The similar unanimous results were observed for sulfate, i.e., high O_2 outflow concentrations were sometimes connected with high reduction of sulfate and on the other hand low O_2 outflow concentrations were not connected with sulfate reduction. We concluded that DO concentration at the outflow from HF CWs does not provide good information about the processes occurring in the filtration beds.

Keywords Ammonia, dissolved oxygen, nitrate, subsurface horizontal flow, sulfate

[1] ENKI, o.p.s., Dukelská 145, 379 01 Třeboň, Czech Republic

[2] Institute of Systems Biology and Ecology, Dukelská 145, 379 01 Třeboň, Czech Republic

($\boxtimes$) Corresponding author: e-mail: vymazal@yahoo.com

J. Vymazal (ed.) *Wastewater Treatment, Plant Dynamics and Management in Constructed and Natural Wetlands,*
© Springer Science + Business Media B.V. 2008

27.1   Introduction

Excess water applied to a permeable soil will rapidly drain from the upper profile through the interconnected pore spaces. Much of this pore space is again filled with gas, which is continuous with the atmosphere, after draining for several hours. When soils are inundated, the pore spaces are filled with water and the rate at which oxygen can diffuse through the soil is drastically reduced. Diffusion of oxygen in an aqueous solution has been estimated at 10,000 times slower than oxygen diffusion through a porous medium such as drained soils (Greenwood, 1961; Greenwood & Goodman, 1964). As a result of prolonged flooding and continued oxygen demand for root and microbial respiration, as well as chemical oxidation of reduced organic and inorganic components, the oxygen content of the soil solution begins an immediate decline and may be depleted within several hours to a few days. The rate at which the oxygen is depleted depends on the ambient temperature, the availability of organic substrates for microbial respiration, and sometimes the chemical oxygen demands from reductants such as ferrous iron.

Oxygen consumption rates have been thought to be a function of microbial respiration. However, Howeler and Bouldin (1971) demonstrated that oxygen consumption rates in some flooded soils can best be described by models including oxygen consumption for both biological respiration and for chemical oxidation of both mobile and nonmobile constituents. Reduced iron and manganese ions were thought to represent the bulk of the mobile reductants while precipitated ferrous iron, manganous manganese, and sulfide compounds, encountered as the oxidized zone increased in thickness, likely comprised much of the nonmobile constituents. Howeler (1972) pointed out that the ratio between biological and chemical oxygen consumption rates may vary widely, depending on the organic matter content of the soil or sediment.

Oxygen diffusion is not the only route for oxygen transport in the flooded soil. It is well documented that aquatic and wetland macrophytes release oxygen from roots into the rhizosphere (e.g., Armstrong & Armstrong, 1988; Sorrell & Boon, 1992). Oxygen release rates from roots depend on the internal oxygen concentration, the oxygen demand of the surrounding medium, and the permeability of the root walls (Sorrell & Armstrong, 1994). Wetland plants conserve internal oxygen because of suberized and lignified layers in the hypodermis and outer cortex (Armstrong & Armstrong, 1988). These stop radial leakage outward, allowing more oxygen to reach the apical meristem. Thus, wetland plants attempt to minimize their oxygen losses to the rhizosphere. Wetland plants do, however, leak oxygen from their roots (Brix, 1998).

Using different assumptions of root oxygen release rates, root dimensions, numbers, permeability, etc., Lawson (1985) calculated a possible oxygen flux from roots of *Phragmites australis* up to 4.3 g m^{-2} day^{-1}. Others, using different techniques, have estimated root oxygen release rates from *Phragmites* to be 0.02 g m^{-2} day^{-1} (Brix, 1990; Brix & Schierup, 1990), 1–2 g m^{-2} day^{-1} (Gries et al., 1990) and 5–12 g m^{-2} day^{-1} (Armstrong et al., 1990). Brix (1998) reported that gas-exchange experiments in Denmark have shown that $4\,g\ O_2\ m^{-2}\ day^{-1}$ is transferred from the atmosphere to the soil. The reed vegetation transport $2\,g\ O_2\ m^{-2}\ day^{-1}$ to the root zone is mainly utilized by the roots and rhizomes themselves.

While oxygen release from the roots of emergent plants such as *P. australis* has been measured, the information on the diffusion of oxygen from the atmosphere to the vegetated beds of constructed wetlands is missing. The objective of this chapter is to find out what are the DO concentrations at the outflow from constructed wetlands with horizontal subsurface flow (HF CWs) in the Czech Republic, and if these concentrations could be related to the processes in filtration beds.

27.2   Methods

During 2004 and 2005, a total of 59 HF CWs across the Czech Republic were visited and inflow and outflow DO concentrations were measured. The systems where chemical parameters, namely COD, BOD, ammonia-N, nitrate-N, or sulfate, were available, have been visited repeatedly. The number of DO measurements varies between 1 and 40 for individual systems. For the measurements, portable device WTW Oxi 340i with oxygen sensor CellOx 325 was used.

27.3   Results and Discussion

The field measurements revealed wide range of outflow DO concentrations (Fig. 27.1) with 33 CWs having DO concentration $> 3\,mg\ l^{-1}$ and 18 CWs having DO outflow concentrations $> 5\,mg\ l^{-1}$. This was a surprising finding as HF CWs are considered as mostly anoxic/anaerobic systems. It is true that for some systems only one measurement is available. However, more detailed studies indicated that the outflow concentration from horizontal systems is quite steady during the year and also during the day.

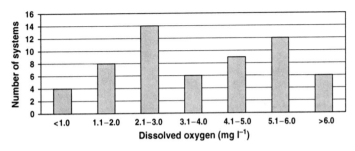

Fig. 27.1 Dissolved oxygen concentration at the outflow from vegetated beds of 59 HF CWs in the Czech Republic. The number of measurements in individual systems varies between 1 and 40

In Fig. 27.2, the DO concentrations at the outflow from CW Břehov during the period June 2004 to December 2005 is presented. The average DO outflow concentration was 4.19 (± 0.95) mg l⁻¹ during this period and fluctuated only slightly. The average inflow concentration (i.e. in the outflow from a septic tank) was 1.90 (± 1.52) mg l⁻¹.

Figure 27.3 shows more detailed study from a CW in Mořina – 37 measurements during the period of 2 months were completed. The study started during the spring snow melting, which resulted in high inflow DO concentrations. The retention time in the Imhoff tank was quite short, and therefore, the DO concentrations did not decrease too much. The outflow DO concentrations were quite steady around 1 mg l⁻¹. In mid-April, water from the adjacent stream leaked to the sewer system. As the

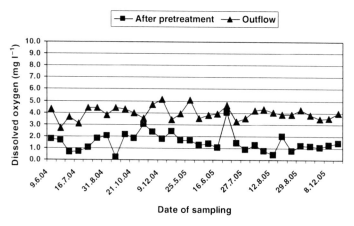

Fig. 27.2 Concentrations of dissolved oxygen in water entering the filtration bed (after pretreatment in a septic tank) and discharged water from the bed in CW Břehov

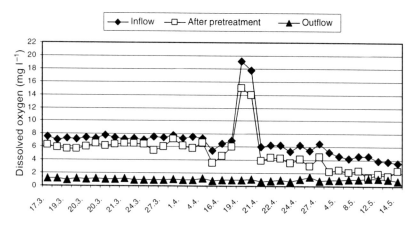

Fig. 27.3 Concentrations of DO in raw wastewater, after pretreatment in an Imhoff tank and at the outflow from the filtration beds in HF CW Mořina during the period March 17–May 14, 2006

stream was fed by water from a village pond where algae started to flourish, the DO concentrations were extremely high. However, there was no change in outflow DO concentrations. When this leakage was stopped and all the snow melted, DO concentrations dropped down to about 4 mg l^{-1}. Again, the outflow DO concentrations did not change. During the 2-month period, the DO in raw and pretreated wastewater varied between 3.6 and 19.2 mg l^{-1} and 1.3 and 15.0 mg l^{-1}, respectively but outflow DO concentrations varied only between 0.71 and 1.36 mg l^{-1}. Also, there is little fluctuation in DO concentration at the outflow during the day (Fig. 27.4).

Table 27.1 summarizes inflow and outflow concentrations of BOD$_5$, ammonia-N, sulfate, and DO in several Czech HF CWs. Ammonia and sulfate were selected because oxidation of ammonia (nitrification) and sulfate reduction proceed under entirely different redox conditions. We tried to relate the extent of mentioned processes to DO concentration at the outflow. However, it seems that DO concentrations at the outflow provide little information about the processes in the bed.

In Břehov, in both years removal of ammonia and sulfate occurred at the same time with quite high DO concentrations at the outflow. In Ondřejov, no removal of ammonia occurred while sulfate was substantially reduced. This is in accordance with the theoretical presumptions but at the same time both inflow and outflow DO

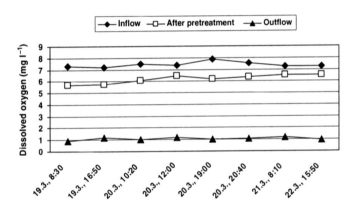

Fig. 27.4 Inflow and outflow DO concentrations during the period March 19–March 22, 2006 in CW Mořina

Table 27.1 Inflow and outflow DO concentrations in five HF CWs in the Czech Republic. All values in mg l^{-1}

Locality	BOD$_5$ in	BOD$_5$ out	NH$_4$-N in	NH$_4$-N out	SO$_4^{2-}$ in	SO$_4^{2-}$ out	DO in	DO out
Břehov, 2004	40.1	7.6	25.9	12.7	46	22	2.5	4.3
Břehov, 2005	109	27	40	24.7	71	35	1.4	4.0
Ondřejov, 1998	104	12	18.3	25.5	86	25	5.5	4.9
T. Dušníky, 2004	716	56	54	27	80	60	0.9	2.0
Čistá, 2002	37	7.3	14.1	12.8	72	68	4.9	3.7
Mořina, 2003	116	27	35.4	32.3	154	73	1.5	0.18

concentrations were high. In extremely highly loaded system in Trhové Dušníky, both oxidation of ammonia and reduction of sulfate occurred with very low DO concentrations at both ends of the filtration beds. In Čistá, where wastewater is diluted by rain and drainage waters, very low nitrification occurred with more or less no sulfate reduction at the same time. The only system which behaved according to the theory was Mořina – low ammonia oxidation, high sulfate reduction under low oxygen concentrations.

We also tried to calculate how much oxygen consumed in the bed is coming in the wastewater. We used as a basic equation for oxygen transfer rate (Cooper, 1999):

$$OTR = \text{flow rate} \times [(BOD_5in - BOD_5out) + 4.3 \, (NH_4 \text{ - Nin} - NH_4 \text{ - Nout})]/area,$$

from which we subtracted (a) nitrogen taken up by plants ($60 \, g \, m^{-2}$) and therefore not available for nitrification, (b) oxygen from sulfate reduction [$0.67 \, (SO_4^{2-}in - SO_4^{2-}out)$] and (c) denitrification [$2.5 \, (NH_4\text{-Nin} - NH_4\text{-Nout})$], provided that all ammonia nitrogen not taken up by plants is nitrified to nitrate and then all nitrate is denitrified. This, however, is an overestimation but the amount of ammonia adsorbed and volatilized is a question mark. However, in HF CWs there is no free water surface and also materials used for filtration beds usually do not provide high sorption capacity. Therefore, volatilzation and adsorption are very limited in HF CWs. Also, the amount of oxygen from manganic and ferric compounds used for reduction of organics is unknown. Using this simple estimation, we calculated that the consumption of DO in the beds in Břehov in 2003 was $2.46 \, g \, O_2 \, m^{-2} \, day^{-1}$ while the DO in the inflow was only $0.037 \, g \, O_2 \, m^{-2} \, day^{-1}$, i.e., 1.5% of the DO consumption in the beds. In Mořina in 2003, the DO consumption was $1.91 \, g \, O_2 \, m^{-2} \, day^{-1}$ with $0.04 \, g \, O_2 \, m^{-2} \, day^{-1}$ coming with the inflowing water (2% of the total DO consumption). These approximate numbers mean, however, that the amount of dissolved oxygen entering the beds in inflowing wastewater is a negligible portion of the DO consumed in the beds.

27.4   Conclusions

1. Concentration of dissolved oxygen at the outflow from the 59 Czech HF CWs varied widely with 33 CWs having DO concentration $> 3 \, mg \, l^{-1}$ and 18 CWs having DO outflow concentrations $> 5 \, mg \, l^{-1}$.
2. The DO concentration at the outflow does not fluctuate too much during the year and it apparently does not depend on the inflow DO concentration.
3. Outflow DO concentration provides little information about the processes in the bed.
4. The amount of DO in the inflowing wastewater forms only a negligible portion (less than 2%) of the DO used in the filtration bed.

Acknowledgements The research was supported by grants No. 206/02/1036 "Processes Determining Mass Balance in Overloaded Wetlands" and No. 206/06/0058 "Monitoring of Heavy Metals and Selected Risk Elements during Wastewater Treatment in Constructed Wetlands" from the Czech Science Foundation.

References

Armstrong, J., & Armstrong, W. (1988). *Phragmites australis* – a preliminary study of soil-oxidizing sites and internal gas transport pathways. *New Phytologist, 108*, 373–382.

Armstrong, W., Armstrong, J., & Beckett, P.M. (1990). Measurement and modelling of oxygen release from roots of *Phragmites australis*. In P.F. Cooper & B.C. Findlater (Eds.), *Constructed wetlands in water pollution control* (pp. 41–51). Oxford: Pergamon.

Brix, H. (1990). Gas exchange through the soil-atmosphere interface and through dead culms of *Phragmites australis* in a constructed wetland receiving domestic sewage. *Water Research, 24*, 259–266.

Brix, H. (1998). Denmark. In J. Vymazal, H. Brix, P.F. Cooper, M.B. Green, & R. Haberl (Eds.), *Constructed wetlands for wastewater treatment in Europe* (pp. 123–156). Leiden, The Netherlands: Backhuys.

Brix, H., & Schierup, H.-H. (1990). Soil oxygenation in constructed reed beds: The role of macrophyte and soil-atmosphere interface oxygen transport. In P.F. Cooper & B.C. Findlater (Eds.), *Constructed wetlands in water pollution control* (pp. 53–66). Oxford: Pergamon.

Cooper, P.F. (1999). A review of the design and performance of vertical flow and hybrid reed bed treatment systems. *Water Science and Technology, 40*(3), 1–9.

Greenwood, D.J. (1961). The effect of oxygen concentration on the decomposition of organic materials in soil. *Plant and Soil, 14*, 360–376.

Greenwood, D.J., & Goodman, D. (1964). Oxygen diffusion and aerobic respiration in soil spheres. *Journal of the Science of Food and Agriculture, 15*, 579–588.

Gries, C., Kappen, L., & Lösch, R. (1990). Mechanism of flood tolerance in reed, *Phragmites australis* (Cav.) Trin. ex Steudel. *New Phytologist, 114*, 589–593.

Howeler, R.H. (1972). The oxygen status of lake sediments. *Journal of Environmental Quality, 1*, 366–371.

Howeler, R.H., & Bouldin, D.R. (1971). The diffusion and consumption of oxygen in submerged soils. *Soil Science Society of America Proceedings, 35*, 202–208.

Lawson, G.J. (1985). Cultivating reeds (*Phragmites australis*) for root zone treatment of sewage: Contract Report ITE (965th ed.). Cumbria, UK: Water Research Centre.

Sorrell, B.K., & Armstrong, W. (1994). On the difficulties of measuring oxygen release by root systems of wetland plants. *Journal of Ecology, 82*, 177–183.

Sorrell, B.K., & Boon, P.I. (1992). Biogeochemistry of billabong sediments. II: Seasonal variations in methane production. *Freshwater Biology, 27*, 435–445.

Chapter 28
Pollutant Transformation Performance and Model Development in African Wetland Systems: Large Catchment Extrapolation

Herbert John Bavor[1](✉) and Michael Thomas Waters[2]

Abstract Two wetland systems in the Lake Victoria Basin, western Kenya, were investigated and monitored to assess buffering capacity and also to develop a model approach to evaluating larger-scale pollutant-buffering capacity of regional wetlands. The Dionosoyiet wetland was located in the highlands of the Rift Valley Province, at approximately 2,000 m altitude. It was located immediately adjacent the Kericho town centre, covered a 34 ha area and was set in a catchment of 23 km[2]. The wetland was located in the upper reaches of the Sondu-Miriu river system which flows into Lake Victoria. The 560 ha Chepkoilel wetland, near Eldoret, had an agricultural catchment area of 210 km[2] with major inflows to the wetland contributed by the Sergoit-Misikuri river system. The catchment drained areas of mild slopes ranging up to 2,160 m above sea level. Water-quality investigations were undertaken in the wetlands from June 2004 to April 2005 for nutrients and suspended solids. Hydrology and water-quality modelling were performed utilising the LAVINKS-WEB model. The model was adapted to incorporate a rainfall-runoff module based on the isochronal histogram technique and a partially stochastic prediction of water quality (TSS, TN and TP) based on incoming flow rates. Using the data gathered from June 2004 to April 2005 for calibration and earlier climatic data, modelling was performed to cover an 11-year period, from January 1994 to December 2004, and indicated that the wetland removed 43% TSS, 41% TP and 20% TN with average areal removal rates of 21.3 TSS, 0.038 TP and 1.03 TN (kg ha^{-1} day^{-1}) for the Dionosoyiet system and considerably greater removal from the Chepkoilel system. The findings and model development show that in addition to being critical ecosystem diversity reservoirs and central community/agricultural-activity resources, the wetlands perform significant functions of water-quality improvement. The preservation of these wetlands and other similar wetlands is important in ensuring sustainable utilisation of water resources in the Lake Victoria Basin.

[1] Centre for Water and Environmental Technology – Water Research Laboratory, University of Western Sydney – Hawkesbury, Locked Bag 1797, Penrith South DC, NSW 1797, Australia

[2] SMEC International, P.O. Box 1052, North Sydney, NSW 2060, Australia

(✉) Corresponding author: e-mail: j.bavor@uws.edu.au

J. Vymazal (ed.) *Wastewater Treatment, Plant Dynamics and Management in Constructed and Natural Wetlands,*
© Springer Science + Business Media B.V. 2008

Keywords Buffering, Kenya, Lake Victoria, modelling, nutrients, runoff, surface flow

28.1   Introduction

Lake Victoria is by surface area the largest freshwater lake in Africa and the second-largest freshwater lake in the world. It has a 6,880,000 ha surface area with an approximately oval shape. The depth profile of the lake is shallow and it is relatively well mixed in the vertical profile.

The Lake:

- Represents a major fishery resource for Kenya, Uganda and Tanzania
- Is a major waterway supporting trade, communication and tourism
- Generates substantial power at the Owen Falls Hydroelectric Station
- Is important in supporting a large and diverse ecological community
- Forms the headwaters of the River Nile, the longest river in the world and a vital water resource for the countries of Uganda, Sudan and Egypt

Approximately 30 million people live in the lake's catchment and the lake is utilised for fish production, transport and power generation. Approximately 2 million people depend directly or indirectly on fish from the lake (Ntiba *et al.*, 2001). Water quality in the lake has declined significantly over recent decades. One of the causes of declining water quality is increasing loadings of nitrogen and phosphorus entering the lake due to anthropogenic inputs (see van der Knaap *et al.*, 2002; Mwanuzi *et al.*, 2003; Verschuren *et al.*, 2002). It has been established that a significant proportion of the nutrients entering the lake arise because of changing patterns of human land-use activities which have exerted huge pressure on environmental resources in recent decades.

Linked with increased nutrient loads, changes in land-use management within the catchment (particularly deforestation) have led to increased erosion, resulting in increased sediment loadings on the lake (Verschuren *et al.*, 2002). A reversal of these processes will require better catchment-management practices, including erosion control and the wise management of the remaining wetlands within the catchment (Ngirigacha, 2000).

These increases in the sediment, phosphorus and nitrogen loads entering the lake are most difficult to address; nutrient load increases result in increased nutrient concentrations within the lake and lead to ecological disturbances such an increased abundance of nuisance plant species in waterways. Modelling of water-quality processes, as carried out in this study, is increasingly recognised as an important aspect of sustainable management of natural resources such as wetlands.

28.2   Methods

Water quality was monitored at the site from June 2004 to the February 2005. Each monitoring event was carried out in two phases as follows:

1. Physical assessments involving on-site measurements for pH, electrical conductivity and total dissolved salts, temperature, turbidity and dissolved oxygen.

These parameters were analysed using either a Hydrolab Data Sonde 4A or YSI 660 Sonde both of which were calibrated regularly for reproducibility of results. Physical observations were also undertaken noting the general environmental conditions at the time of the sampling and measurements. Among the observations made were rainfall, prevailing activities around the monitoring stations and any relevant observations of the nature of water flowing in the stream – odour, colour, etc.

2. Chemical parameters including nitrogen, phosphorous and total suspended solids were analysed using standard methods. Intensive monitoring was carried out at the wetland from October to November 2004 and again from March to April 2005. The primary aim of the intensive monitoring was to characterise the hydrology and water-quality trends in the wetland through wet-weather events. It was also necessary to undertake detailed sampling of soils and vegetation to characterise the nutrient and heavy metals status of these compartments of the wetland in order to more fully characterise the buffering processes taking place.

Measurements were made in coincidence with hydrologic assessment of flow rates for the streams. All water samples were collected in pre-rinsed plastic bottles for the above-mentioned parameters and held in cooler boxes for transport to the laboratory for analyses.

The base modelling framework adopted here utilised the LAVINKS-WEB model derived from a Cooperative Research Centre for Freshwater Ecology (CRCFE) model denoted by the acronym 'POND' (**PO**llutant **N**utrient **D**ynamics). This model has been most widely used in Australia, but is written generally enough so that other applications across a range of climatic and landscape conditions are possible. The model has also been used successfully in countries with tropical climates such as Thailand. The background and details for the model are given in Lawrence and Breen (1998).

Basic variables used in the model can be considered as inflow variables, variables within the wetland water column, variables within the wetland bed sediments and outflow variables as described below.

- Inflow variables: daily rainfall, inflow estimate, incoming water-quality concentrations for SS, TP, TN and TDS.
- Variables within the wetland water column: volume, surface area, depth and total masses for SS, TP, TN, TDS, N remobilised from bed, inorganic N, organic N, DO, total organic store, plant growth rate, P uptake due to plant growth, settling velocity for SS, SS interception, TP interception, TN interception, organic store interception and uptake rates for SS, TP and TN.
- Variables within the wetland bed sediments: total masses for cumulative and daily organic stores, sediment oxygen content, residual oxygen store, NO_3, $Fe(III)$, SO_4, H_2S, $Fe(II)$, NH_4, PO_4, respiration rate.
- Outflow variables: discharge, concentrations and loads and reduction rates for SS, TN, TP and TDS and N:P ratios.
- The model sheets of the LAVINKS-WEB model, modified from the CRCFE model, fall into three classes as shown in Fig. 28.1:

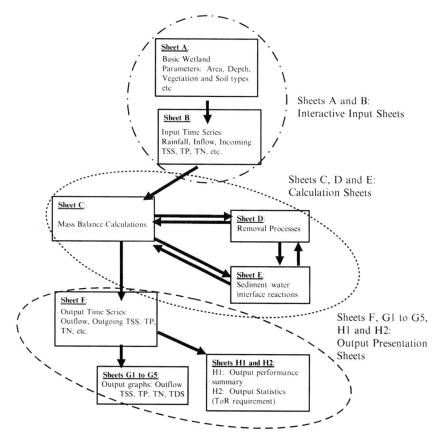

Fig. 28.1 Modelling framework for LAVINKS-WEB model

- Input sheets A and B where the user enters the basic parameters describing the wetland (sheet A) and the time series inputs (sheet B)
- Calculations sheets C, D and E, where the processes taking place are modelled, and which the user does not generally need to interact with
- Output sheets F, G1 to G5, H1 and H2, which present the model results

28.3   Results and Discussion

It has long been considered that wetlands play a significant role in reducing the loads of pollutants entering Lake Victoria; however, the extent to which this takes place has until now been very poorly understood. No attempts are known to have been made to quantify the buffering capacity of the Kenyan part of the Lake Victoria watershed. The term 'buffering capacity' of wetlands refers to the net improvement in water quality from inflows to outflow that takes place as water flows through a wetland. This is most apparent as reductions in suspended solid and nutrient loads are transmitted through the wetland.

Given the extent of data collection for the pilot wetlands, the mapping activities that have taken place, and the development of a model for wetland-buffering determination, it is now possible to give some preliminary estimation of the extent of buffering capacity that takes place in wetlands in the basin (Bavor *et al.*, 2001). This is done by firstly considering briefly some issues associated with scaling, then by describing the data available to assist in scaling, followed by a description of the estimation methodology and finally by presenting the results.

In up-scaling from pilot scale to catchment scale, factors accounted for included:

- The net long-term average runoff rate, SS, TP and TN loads entering Lake Victoria from the Kenyan portion of the basin.
- Differences in rainfall between the two pilot wetland sites and the whole basin area. Daily rainfall data was obtained from a Microsoft Access format climate database maintained by the LVEMP Water Quality Component in Kisumu.
- Differences in water quality between the pilot wetland sites and across the whole lake basin.
- For rainfall data on record, the period from 1 January 1983 to 31 December 1990 was the longest contiguous period over which rainfall data was available with a minimum number of data gaps at stations which covered the major part of the Kenyan portion of the Lake basin.

Stations from which rainfall data was used were Bungoma Water Supply, Kisii Water Supply, Miwani, European quarters, Narok Keekorok Game Lodge, Eldoret Experimental Farm, Turbo Forest Nursery, Kibos Cotton Experiment Station, Ahero Kano Irrigation Station and Kericho Tumbilil. Uniform weighting was applied to the rainfall from each station in order to estimate the basin-wide average rainfall. A summary of average annual rainfall from these stations is presented in Table 28.1.

Long-term average river discharges into the Lake were determined by COWI (2002). These estimates are based on rainfall-runoff modelling using the NAM and

Table 28.1 Rainfall summary from weather stations across the Lake Victoria Basin, 1983–1990

Station	Average yearly rainfall (mm)
Bungoma Water Supply	1,564
Kisii Water Supply	1,796
Miwani, European quarters	943
Narok Keekorok Game Lodge	1,038
Eldoret Experimental Farm	982
Turbot Forest Nursery	1,238
Kibos Cotton Experiment Station	1,231
Ahero Kano Irrigation station	1,200
Kericho Tumbilil	2,042
Basin average rainfall estimate	1,345

SMAP models for gauged rivers and from the modified rational method for ungauged catchments. Results from that study reported a total average discharge of $292\,m^3\,s^{-1}$.

Buffering-capacity estimation was performed using the basin default LAVINKS-WEB model with the entire wetlands in the Kenyan portion of the basin by accounting for the total area of wetlands in the basin, estimated from Landsat imagery and detailed ground verification to be 216,800 ha. The basin wide averages/totals of rainfall, runoff and water-quality data were also required as discussed below.

With the estimate of the long-term average discharge and knowing the average rainfall on the catchment, inflow rates in the model were determined by assuming a simple rational method of rainfall-runoff relationship and by simply changing the runoff coefficient until the net average outflow from the wetland matched the estimated discharge entering the lake. With this approach, a runoff coefficient of 0.177 was adopted as a reasonable value for an area such as the Lake Victoria Basin.

Concentrations coming into the wetlands from the catchment for TSS, TP and TN were determined by taking the average concentrations at the inlet of the two pilot wetlands for each of these parameters and then applying a scaling parameter to account for the much larger catchment size over the whole basin compared to the two wetlands individually. Adjusting plant uptake rates and relative cover factors only had a very minor effect on the basin-wide net wetland model, so these rates were left at default values.

After adjusting the model by averaging phosphorus and nitrogen loads, the basin-wide loads for the Kenyan portion into Lake Victoria were estimated. Model outputs from July 1984 to December 1990 were analysed. For this period, model loads were found to closely correlate with loads estimated as entering the Lake by COWI (2002). The model was therefore considered to be running under acceptable conditions. Under the conditions described above, load reduction rates of 47% for TSS, 48% for TP and 27% for TN were determined from the modelling results run for July 1984 to December 1990.

The lower removal rate for nitrogen is to be expected as:

- Generally the wetlands in the basin would be expected to be phosphorus limited and, therefore, plants and periphyton would be expected to uptake phosphorus very efficiently but nitrogen less so.
- As discussed earlier, phosphorus is usually much more strongly associated with sediments especially at long timescales and is therefore more effectively removed by sedimentation processes than nitrogen.
- Timescales for nitrogen cycling mean there are significant delays in transformation and removal of nitrogen from the water column.

Summary statistics for the basin-wide buffering-capacity estimation are contained in Table 28.2 and graphically presented in Fig. 28.2.

Table 28.2 Summary of buffering-capacity effectiveness for wetlands of the Kenyan portion of the Lake Victoria Basin

Inlet water quality

	TSS	TP	TN	TDS
Maximum	346.79	1.11	4.86	0.08
Average	112.16	0.39	1.38	0.03
Minimum	0.98	0.00	0.11	-0.01
Unit	mg l^{-1}	mg l^{-1}	mg l^{-1}	mg l^{-1}

Load in

	Discharge	TSS	TP	TN	TDS
Average	25,140	2,799,162	9,721	34,442	888
Unit	m$^3 \times 10^3$ day^{-1}	kg day^{-1}	kg day^{-1}	kg day^{-1}	kg day^{-1}

Loading Rates

	Discharge	TSS	TP	TN	TDS
Average	0.0116	12.9	0.0448	0.1588	0.0041
Unit	m day^{-1}	kg ha^{-1} day^{-1}	kg ha^{-1} day^{-1}	kg ha^{-1} day^{-1}	kg ha^{-1} day^{-1}

Areal removal rate (kg ha^{-1} day^{-1})

	TSS	TP	TN	TDS
Average	5.96	0.02	0.04	5.96

Outlet water quality

	TSS	TP	TN	TDS	N/P ratio
Maximum	78.14	0.28	1.20	0.04	7.94
Average	58.81	0.20	0.99	0.04	5.10
Minimum	30.14	0.10	0.70	0.03	3.47
Unit	mg l^{-1}	mg l^{-1}	mg l^{-1}	mg l^{-1}	–

Load out

	Discharge	TSS	TP	TN	TDS	N/P ratio
Average	25,249	1,517,103	5,154	25,220	891	–
Unit	m$^3 \times 10^3$ day^{-1}	kg day^{-1}	kg day^{-1}	kg day^{-1}	kg day^{-1}	

Load removed

	TSS	TP	TN	TDS
Average	1,292,138	4,566	9,482	18
Unit	kg day^{-1}	kg day^{-1}	kg day^{-1}	kg day^{-1}

Relative reduction (%) (load removed/incoming load)

	TSS	TP	TN	TDS
Average	47	47	27	5

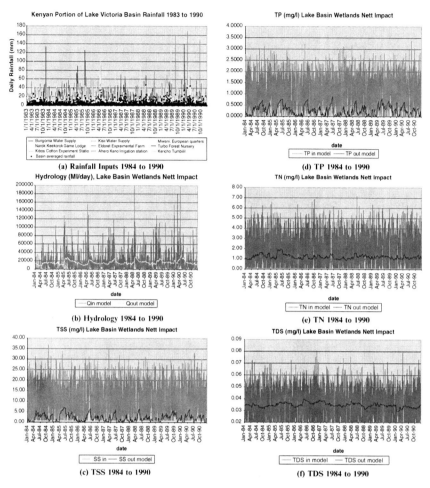

Fig. 28.2 Basin-wide wetland model inputs and expected net modelled impacts in buffering of nutrient inputs to Lake Victoria, 1984–1990. (a) rainfall inputs, (b) hydrology, (c) TSS, (d) TP, (e) TN, (f) TDS

28.4   Conclusions

From the results, it can be seen that wetlands are a highly significant component of water-quality control for the Kenyan portion of the Lake Victoria Basin. Following determination of the extent of wetlands across the basin from mapping activities, modelling was then extended to a basin-wide level to examine the buffering potential of wetlands across the Kenyan portion of the basin. At a basin-wide level it was estimated that wetlands were responsible for buffering approximately 50% of the suspended solids and phosphorus leaving the catchment and approximately 30% of nitrogen leaving the catchment. These estimates should be considered as indicative but valuable 'order of magnitude' determinations, as it must be recognised that such estimates are subject to a high degree of uncertainty and much more work will be required over many years to more fully validate the results presented here.

Ecological studies of tropical wetlands have rarely included mass balances of sediment and nutrients over prolonged periods with consistent, continuous monitoring including storm events because, often, conditions for compiling water/nutrient balances are not suitable. The Kericho-Dionosoyiet and Eldoret wetland site infrastructure and the equipment made available through LVEMP provided an excellent opportunity to fill this gap in the current understanding of tropical wetland ecology. Wetlands preservation, restoration management and rehabilitation are very necessary activities in maintaining a healthy ecosystem in the Lake Victoria Basin. Furthermore, it is vitally necessary that water-quality monitoring for the basin address the issue of wetlands-buffering capacity. Simply measuring the concentrations of pollutants entering the lake and modelling in-lake processes is insufficient to develop effective management strategies for the lake and its basin.

References

Bavor, H.J., Davies, C.M., & Sakadevan, K. (2001). Removal of storm water associated nutrients and bacteria in constructed wetland and water pollution control pond systems. In J. Vymazal (Ed.), *Transformation of nutrients in natural and constructed wetlands* (pp. 483–495). Leiden, Netherlands: Backhuys.

COWI. (2002) Consultants to the Water Quality Component, Annual Report of Lake Victoria Environment Management Programme (LVEMP): Water Quality Component, Kisumu, Kenya.

Lawrence, I., & Breen, P. (1998). *Design guidelines: Stormwater pollution control ponds and wetlands*. Canberra, Australia: CRC for Freshwater Ecology.

Mwanuzi, F., Aalderink, H., & Mdamo, L. (2003). Simulation of pollution buffering capacity of wetlands fringing Lake Victoria. *Environment International, 29*, 95–103.

Ngirigacha, H.W. (2000): The potential of *Cyperus immensus, Cyperus papyrus, Typha domingensis* and *Phragmites mauritianus* for pulp and paper mill waste water treatment. MSc Thesis, DEW 130 IHE Delft, The Netherlands.

Ntiba, M.J., Kudoja, W.M., & Mukasa, C.T. (2001). Management issues in the Lake Victoria watershed. *Lakes Reservoirs Research and Management, 6*(3), 211–216.

Van der Knaap, M., Ntiba, M.J., & Cowx, I.G. (2002). Key elements of fisheries management of Lake Victoria. *Aquatic Ecosystem Health and Management, 593*, 245–254.

Verschuren, D., Johnson, T.C., Kling, H.J., Edgington, D.N., Leavitt, P.R., Brown, E.T., Talbot, M.R., & Hecky, R.E. (2002). History and timing of human impact on Lake Victoria, East Africa. *Proceedings of the Royal Society of London, Series B, 269*, 289–294.

Chapter 29
Sulfur Cycling in Constructed Wetlands

Paul J. Sturman[1], **Otto R. Stein**[1,2](✉), **Jan Vymazal**[3,4],
and Lenka Kröpfelová[3]

Abstract Constructed wetlands (CWs) have been successfully employed in both mining and domestic wastewater applications, yet the fundamental processes responsible for treatment are poorly quantified. Sulfur is common in CW influent streams and is highly reactive, redox-sensitive, and microbially active; therefore, it plays an important role in both desirable and deleterious processes in CWs. In this chapter we review the major sulfur transformations likely occurring in CWs, their interactions with other important processes, and their role in the treatment process. We also present two case studies on the influence of sulfate-reducing bacteria and sulfur-oxidizing bacteria on the performance of CW systems designed to treat mining-contaminated and municipal wastewater, respectively. In both cases there is a feedback between these microbial consortia and other microbes responsible for treatment. A better understanding of the important sulfur transformations in CWs will lead to better design and more confident performance expectations.

Keywords Bacteria, chemical precipitation, mining, sulfate, sulfide, treatment wetlands

29.1 Introduction

Constructed wetlands (CWs) have been successfully employed to treat a wide range of municipal, industrial, and mining wastewaters. In addition to their low operational costs, an attractive feature of CWs is their versatility in treating waste streams which

---

[1] Center for Biofilm Engineering, Montana State University, Bozeman, MT 59717, USA

[2] Department of Civil Engineering, Montana State University, Bozeman, MT 59717, USA

[3] ENKI, o.p.s., Dukelská 145, 379 01 Třeboň, Czech Republic

[4] Institute of Systems Biology and Ecology, Dukelská 145, 379 01 Třeboň, Czech Republic

(✉) Corresponding author: email: ottos@ce.montana.edu

J. Vymazal (ed.) *Wastewater Treatment, Plant Dynamics and Management in Constructed and Natural Wetlands,*
© Springer Science + Business Media B.V. 2008

may vary significantly in organic carbon and nutrient loading. Constructed wetlands are adaptable to organic carbon removal from high strength (copiotrophic) domestic and municipal wastes as well as contaminant removal from mesotrophic and oligotrophic waste streams, such as nutrient removal in agricultural runoff and acid neutralization and precipitation of metals from acid mine drainage (Kadlec & Knight, 1996). This versatility is, at least in part, attributable to the wide range of bacterial niches present in CWs. Oxidation-reduction (redox) state is known to vary spatially along the flow path, with distance from the air–water interface, and/or distance from plant tissues, especially roots (Kadlec & Knight, 1996; Allen *et al.*, 2002; Garcia *et al.*, 2003). Spatial gradients lead to a variety of ecologically distinct zones which may be inhabited by obligate aerobic microorganisms, facultative anaerobes, iron- and manganese-oxidizing organisms, sulfate-reducing bacteria, methane-producing bacteria, and fermentative organisms. The extent to which members of these groups thrive is dependant on characteristics of the wastewater, temperature, season, wetland plant phenology, and wetland design.

Gaining a comprehensive understanding of the interplay between the biotic and abiotic reactions in wetlands has been a major challenge in CW engineering and operation. The success of the traditional "black box" approach to CW operation is testament to the above-mentioned innate versatility of these systems, yet a broader understanding of microscale wetland processes would clearly assist their design. Research over the past decade has shed light on the complexity of wetlands microbiology and geochemistry, yet, many questions remain. Several recent investigators have identified sulfur-related processes within wetlands as both poorly understood and of prime importance in advancing a broader understanding of wetland function (Whitmire & Hamilton, 2005; Wiessner *et al.*, 2005). Sulfur can occur in four valence states, $-2(H_2S)$, $0(S^0)$, $+2(S_2O_3^{2-})$ and $+6(SO_4^{2-})$; thus, it is reactive under both oxidized and reduced conditions as well as in both biotic and abiotic settings. It can be an electron donor or electron acceptor in energy-producing microbial reactions and reacts with virtually all metals (except gold and platinum) to form metal sulfides. Sulfur is also a macronutrient for microbial and plant growth. It is typically abundant in CW influent streams, including municipal and industrial wastewater and especially, acid rock drainage. Sulfur's high reactivity, redox sensitivity, and microbial activity, when combined with the range of conditions found in many CWs, leads to complex geomicrobial interactions. These interactions are often an integral part of the treatment process, e.g., metals removal; may inhibit a desired process, e.g. plugging due to precipitate formation; and at the least are indicators of dominant pathways for, and/or the relative importance of other important removal mechanisms.

In this chapter, we review the various biologically catalyzed and abiotic sulfur fate and transport pathways that are likely active within CWs. To keep application as broad as possible, we do not specify the type of CWs, e.g., free water surface (FWS) or subsurface flow (SF), nor do we specify the type of wetland, scales, and other considerations – such as influent water chemistry, macrophytic plant species selection, temperature, and season – that will influence the magnitudes of the sulfur transformations that might occur in a specific CW. Finally, we present two case examples of sulfur transformations in operating and model CWs.

29.2 Conceptual Model of Sulfur Fate and Transport in Constructed Wetlands

Influent sulfur to wetlands is typically in the form of sulfate in oxidized environments and sulfide in reduced environments, although other sulfur compounds representing intermediate valence states, including thiosulfate, other polythionates, and organic sulfur can also occur in CW influent. Sulfate is highly soluble under all temperature and pH conditions, whereas sulfide solubility is pH-dependant. Sulfide solubility increases tenfold between pH 6–8. At acidic pH (<6) sulfide will be present as H_2S, which has a much lower solubility than the deprotonated form, HS^-, which predominates above neutral pH ($pK_a = 7.04$) (Stumm & Morgan, 1996). Lower pH systems thus have the propensity to offgas H_2S, with the accompanying rotten-egg odor, whereas pH-neutral and above systems maintain higher sulfide concentrations in solution.

In well-aerated systems, or systems with relatively low levels of assimilable organic carbon (AOC), the entire water column (in the case of FWS) and the upper reaches of the sediment strata or sediment within the plant rhizosphere (in the case of FWS and SF) may be aerobic. Sulfate concentrations will be lowered under these conditions mainly through abiotic mineral precipitation (such as gypsum, $CaSO_4$) or biological assimilation into plant or microbial tissue (Fig. 29.1). Bacterial assimilation of sulfur occurs through the reduction of sulfate to the amino acid cysteine ($C_3H_7NO_2S$), thus creating organic sulfur (Le Faou *et al.*, 1990).

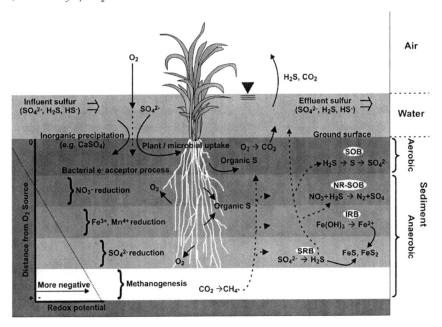

Fig. 29.1 Major biotic and abiotic sulfur transformations in constructed wetlands and their relation to redox potential

Sulfide dissolved in influent water or diffusing from anoxic sediment strata may be subject to oxidation in more aerobic zones through either abiotic processes or via reactions catalyzed by sulfur-oxidizing bacteria (SOB). Under anoxic conditions, sulfide may also precipitate with metals to form metal sulfides, such as iron sulfide (FeS). By far the most important biotic reactions influencing sulfur are those catalyzed by dissimilatory sulfate-reducing bacteria (SRB), which generate energy from the transfer of electrons from organic substrates to sulfate, thereby reducing it to sulfide. Sulfide thus generated may precipitate (as above), diffuse into the water column and offgas, or be oxidized back to sulfate through biotic or abiotic reactions. A variety of biogenic organic sulfur compounds may also exist in equilibrium in the aqueous phase or in transition to the vapor phase. These compounds may be produced by either sediment-associated microbial activity, or as a result of plant activity.

Microbially catalyzed reactions in CW sediments occur in zones which reflect the sequential consumption of electron acceptors based on the energy available from each process. Oxygen diffusing into sediments is rapidly consumed by aerobic or facultative organisms, as it has the highest available energy of potential electron acceptors. In anoxic sediments, nitrate (if present) is often the next most energetic electron acceptor, followed by mineral phase ferric iron or manganese (Mn^{+4}), sulfate, and methanogenesis. The conceptual location of these electron acceptor zones in relation to an oxygen source (atmosphere and/or plant roots) is shown in Fig. 29.1. In all but the most oligotrophic wetlands environments, AOC is present at concentrations sufficient to deplete available dissolved oxygen and overwhelm the oxygen supply rate, leading to the sequential consumption of the less-energetic electron acceptors. Sulfate reduction has been found to be the dominant terminal electron acceptor process in high organic carbon-containing groundwater and open water systems (Wiedemeier *et al.*, 1999), as well as in low-AOC systems containing high concentrations of sulfate (D'Hondt *et al.*, 2002). The abundance of both AOC and sulfate in many CW environments therefore suggests that sulfate reduction will likely account for a large proportion of the total microbial activity. The fate and transport of influent sulfur in a CW depends on the interplay between these bacterially catalyzed reactions, which are discussed in detail below.

29.2.1 Bacterially Catalyzed Sulfur Reactions

29.2.1.1 Bacterial Sulfate Reduction

Sulfur participates in a variety of bacterially catalyzed oxidation and reduction reactions which may impact its mobility and fate in CWs. The most familiar of these is the dissimilatory sulfate-reduction reaction catalyzed by SRB in anoxic water and sediment. SRB gain energy by coupling the oxidation of organic compounds or H_2 to sulfate reduction and liberate inorganic carbon and sulfide (H_2S, HS^-, or S^{2-}, depending on pH), as the primary end products of organic matter mineralization (Megonikal *et al.*, 2004). The process is of major importance in wetlands due to the propensity of produced sulfide to form insoluble metal sulfides. Dissimilatory sulfate reduction is thought to be among the oldest metabolic processes of life on Earth and is found among a wide range of both gram-positive and gram-negative bacterial genera. SRB are a phylogenetically diverse group of δ-Proteobacteria encompassing over 20 genera and utilizing a range of organic electron donors, including H_2, volatile fatty acids (VFAs), and some primary alcohols. SRB are divided into two main groups: (1) incomplete oxidizers (*Desulfovibrio, Desulfomicrobium*), which utilize VFAs such as pyruvate, formate, and butyrate and produce acetate; and (2) complete oxidizers (*Desulfobacter, Desulfobacterium*), which utilize fatty acids, including acetate, and produce carbon dioxide (Widdel, 1988). SRB were once thought to be capable of growth using only sulfate as the electron acceptor, and only in the absence of dissolved oxygen. Recent research has shown that some SRB are capable of growth using more energetic electron acceptors, particularly nitrate (Itoh *et al.*, 2002; Lopez-Cortes *et al.*, 2006). In addition, some SRB have been shown to both tolerate low concentrations of oxygen and possess mechanisms for oxygen detoxification (Vasconcelos & McKenzie, 2000). Such attributes generally insure the survival of SRB populations in sediments that are periodically (or seasonally) exposed to oxygen or other electron acceptors.

The catabolic bacterial sulfate-reduction reaction generates one mole of sulfide per mole of sulfate utilized, as illustrated in the following stoichiometry with acetate as the electron donor:

$$CH_3COO^- + H^+ + SO_4^{2-} \rightarrow H_2S + 2HCO_3^-$$

SRB activity also results in the generation of alkalinity, which may raise the pH of acidic systems. It is important to note that sulfate reduction does not occur in isolation, but in concert with other microbial reactions (Fig. 29.1), including fermentation and methanogenesis. The use of organic acids as electron donors by SRB implies a close relationship between SRB activity and the activity of fermentative organisms which generate VFAs as a product of metabolism. These reactions create a highly reducing environment, which in sediments may accumulate reduced inorganic species such as Fe^{2+}, Mn^{2+}, NH_4^+, and CH_4, in addition to sulfide and bicarbonate. Biogenic sulfide may undergo further biotic and abiotic reactions with these compounds.

Dissimilatory sulfate reduction can account for half or more of the total organic carbon mineralization in many environments (Jørgensen, 1982). In freshwater environments, SO_4^{2-} reduction can account for a significant portion of anaerobic mineralization processes, and in some instances can be the dominant pathway (Bak & Pfennig, 1991; Urban *et al.*, 1994; Holmer & Storkholm, 2001). In wetlands that do not have significant dissolved metal concentrations or iron-containing sediments, biogenic sulfide production may exceed the capacity of the precipitation mechanisms mentioned above and diffuse out of anoxic sediments into either oxygen-containing sediments or the water column. Under these conditions, H_2S may offgas to the atmosphere, or more commonly, reoxidize back to thiosulfate

(S_2O_3) or sulfate through reaction with ferric iron, manganese dioxide, nitrate, or oxygen. Sulfide can also be oxidized back to elemental sulfur or sulfate by chemolithotrophic bacteria under aerobic conditions (Nelson *et al.*, 1986). Sulfur-oxidizing bacteria (SOB) are typically active in a relatively narrow ecological zone where oxygen diffusing in one direction occurs concurrently with sulfide diffusing in the other, and rates of bacterial sulfur oxidation are highest where oxygen concentrations are limited. Where oxygen is plentiful, abiotic sulfide oxidation accounts for the majority of reoxidation to sulfate.

29.2.1.2 Bacterial Iron Reduction

Ferric oxyhydroxide minerals are very common in wetland sediments, and ferrous iron (Fe^{2+}) results from bacterial reduction of these minerals under anoxic conditions. Solution phase Fe^{2+} reacts rapidly with biogenic sulfide to form amorphous iron(II) monosulfide (FeS), which typically precipitates as a black solid. Newly formed FeS is noncrystalline and thus does not have a repeating structure; however, amorphous FeS further reacts with reduced sulfur species to form more sulfur-enriched crystalline solids, such as greigite (Fe_3S_4) and, ultimately, pyrite (FeS_2). While newly formed FeS is subject to re-dissolution under some conditions (such as low pH), the evolved crystalline solids are more thermodynamically stable and resistant to dissolution (Sweeney & Kaplan, 1973). Immobilization of precipitated metal sulfides is an important metal- and sulfur-removal process in CWs and other bioreactor systems designed to treat metal-rich wastewater. Furthermore, rapid consumption of biogenic sulfide via FeS precipitation acts as a detoxification mechanism to prevent sulfide accumulation and toxicity to SRB (Reis *et al.*, 1992), thereby allowing further sulfate consumption.

29.2.1.3 Bacterial Manganese Reduction

Like ferrous iron, solution-phase divalent manganese results from the bacterial reduction of manganese minerals, such as manganese dioxide (MnO_2), under anoxic conditions. Because manganese sulfides are highly soluble, Mn(II) typically does not precipitate as MnS. However, where SRB are active, MnO_2 may react directly with H_2S to form solution phase Mn^{2+} and elemental sulfur (S^0). Sulfur may then undergo bacterial disproportionation to sulfide and sulfate (Thamdrup *et al.*, 1993). Sulfide may then be reoxidized to S to repeat this cycle, and sulfate may be reduced by SRB or diffuse into the water column.

29.2.1.4 Bacterial Nitrate Reduction

As the next most energetic electron acceptor after oxygen, nitrate is usually rapidly consumed by heterotrophic nitrate-reducing bacteria (NRB) in anoxic zones of CWs.

The first step in nitrate reduction produces nitrite (NO_2^-), which is actively inhibitory to SRB (Sturman *et al.*, 1999). Nitrite is then typically further reduced to nitric oxide (NO), nitrous oxide (N_2O), and ultimately to di-nitrogen (N_2). Where biogenic sulfide diffusing from the SRB-active zone is present concurrently with nitrate, chemolithotrophic bacteria can couple the reduction of nitrate with the oxidation of sulfide, as shown below.

$$5HS^- + 8NO_3^- \rightarrow 5\ SO_4^{2-} + 4N_2 + 3OH^- + H_2O$$

These so-called nitrate-reducing – sulfur-oxidizing bacteria (NR-SOB) are typified by members of the genus *Thiomicrospira*, and their activity has been noted to inhibit SRB through the production of the intermediate species nitrite (NO_2^-) during nitrate reduction (Haveman *et al.*, 2005). Therefore, in a CW, NR-SOB bacteria would likely be located in anoxic regions near SRB activity, but not concurrent with them. Sulfate produced by NR-SOB would either reenter the water column or diffuse into the SRB-active zone.

29.2.1.5   Bacterial Sulfur Oxidation

In the presence of available electron acceptors, sulfide, elemental S, thiosulfate, and tetrathionate are oxidized by both chemical and biological pathways (Wainwright, 1984; Paul & Clark, 1996):

$$SH^- \rightarrow S^0 \rightarrow S_2O_3^{2-} \rightarrow S_4O_6^{2-} \rightarrow SO_3^{2-} \rightarrow SO_4^{2-}$$

Sulfur oxidizing bacteria include primarily chemolithotrophic genera, but also phototrophic genera. Photosynthetic SOB couple the oxidation of reduced sulfur (H_2S, S^{2-}, S^0) with CO_2 reduction. They typically occupy anaerobic zones where light penetrates and sulfide is abundant, and accumulate elemental sulfur. So-called purple sulfur bacteria (Thiorhodaceae; e.g., *Chromatium*) generally deposit sulfur internally, whereas green sulfur bacteria (Chlorobacteriaceae, e.g., *Chlorobium*) accumulate sulfur extracellularly. In both cases, accumulated sulfur may be further oxidized to sulfate under conditions of sulfide limitation (Madigan *et al.*, 2000; Wetzel, 2001). Both forms are commonly found in mud and stagnant waters containing H_2S and exposed to light. They reoxidize H_2S, coming from lower anaerobic layers. They require light as an energy source and H_2S as an electron donor in the photosynthetic reduction of CO_2 (Trudinger, 1979; Paul & Clark, 1996; Wetzel, 2001).

Aerobic chemolithotrophic SOB can catalyze the oxidation of reduced sulfur to sulfate where sulfide and oxygen occur concurrently. Also known as colorless sulfur bacteria, these genera most commonly associated with acidic conditions, such as would be associated with mine waste, but some genera are capable of growth under neutral pH conditions as well. The most common SOB genera in low pH mine waste streams are *Acidithiobacillus*, *Acidiphilium*, and *Sulfobacillus* where they catalyze the transformation of thiosulfate ($S_2O_3^{2-}$), elemental sulfur (S^0), or polysulfide (H_2S_n) from the immediate vicinity of active pyrite (or other metal sulfide) dissolution (Johnson, 1998; Fowler & Crundwell, 1999). Many acidophilic SOB are also capable of iron oxidation, and some species are also capable of heterotrophic growth utilizing organic carbon sources in addition to CO_2 (Johnson, 1998).

In neutral streams chemolithotrophic SOB typically occupy microaerophilic zones where they catalyze the oxidation of H_2S to sulfate. *Beggiatoa*, a long filamentous gliding bacterium, and *Thiothrix* are common bacteria that oxidize H_2S with deposition of sulfur intracellularly (Kowallik & Pringsheim, 1966; Shively, 1974; Strohl & Larkin, 1978). Colorless sulfur bacteria of the genus *Beggiatoa* are among the largest and most conspicuous of all bacteria. In nature, the filaments grow only where both H_2S and O_2 are present (Jørgensen, 1977; Kuenen & Beudeker, 1982). Since H_2S is not stable in oxic waters due to autocatalytic oxidation by O_2, the habitat of *Beggiatoa* is restricted to the transition zone between oxic and anoxic environments where O_2 and H_2S are continuously supplied by diffusion along opposite gradients. Where these gradients are steep, *Beggiatoa* and other types of colorless sulfur bacteria may form white patches of dense cell masses (Jørgensen, 1977; Whitcomb *et al.*, 1989). Oxidation of sulfide to sulfate, via S^0 intermediate, was described for *Beggiatoa* more than 100 years ago by Winogradsky (1887, 1888). These so-called gradient organisms (Konhauser, 2007) occupy a relatively narrow zone of low dissolved oxygen, taking advantage of the energy available in reduced sulfur before it can diffuse into more oxidized zones where sulfide is more likely to be oxidized abiotically. Because the zone of sulfide and oxygen overlap may vary temporally, many colorless sulfur bacteria are capable of storing partially oxidized sulfur (in the form of elemental sulfur) intracellularly, thereby insuring a source of sulfur if sulfide becomes limiting. In the wetlands context, it has been observed that *Beggiatoa* growing in association with plant roots serves to detoxify sulfide in the root zone, utilizing oxygen exuded by wetlands plants (Joshi & Hollis, 1977).

29.2.1.6 Methanogenesis (CH_4)

Methane is produced under anaerobic conditions through the activity of methanogenic bacteria. Methanogens utilize hydrogen and CO_2 (and in some cases simple organic molecules) as substrates to form methane. Methanogenic respiration yields the least energy of the common electron acceptor processes ($O_2 > NO_3^- > Mn^{4+}, Fe^{3+} > SO_4^{2-} >$ methanogenesis) and therefore methanogenesis typically occurs in sediment strata most isolated from atmospheric or dissolved oxygen. Methane is a highly energetic compound, of course, and may be utilized by SRB (in symbiotic association with some *Archaea*) or other heterotrophic bacteria as it diffuses away from methanogenic activity following production (Niewöhner *et al.*, 1998).

29.3   Sulfur Transformations in Constructed Wetlands for Mining Applications

Mining wastewater is typically high in dissolved metals and sulfate, and can range in pH from highly acidic (pH 1–3) to circumneutral, depending on the mineralogy of the mine and buffering capacity of subsequently encountered rock. Recognizing that this chemistry can be ideal for SRB activity and the subsequent precipitation of dissolved metals with the produced sulfide (provided organic carbon is available), CWs have been successfully employed to treat mine wastewater since the 1980s. However, evidence suggests that rates of sulfate reduction in wetlands are extremely variable and depend on many factors including pH, redox potential, type and quantity of available organic matter, and the ratio of organic carbon to sulfur (Westrich & Berner, 1988; Webb et al., 1998; Lyew & Sheppard, 1999). Because SRB activity is essential to successful metals removal in CWs, the wetland should be designed to provide: (1) anaerobic conditions, (2) adequate organic carbon for SRB growth, and (3) some means of preventing sediment plugging that could result from the precipitation of metal sulfide solids. Maintaining adequate permeability to insure proper treatment is largely an engineering challenge, and is accomplished through either periodic solids removal or adequate initial treatment volume to insure the necessary life-span.

As noted in Section 29.2, SRB can survive periodic exposure to oxidized conditions, but will not actively reduce sulfate unless more energetic electron acceptors are absent. Since wetland plants add organic carbon necessary for consumption of more energetic electron acceptors (such as oxygen, nitrate, and ferric iron), but also oxygen, the most favorable electron acceptor, their effect on CW redox potential and microbial processes is important, site-specific, and poorly understood (Stein & Hook, 2005). To insure that energetically more-favorable electron acceptors do not overwhelm the desired SRB activity, dissolved organic carbon is usually added to the CW system (Lloyd et al., 2004). However, the quantity required is likely influenced by specific influent chemistry, plant species selection, temperature, and season.

Utilizing a year-long cycle of varying temperature simulating seasonal variation under greenhouse conditions, Stein et al. (2007) compared the influence of two plant species (and unplanted CW) and two influent organic carbon concentrations on redox potential, sulfate reduction, and subsequent zinc precipitation. Results indicated that temperature, season, and plant species had significant interacting affects on redox potential, quantity of sulfate utilized, and the relative influence of sulfate reduction on organic carbon utilization. At identical organic carbon concentrations, redox potential was universally lowest and sulfate reduction was typically highest in unplanted CW, indicating that the net influence of plants is inhibitory for sulfate reduction. Across all plant treatments, sulfate reduction was least at 4°C in winter, but winter inhibition was greater in the planted CW, especially those planted with bulrush (Schoenoplectus acutus), which also displayed increased winter redox levels indicating oxygen was being utilized over sulfate for removal of organic carbon.

Higher influent organic carbon concentrations in bulrush treatments increased sulfate reduction in all seasons and dampened the observed increase in redox during winter. Similar patterns of zinc removal were observed; but variation due to temperature, season, and plant species was typically dampened.

The above-mentioned results clearly demonstrate that plant species selection and season can influence sulfate reduction in CWs by influencing root-zone oxygen release (Stein *et al.*, 2007). Because utilization of the influent organic carbon (as measured by chemical oxygen demand, COD) was virtually complete, regardless of temperature, season, and plant species – variation in sulfate removal is an indication of the competition between aerobic heterotrophs, methanogens and SRB in CW systems (no other electron acceptors were present). Results reinforce conclusions of previous studies (Callaway & King, 1996; Moog & Brüggemann, 1998) that roots of some plant species (but not others) release oxygen in winter. Increased winter oxygen availability increases aerobic respiration over other less-favorable metabolic pathways including sulfate reduction (Allen *et al.*, 2002; Stein & Hook, 2005). Thus, the quantity of organic carbon required to optimize a CW for sulfate reduction and the removal of dissolved metals will vary depending on operating temperature, season, and plant species (Fig. 29.2).

An unreported result of the above study was the evidence of purple photosynthetic SOB growing on the inside walls of the clear influent tubing. Some sulfate reduction occurred in the holding tanks and, due to the presence of sunlight in the connecting lines, these bacteria were able to utilize the produced sulfide and available organic carbon for growth. It is unknown whether these bacteria produced

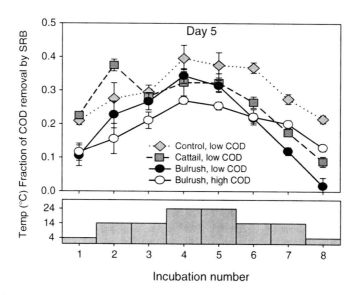

Fig. 29.2 Seasonal variation in the fraction of assimilable organic carbon removal attributable to the activity of sulfur-reducing bacteria (From Stein *et al.*, 2007. With permission from Elsevier)

elemental sulfur or if the sulfide was complexly oxidized to sulfate and then available for sulfate reduction once again in the CW, but their existence indicates that sulfur can cycle between oxidized, reduced, and back to oxidized states over relatively short spatial and temporal scales.

29.4   Sulfur Transformations in Constructed Wetlands for Domestic Wastewater Applications

Wastewater from domestic sources is rich in organic carbon and typically has sulfur concentrations 5–20 mg l^{-1} higher than the original water source, which can regionally have widely varying sulfur concentrations (Crites & Tchobanoglous, 1998). Significant industrial inputs can increase CW influent sulfur concentrations even more; thus, sulfur cycling can be an important component in domestic wastewater treatment CWs. As with mining applications, the most important biologically catalyzed sulfur transformation is sulfate reduction by SRB as the copious organic carbon concentrations typically overwhelm any oxygen supply and transfer mechanisms. Hook *et al.* (2003) observed that temperature, season, and plant species effects on sulfate reduction and redox potential at domestic wastewater influent concentrations were similar to, but often even more dramatic than, those subsequently observed at mining wastewater concentrations (Stein *et al.*, 2007). Thus interactions between plant-mediated oxygen transfer and SRB activity may be more dramatic in CWs treating domestic wastewater.

Vymazal and Kröpfelová (2005) noted that a few CWs for domestic wastewater treatment in the Czech Republic precipitated what proved to be elemental sulfur within the effluent conduits and/or immediately upon contact with the receiving stream (Fig. 29.3), but most seemingly similar CWs did not. Treatment plant operators and local inhabitants equated the presence of these deposits with CW failure despite good performance for traditional parameters such as suspended solids and BOD$_5$ which met the discharge limits. Presumably the elemental sulfur deposits are evidence of SOB activity at the anoxic–oxic transition at the tail end of the CWs; however, there was no visual evidence of photosynthetic SOB anywhere in the system and the deposits appear to be formed extracellularly, while most chemolithotrophic SOB deposit sulfur intracellularly. The formation could be an abiotic process of unknown type. Regardless, the necessary requirement for elemental sulfur deposition is the reduction of influent sulfate to sulfide within the CW bed by SRB activity and consequent oxidation of sulfide upon release to oxic conditions.

A collection of limited water-quality data (Table 29.1) has not revealed a method to successfully predict the formation of white elemental sulfur patches in the CW effluent. The initial assumption was that systems with large reductions in sulfate concentration might lead to sulfur deposition due to high concentrations of effluent sulfide. However, the data revealed that at some systems with substantial sulfate concentration reductions (Chmelná, Břehov), white patches do not occur while massive patches occur even in systems with a very little evidence of sulfate reduction

Fig. 29.3 White patches of elemental sulfur in a stream receiving the outflow from the constructed wetland Mořina (Photo Jan Vymazal)

(Obecnice, Trhové Dušníky). Perhaps the sulfide produced in locations such as Chmelná and Břehov was retained in the CW by precipitation of metal sulfides (or outgassed at low pH), but this would not explain why some systems with poor sulfate reduction (Obecnice, Trhové Dušníky) created elemental sulfur deposits.

Table 29.1 Data from horizontal flow constructed wetlands treating municipal sewage in the Czech Republic

Locality	BOD_5 in (kg ha^{-1} day^{-1})	BOD_5 out (mg l^{-1})	BOD_5 removal (%)	NH_4-N removal (%)	Total N removal (%)	SO_4 removal (%)	Elemental sulfur deposits
Chlístovice	6	4	25	41	46	12	NO
Onšov	7	6	46	37	26	1	NO
Doksy	9	6	84	66	44	2	NO
Mořina 2002	16	4	90	28	35	28	NO
Chmelná	26	5	76	17	21	49	NO
Obecnice	27	11	93	28	33	29	MASSIVE
Břehov	30	11	78	33	35	56	NO
Mořina 2003–2004	36	45	68	15	17	55	MASSIVE
Koloděje	54	10	93	35	38	23	NO
Čistá	59	7	81	17	21	15	NO
Ondřejov	73	11	92	15	20	29	MILD
Trhové Dušníky	145	56	92	50	51	20	MASSIVE

In general, elemental sulfur precipitation occurred in systems with higher organic loads, but not all heavily loaded systems exhibited the formation of elemental sulfur. There is also a mild correlation between sulfur deposits and higher outflow BOD_5 concentrations, but in Obecnice massive deposition occurred with BOD_5 concentration as low as 11 mg l^{-1}. More detailed water-quality monitoring and/or microbial assays in the vicinity of the sulfur deposits will be required to determine the cause of their formation.

29.5 Conclusions

Sulfur transformations play an important role in many biogeochemical reactions occurring in CWs. Most important of these is the reduction of sulfate to sulfide, catalyzed by the ubiquitous SRB. The subsequent precipitation of metal sulfides in systems with high dissolved metal concentrations makes this the dominant mechanism for removal of metals in CW treating mining wastewater. Because most CWs treating domestic wastewater have high concentrations of assimilable organic carbon, thereby making the CWs largely anaerobic, SRB activity is also likely an important mechanism for organic carbon removal in these systems. However, oxygen release by plants under some conditions can interfere with the activity of SRB.

There is clear evidence that SOB are also active in CWs. These bacteria are most likely active at the oxic–anoxic interface and can cycle sulfide back to sulfate which can be subsequently utilized by SRB and/or can lead to deposits of elemental sulfur in the exit region of the CWs. A better understanding of sulfur transformations in CWs, and their spatial and temporal variation, will shed light on the relative magnitudes of the microbially catalyzed reactions occurring in CWs that are important in both domestic and mining applications. Also, better understanding sulfur cycling in CWs is necessary to explain the occasional formation of unwanted sulfur deposits at discharge points.

Acknowledgement The study was partially supported by grants No. 206/06/0058 "Monitoring of Heavy Metals and Selected Risk Elements during Wastewater Treatment in Constructed Wetlands" from the Czech Science Foundation and No. 2B06023 "Development of Mass and Energy Flows Evaluation in Selected Ecosystems" from the Ministry of Education, Youth and Sport of the Czech Republic.

References

Allen, W.C., Hook, P.B., Biederman, J.A., & Stein, O.R. (2002). Temperature and wetland plant species effects on wastewater treatment and root-zone oxidation. *Journal of Environmental Quality, 31*, 1011–1016.

Bak, F., & Pfennig, N. (1991). Microbial sulfate reduction in littoral sediments of lake Constance. *FEMS Microbiology Ecology, 85*, 31–42.

Callaway, R.M., & King, L. (1996). Temperature-driven variation in substrate oxygenation and the balance of competition and facilitation. *Ecology, 77*, 1189–1195.

Crites, R., & Tchobanoglous, G. (1998). *Small and decentralized wastewater management systems*. Boston, MA: WCB McGraw-Hill.

D'Hondt, S., Rutherford, S., & Spivack, A.J. (2002). Metabolic activity of subsurface life in deep-sea sediments. *Science, 295*, 2067–2070.

Fowler, T.A., & Crundwell, F.K. (1999). Leaching of zinc sulfide by *Thiobacillus ferrooxidans*: Bacterial oxidation of the sulfur product layer increases the rate of zinc sulfide dissolution at high concentrations of ferrous ions. *Applied Environmental Microbiology, 65*, 5285–5292.

Garcia, J., Ojeda, E., Sales, E., Chico, F., Piriz, T., Aguirre, P., & Mujeriego, R. (2003). Spatial variations of temperature, redox potential and contaminants in horizontal flow reed beds. *Ecological Enginnering, 21*, 129–142.

Haveman, S.A., Greene, E.A., & Voordouw, G. (2005). Gene expression analysis of the mechanism of inhibition of *Desulfovibrio vulgaris* Hildenborough by nitrate-reducing, sulfide-oxidizing bacteria. *Environmental Microbiology, 7*, 1461–1465.

Holmer, M., & Storkholm, P. (2001). Sulphate reduction and sulphur cycling in lake sediments: A review. *Freshwater Biology, 46*, 431–451.

Hook, P.B., Stein, O.R., Allen, W.C., & Biederman, J.A. (2003). Plant species effects on seasonal performance patterns in model subsurface wetlands. In Ü. Mander & P.D. Jenssen (Eds.), *Constructed wetlands for wastewater treatment in cold climates* (pp. 87–106). Southampton, UK: WTI.

Itoh, T., Nielsen, J.L., Okabe, S., Watanabe, Y., & Nielsen, P.H. (2002). Phylogenetic identification and substrate uptake patterns of sulfate-reducing bacteria inhabiting an oxic-anoxic sewer biofilm determined by combining microautoradiography and fluorescent in situ hybridization. *Applied Environmental Microbiology, 68*, 356–364.

Johnson, D.B. (1998). Biodiversity and ecology of acidophilic microorganisms. *FEMS Microbiology Ecology*, 27, 307–317.

Jørgensen, B.B. (1977). Distribution of colorless sulfur bacteria (*Beggiatoa* spp.) in a coastal marine sediments. *Marine Biology*, 41, 19–28.

Jørgensen, B.B. (1982). Mineralization of organic matter in the sea bed: The role of sulphate reduction. *Nature*, 296, 643–645.

Joshi, M.M., & Hollis, J.P. (1977). Interaction of *Beggiatoa* and rice plant: Detoxification of hydrogen sulfide in the rice rhizosphere. *Science*, 195, 179–180.

Kadlec, R.H., & Knight, R.L. (1996). *Treatment wetlands*. Boca Raton, FL: CRC Press.

Konhauser, K. (2007). *Introduction to geomicrobiology*. Malden, MA: Blackwell.

Kowallik, U., & Pringsheim, E.G. (1966). The oxidation of hydrogen sulfide by *Beggiatoa*. *American Journal of Botany*, 53, 801–806.

Kuenen, J.G., & Beudeker, R.F. (1982). Microbiology of thiobacilli and other sulphur-oxidizing autotrophs, mixotrophs and heterotrophs. *Philosophical Transactions of the Royal Society of London*, 298, 473–497.

Le Faou, A., Rajagopal, B.S., Daniels, L., & Fauque, G. (1990). Thiosulfate, polythionates and elemental sulfur assimilation and reduction in the bacterial world. *FEMS Microbiological Reviews*, 75, 351–382.

Lloyd, J.R., Klessa, D.A., Parry, D.L., Buck, P., & Brown, N.L. (2004). Stimulation of microbial sulfate reduction in a constructed wetland: Microbiological and geochemical analysis. *Water Research*, 38, 1822–1830.

Lopez-Cortes, A., Fardeau, M.L., Fauque, G., Joulian, C., & Ollivier, B. (2006). Reclassification of the sulfate- and nitrate-reducing bacterium *Desulfovibrio vulgaris* subsp. *oxamicus* as *Desulfovibrio oxamicus* sp. nov., comb. nov. *International Journal of Systematic Evolutionary Microbiology*, 56, 1495–1499.

Lyew, D., & Sheppard, J. (1999). Sizing considerations for gravel beds treating acid mine drainage by sulfate reduction. *Journal of Environmental Quality*, 28, 1025–1030.

Madigan, M.T., Martinko, J.M., & Parker, J. (2000). *Biology of microorganisms* (9th ed.). Upper Saddle River, NJ: Prentice Hall.

Megonikal, J.P., Hines, M.E., & Visscher, P.T. (2004). Anaerobic metabolism: Linkages to trace gases and aerobic processes. In W.H. Schlesinger (Ed.), *Biogeochemistry* (pp. 317–424). Oxford: Elsevier-Pergamon.

Moog, P.R., & Brüggemann, W. (1998). Flooding tolerance of *Carex* species. II: Root gas-exchange capacity. *Planta*, 207, 199–206.

Nelson, D.C., Jørgensen, B.B., & Revsbech, N.P. (1986). Growth pattern and yield of a chemoautotrophic *Beggiatoa* sp. in oxygen-sulfide microgradients. *Applied Environmental Microbiology*, 52, 225–233.

Niewöhner, C., Hensen, C., Kasten, S., Zabel, M., & Schulz, H.D. (1998). Deep sulfate reduction completely mediated by anaerobic methane oxidation in sediments of the upwelling area off Namibia. *Geochimica Cosmochimica Acta*, 62, 455–464.

Paul, E.A., & Clark, F.E. (1996). *Soil microbiology and biochemistry* (2nd ed.). San Diego, CA: Academic.

Reis, M.A.M., Almeida, J.S., Lemos, P.C., & Carrondo, M.J.T. (1992). Effect of hydrogen sulfide on growth of sulfate-reducing bacteria. *Biotechnology Bioengineering*, 40, 593–600.

Shively, J.M. (1974). Inclusion bodies in prokaryotes. *Annual Review of Microbiology*, 28, 167–187.

Stein, O.R., & Hook, P.B. (2005). Temperature, plants and oxygen: How does season affect constructed wetland performance?. *Journal of Environmental Science and Health*, 40A, 1331–1342.

Stein, O.R., Borden, D.J., Hook, P.B., &Jones, W.L. (2007). Seasonal influence on sulfate reduction and metal sequestration in sub-surface wetlands. *Water Research*, 41, 3440–3448.

Strohl, W.R., & Larkin, J.M. (1978). Enumeration, isolation, and characterization of *Beggiatoa* from fresh water sediments. *Applied Environmental Microbiology*, 36, 755–770.

Stumm, W., & Morgan, J.J. (1996). *Aquatic chemistry*. New York: Wiley.

Sturman, P.J., Goeres, D.M., & Winters, M.A. (1999). *Control of hydrogen sulfide in oil and gas wells with nitrite injection.* Society of Petroleum Engineers Paper #56772, Proceedings SPE Annual Technical Conference, Houston TX, 3–6 October 1999.

Sweeney, R.E., & Kaplan, I.R. (1973). Pyrite framboid formation: Laboratory synthesis and marine sediments. *Economic Geology, 68,* 618–634.

Thamdrup, B., Finster, K., Hansen, J.W., & Bak, F. (1993). Bacterial disproportionation of elemental sulfur coupled to chemical reduction of iron or manganese. *Applied Environmental Microbiology, 59,* 101–108.

Trudinger, P.A. (1979). The biological sulfur cycle. In P.A. Trudinger & D.J. Swaine (Eds.), *Biogeochemical cycling of mineral-forming elements* (pp. 293–313). Amsterdam: Elsevier.

Urban, N.R., Brezonik, P.L., Baker, L.A., & Sherman, L.A. (1994). Sulfate reduction and diffusion in sediments of Little Rock Lake, Wisconsin. *Limnology and Oceanography, 39,* 797–815.

Vasconcelos, C., & McKenzie, J.A. (2000). Biogeochemistry. Sulfate reducers: Dominant players in a low-oxygen world?. *Science, 290,* 1744–1747.

Vymazal, J., & Kröpfelová, L. (2005). Sulfur deposits at the outflow from horizontal sub-surface flow constructed wetlands – what does it indicate? In *Proceedings of the International Symposium on Wetland Pollutant Dynamics and Control* (pp. 200–201). Ghent, Belgium: University of Ghent.

Wainwright, M. (1984). Sulfur oxidation in soils. *Advance in Agronomy, 37,* 349–396.

Webb, J.S., McGinness, S., & Lappin-Scott, H.M. (1998). Metal removal by sulphate-reducing bacteria from natural and constructed wetlands. *Journal of Applied Microbiology, 84,* 240–248.

Westrich, J.T., & Berner, R.A. (1988). The effect of temperature on rates of sulfate reduction in marine sediments. *Geomicrobiology Journal, 6,* 99–117.

Wetzel, R.G. (2001). *Limnology: Lake and river ecosystems* (3rd ed.). San Diego, CA: Academic.

Whitcomb, J.H., DeLaune, R.D., & Patrick, W.H. Jr. (1989). Chemical oxidation of sulfide to elemental sulfur: Its possible role in marsh energy flow. *Marine Chemistry, 26,* 205–214.

Whitmire, S.L., & Hamilton, S.K. (2005). Rapid removal of nitrate and sulfate in freshwater wetlands sediments. *Journal of Environmental Quality, 34,* 2062–2071.

Widdel, F. (1988). Microbiology and ecology of sulfate- and sulfur reducing bacteria. In A.J.B. Zehnder (Ed.), *Biology of anaerobic microorganisms* (pp. 469–585). New York: Wiley.

Wiedemeier, T.H., Rifai, H.S., Newell, C.J., & Wilson, J.T. (1999). *Natural attenuation of fuels and chlorinated solvents in the subsurface.* New York: Wiley.

Wiessner, A., Kappelmeyer, U., Kuschk, P., & Kästner, M. (2005). Sulphate reduction and the removal of carbon and ammonia in a laboratory-scale constructed wetland. *Water Research, 39,* 4643–4650.

Winogradsky, S. (1887). Über Schwefelbacterien. *Botanische Zeitung, 45,* 489–610.

Winogradsky, S. (1888). *Beiträge zur Morphologie und Physiologie der Bakterien, Heft I: Zur Morphologie und Physiologie der Schwefelbakterien.* Leipzig, Germany: Arthur Felix.

Index

Printed in the United States
124440LV00002B/70-78/P